11048665

REPRODUCTION OF MARINE INVERTEBRATES

Volume V

Molluscs: Pelecypods and Lesser Classes

ADVISORY BOARD

REPRODUCTION OF MARINE INVERTEBRATES

Volume V

Molluscs: Pelecypods and Lesser Classes

Edited by

Arthur C. Giese
*Department of Biological Sciences and
Hopkins Marine Station
Stanford University
Stanford, California*

John S. Pearse
*Division of Natural Sciences
University of California
Santa Cruz, California*

ACADEMIC PRESS New York San Francisco London 1979
A Subsidiary of Harcourt Brace Jovanovich, Publishers

ACADEMIC PRESS, INC.
111 Fifth Avenue, New York, New York 10003

United Kingdom Edition published by
ACADEMIC PRESS, INC. (LONDON) LTD.
24/28 Oval Road, London NW1 7DX

Library of Congress Cataloging in Publication Data

Giese, Arthur Charles, Date
    Reproduction of marine invertebrates.

    Includes bibliographies.
    CONTENTS: v. 1. Acoelomate and pseudocoelomate
metazoans.--v. 2. Entoprocts and lesser coelomates.
--v. 3. Annelids and echiurans.   [etc.]
    1. Marine invertebrates--Physiology. 2. Marine
biology. 3. Reproduction. QL364 G455
QL364.15.G43   592'.01'6   72-84365
ISBN 0-12-282505-5 (v. 5)

PRINTED IN THE UNITED STATES OF AMERICA

79 80 81 82   9 8 7 6 5 4 3 2 1

087037

CONTENTS

LIST OF CONTRIBUTORS vii

PREFACE ix

OBITUARY—DONALD P. COSTELLO xiii

CONTENTS OF OTHER VOLUMES xv

CHAPTER 1   APLACOPHORA

MICHAEL G. HADFIELD

1.1 Introduction 1
1.2 Asexual Reproduction 5
1.3 Sexual Reproduction 5
1.4 Development 12
1.5 Conclusions 22
    References 23

CHAPTER 2   POLYPLACOPHORA

JOHN S. PEARSE

2.1 Introduction 27
2.2 Asexual Reproduction 28
2.3 Sexual Reproduction 28
2.4 Development 64
    References 79

CHAPTER 3   MONOPLACOPHORA

JEFFERSON J. GONOR

3.1 Introduction 87
3.2 Asexual Reproduction 88
3.3 Sexual Reproduction 88
3.4 Postembryonic Development 92
    References 93

CHAPTER 4   SCAPHOPODA

M. McFADIEN-CARTER

4.1 Introduction 95
4.2 Asexual Reproduction 96
4.3 Sexual Reproduction 96
4.4 Development 104
    References 110

CHAPTER 5   PELECYPODA (EXCLUDING OSTREIDAE)

A. N. SASTRY

5.1 Introduction 113
5.2 Asexual Reproduction 114
5.3 Sexual Reproduction 114
5.4 Development 217
    References 265

CHAPTER 6   PELECYPODA: OSTREIDAE

JAY D. ANDREWS

6.1 Introduction 293
6.2 Sexual Reproduction 298
6.3 Development 307
6.4 Setting (Spatfall or Settlement) 322
    References 335

AUTHOR INDEX 343

SUBJECT INDEX 353

TAXONOMIC INDEX 363

LIST OF CONTRIBUTORS

Numbers in parentheses indicate the pages on which the authors' contributions begin.

Jay D. Andrews (293), *Virginia Institute of Marine Science and School of Marine Science, College of William and Mary, Gloucester Point, Virginia 23062*

Jefferson J. Gonor (87), *Marine Science Center, Oregon State University, Newport, Oregon 97365*

Michael G. Hadfield (1), *Kewalo Marine Laboratory, Pacific Biomedical Research Center, University of Hawaii, Honolulu, Hawaii 96813*

M. McFadien-Carter (95), *College of Marine Studies, University of Delaware, Newark, Delaware 19711*

John S. Pearse (27), *Center for Coastal Marine Studies, Division of Natural Sciences, University of California, Santa Cruz, California 95064*

A. N. Sastry (113), *Graduate School of Oceanography, University of Rhode Island, Kingston, Rhode Island 02881*

PREFACE

More marine species of Mollusca have been described than for any other phylum; most of them are gastropods. Volume IV covered gastropods and cephalopods. Because of their economic importance, a great deal of research has been done on pelecypods, especially oysters. The present volume covers pelecypods and the remaining four smaller classes.

The small classes of molluscs include those usually considered to be the most archaic and primitive in the phylum. Members of three of the classes—aplacophorans, polyplacophorans, and monoplacophorans—have been placed near the very base of the phylum Mollusca by different authorities; knowledge of these groups may therefore provide valuable insights into the early evolution of molluscan reproductive biology. Moreover, each of these archaic classes is distinct from the others, reflecting adaptation to different habitats. Scaphopods, comprising a fourth small and distinct class specialized for infaunal life, provide a further perspective on the remarkable adaptability of the molluscan body plan.

Some species of pelecypods are important sources of human food; oysters, in particular, have been cultivated during most of historical time. This activity has provided a wealth of information of both academic and practical importance. Oysters continue to be the most important item of mariculture, and their biology has been more intensively investigated than that of any other marine organism. One chapter in this volume, devoted to oyster reproductive biology, provides a broad overview of current knowledge and problems. Pelecypods other than oysters, such as clams, mussels, and scallops, include food species potentially important for mariculture. Destructive wood-boring species are also of economic importance, as well as numerous species forming major components of many marine communities. The foregoing topics have been subjects of diverse investigation, resulting in an enormous body of information which is comprehensively organized in the most extensive chapter in this volume.

We announce with regret the passing of Donald P. Costello, one of the leading workers in reproductive biology of marine invertebrates. As a member of our Advisory Board, he provided us with helpful and wise counsel. In appreciation this volume is dedicated to his memory.

We are indebted to our Advisory Board for suggestions on the scope and organization of the treatise, to the Board and to a larger community of biologists for encouragement and suggestions for additional prospective authors, to all the authors who graciously assumed responsibility for chapters that demanded of them much effort and time, and to Catherine Henley for the account of Dr. Costello's contribution to science and for the photograph. We are grateful to Dr. R. Andrew Cameron for preparing the Subject and Taxonomic Indexes. Finally, we are indebted to Dr. Vicki Buchsbaum Pearse for her painstaking editorial assistance and to the staff of Academic Press for their help with the development of the treatise.

ARTHUR C. GIESE
JOHN S. PEARSE

Donald P. Costello (1909–1978)

DONALD P. COSTELLO (1909–1978)

Donald Paul Costello, Kenan Professor, Emeritus, at the University of North Carolina, died in Woods Hole, Massachusetts, on February 6, 1978, after a long illness with a heart ailment. He was 68 years old.

Donald Costello could never have been characterized as indecisive in his views or compromising, especially where the pursuit of excellence was concerned. However, there was a striking series of contrasting and paradoxical aspects to his personality and to his scientific career. He was, for example, a traditionalist, steeped in the lore of the classical biologist. But there was also a fascination with the possibilities offered by new techniques (electron microscopy was one example) for answering old questions. He was a talented artisan, whether carving wooden gears for his antique clocks or fashioning jewelry; thin-sectioning or making free-hand blastomere isolations; painting or drawing; or roofing his house. There was also a soaring imagination that carried him over mundane practicalities. This farsighted visionary could, and did, spend hours redoing a photographic enlargement, a scientific drawing, or an iron hematoxylin preparation until they met his exacting standards.

His early training, under the late L. V. Heilbrunn, was as a cell physiologist using marine embryological materials. This later led him to wide-ranging interests in the developmental and reproductive biology of marine invertebrates and gave him an extraordinarily varied, though detailed, acquaintance with such aspects as they exist in "simpler" animals. Dr. Costello's book, "Methods for Obtaining and Handling Marine Eggs and Embryos," was conceived as a preservation, extension, and updating of information contained in an earlier work by E. E. Just: "Basic Methods for Experiments in Eggs of Marine Animals." It was a direct outgrowth, also, of his long participation in the embryology course at Woods Hole, during which he wrote many of the directions for laboratory exercises.

This change in scientific direction, from cell physiologist to developmental and cell biologist, resulted in his devoting a very large proportion of his subsequent work to various aspects of the reproductive biology of marine invertebrates, especially those of the free-living flatworms and the annelids. In collaboration with his wife, Dr. Helen Miller Costello, he described in 1938 a new species of the acoel genus *Polychoerus* from waters adjacent to the Hopkins Marine Station at Pacific Grove, California. In the following years, this beautiful cytological material was extensively investigated, at the levels of both light and electron micro-scopy. The striking resting first-cleavage metaphase of the *Polychoerus* egg, with its huge centrioles and chromosomes arranged on a central spindle, inspired him to set forth at least two important theories in de-velopmental biology: the postulated role of centriole replication in the orientation of dividing cells, and the linear order of chromosomes in the very long and slender spermatozoon of this form. Observations utilizing electron microscopy on male gametes of a number of marine inverte-brates resulted in a provocative theory concerning mechanisms of mo-tility in flagella of these forms and, by extension, in cilia as well.

Perhaps Donald Costello's greatest fame resides in his classical 1945 paper demonstrating the developmental potencies in isolated blasto-meres of the annelid *Nereis limbata*. These meticulously planned and executed experiments were the modern capstones to earlier studies of this sort by, among others, Wilson, Lillie, Whitman, and Conklin.

In what was once described by a senior colleague at the University of North Carolina as a quiet and steadfast rise to eminence, Dr. Costello was reticent about the many honors which came to him beginning early in his career. He was, for instance, recipient of a National Research Council Fellowship in 1934 and of a Rockefeller Fellowship seven years later. At the University, he was named to a prestigious Kenan Professor-ship in 1949 and to the chairmanship of the Department of Zoology in 1947. As a leading research scientist in developmental and cell biology, he served on review and advisory panels for several granting agencies and was elected a Founding Member of the Institut Internationale d'Em-bryologie. Besides serving on the editorial boards of a number of journals, Dr. Costello was Managing Editor of *The Biological Bulletin* for eighteen years. He was an outstanding teacher and, in addition, maintained an active research program, presenting his results in a series of about 100 papers.

The complex personality of this gentle, sensitive, and highly intelligent man cannot adequately be summarized in words. An approach to such a summary might be found, however, in the words of one of his former graduate students: "He represents an ideal toward which I always strive, however ineffectually. His integrity, compassion, and commitment to ex-cellence gave those of us who try to follow his path a standard by which to judge and be judged."

CATHERINE HENLEY

CONTENTS OF OTHER VOLUMES

VOLUME I—ACOELOMATE AND PSEUDOCOELOMATE METAZOANS

Introduction: General Principles
  ARTHUR C. GIESE and JOHN S. PEARSE
Porifera
  PAUL E. FELL
Cnidaria
  RICHARD D. CAMPBELL
Ctenophora
  HELEN DUNLAP PIANKA
Platyhelminthes (Turbellaria)
  CATHERINE HENLEY
Gnathostomulida
  WOLFGANG STERRER
Nemertinea
  NATHAN W. RISER
Nematoda
  W. D. HOPE
Rotifera
  ANNE THANE
Gastrotricha
  WILLIAM D. HUMMON
Kinorhyncha
  ROBERT P. HIGGINS
Author Index—Subject Index—Taxonomic Index

VOLUME II—ENTOPROCTS AND LESSER COELOMATES

Entoprocta
  RICHARD N. MARISCAL
Tardigrada
  LELAND W. POLLOCK
Priapulida
  JACOB VAN DER LAND
Sipuncula
  MARY E. RICE
Pogonophora
  EVE C. SOUTHWARD
Chaetognatha
  M. R. REEVE and T. C. COSPER

Hemichordata
    MICHAEL G. HADFIELD
Chordata: Tunicata
    N. J. BERRILL
Chordata: Acrania (Cephalochordata)
    JOHN H. WICKSTEAD
Author Index–Subject Index–Taxonomic Index

VOLUME III–ANNELIDS AND ECHIURANS

Annelida: Polychaeta
    PAUL C. SCHROEDER and COLIN O. HERMANS
Annelida: Clitellata
    PIERRE LASSERRE
Echiura
    MEREDITH GOULD-SOMERO
Author Index–Subject Index–Taxonomic Index

VOLUME IV–MOLLUSCS: GASTROPODS AND CEPHALOPODS

Gastropoda: Prosobranchia
    HERBERT H. WEBBER
Gastropoda: Opisthobranchia
    ROBERT D. BEEMAN
Gastropoda: Pulmonata
    A. J. BERRY
Cephalopoda: Nautiloidea
    NORINE HAVEN
Cephalopoda: Decapoda
    JOHN M. ARNOLD and LOIS D. WILLIAMS-ARNOLD
Cephalopoda: Octopoda
    M. J. WELLS and J. WELLS
Author Index–Subject Index–Taxonomic Index

REPRODUCTION OF MARINE INVERTEBRATES

Volume V

Molluscs: Pelecypods and Lesser Classes

Chapter 1

APLACOPHORA

Michael G. Hadfield

1.1	Introduction	1
1.2	Asexual Reproduction	5
1.3	Sexual Reproduction	5
	1.3.1 Sexual Dimorphism and Hermaphroditism	5
	1.3.2 Anatomy of the Reproductive System	6
	1.3.3 Origin of the Germ Cells	9
	1.3.4 Cytodifferentiation of the Gametes	9
	1.3.5 Gametogenic Cycles within Populations	9
	1.3.6 Factors Influencing Gametogenesis	11
	1.3.7 Reproductive Behavior and Spawning	11
1.4	Development	12
	1.4.1 Fertilization	12
	1.4.2 Cleavage and Early Embryology	13
	1.4.3 Gastrulation and Germ Layer Formation	13
	1.4.4 Developmental Patterns in the Neomenioidea	17
	1.4.5 Organogenesis	19
	1.4.6 Larvae and Metamorphosis	20
1.5	Conclusions	22
	References	23

1.1 Introduction

Reproduction and developmental biology of the Aplacophora have received scant attention in recent years. The best studies of reproductive anatomy remain those in older expedition reports and similar turn-of-the-century monographs (Hubrecht, 1881; Pruvot, 1891; Wirén, 1892a,b; Pelseneer, 1901, 1903; Nierstrasz, 1902; Heath, 1905, 1911, 1918; Thiele, 1913). Hyman (1967) reviewed these topics in Volume VI of her treatise "The Invertebrates," and summarized information available through the mid-1960s. Fischer-Piette and Franc (1960) extensively discussed repro-

1

ductive anatomy and biology in their section on the Aplacophora in "Traité de Zoologie," but did not include embryology.

The major recent works on the Aplacophora are those of Salvini-Plawen (1968c, 1972a), who has reinvestigated many anatomical problems in the group and has extensively revised aplacophoran systematics. His position that the two major aplacophoran subgroups, the neomenioid and the chaetodermatoid forms (for examples see Fig. 1), should be elevated to the level of separate classes is worthy of consid-

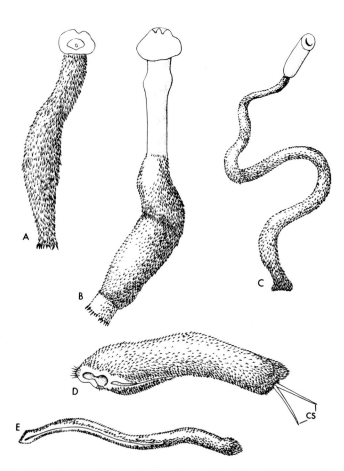

Fig. 1. Aplacophorans. Chaetodermatoidea: (A) *Chaetoderma nanula*, (B) *C. scabra*, (C) *C. attenuata*; Neomenioidea: (D) *Proneomenia vagans*, (E) *P. hawaiiensis*. CS, Copulatory spicules. (A, B, C, E, after Heath, 1911; D, after Pruvot, 1891.)

eration. This classification is not, however, followed here because (1) the differences between the two groups stressed by Salvini-Plawen seem no greater (and perhaps are less) than differences between groups combined in other molluscan classes (consider, for example, the great morphological differences between the two gastropod types represented by marine limpets and terrestrial slugs), and (2) it seems useful to combine a discussion of these two, otherwise small, vermiform molluscan groups. Neomenioids and chaetodermatoids share a common "ground plan" with the following characteristics: they are vermiform and shell-less and have an epidermal cuticle heavily invested with calcareous spicules; their mouths are anterior, and they have a posterior pallial cavity receiving gonadal and fecal products; animals of both groups have posterodorsal pits, probably sensory, called dorsoterminal sensory organs; they share similarities in the layout of their nervous, digestive, and reproductive systems (Figs. 1 and 2). Major differences occur in the presence or absence of gills (similar to prosobranch-pulmonate or septibranch-eulamellibranch differences), presence of monoecious or dioecious individuals, and presence or absence of a "foot groove" (compare the foot of an abalone, *Haliotis*, with that of a vermetid gastropod, *Vermetus*, or a parasitic gastropod such as *Entocolax*). Chaetodermatoids burrow in soft, benthic muds; neomenioids, in contrast, are mostly epibenthic, living on muds or on branching cnidarians which serve as their prey. Anatomical differences between the two aplacophoran groups might be better explained as adaptations to different habitats than as ancient evolutionary divergences from a preaplacophoran ancestor. Scheltema (1978) has argued for the same conclusion.

In light of the preceding discussion, the following discourse considers that vermiform molluscs belong to a common class, the Aplacophora, as two orders, Neomenioidea (with a foot groove and without gills) and Chaetodermatoidea (without a foot groove and with gills). This is in keeping with the usage of Hyman (1967), reflects the decision of the International Commission on Zoological Nomenclature that *Chaetoderma* has legitimate precedence over *Crystallophrisson* (Opinion No. 764, Bull. Zool. Nomen., 1966), avoids the confusion over whether "soleno-gasters" refers to all or only the neomenioid part of the class, and deletes unnecessary new terms, e.g., "Caudofoveata."

Recent works on other aspects of aplacophoran taxonomy and biology are those of Schwabl (1955, 1961a,b, 1963), Thompson (1960), Salvini-Plawen (1967a,b, 1968a,b,c, 1969a,b, 1970a,b, 1971, 1972a,b, 1973), Beedham and Trueman (1968), and Scheltema (1972, 1973, 1978).

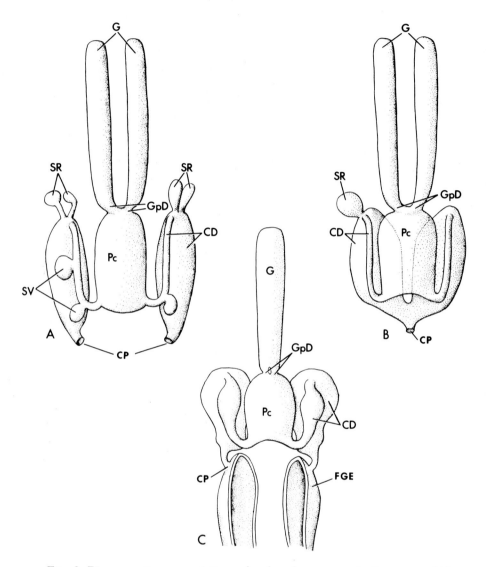

Fig. 2. Diagrammatic representations of aplacophoran reproductive tracts. (A) Neomenioidea with paired coelomopores; (B) Neomenioidea with ventromedially fused coelomoducts and a single coelomopore; (C) Chaetodermatoidea, female. All figures represent dorsal views with anterior toward the top of the page. Numbers, shapes, and positions of seminal receptacles and seminal vesicles are composite and are not intended to represent the condition in any known species. CD, Coelomoduct; CP, coelomopore; FGE, columnar glandular epithelium of the female mantle cavity; G, gonads; GpD, gonopericardial ducts; Pc, pericardium; SR, seminal receptacles; SV, seminal vesicles. (C styled after Hoffman, 1949.)

1.2 Asexual Reproduction

There are no reports of asexual propagation occurring in the Aplacophora. Baba's (1940) observations on regeneration of the posterior
end of *Epimenia verrucosa* may point toward a species with a potential
for vegetative reproduction.

1.3 Sexual Reproduction

1.3.1 Sexual Dimorphism and Hermaphroditism

All neomenioid aplacophorans are hermaphroditic, precluding sexual
dimorphism. A slight delay in the ripening of ova, noted by Salvini-
Plawen (1971), is probably not true protandry. Schwabl (1961b) reported protandry in *Plathymenia branchiosa*. Her evidence was based
on a single animal whose gonads and pericardium were filled with developing oocytes, while spermatozoa were found only in the most caudal
gonadal sacculations.

In the Chaetodermatoidea sexes are separate, but gross external morphological sexual distinctions are lacking. Salvini-Plawen (1972a) and
Scheltema (1973) reported sexually related differences in the epithelium
surrounding the coelomoduct (gonoduct) openings into the mantle
cavity. In *Falcidens* and *Chaetoderma*, the epithelium around the female
coelomopore is built of tall glandular cells (often called a shell gland),
while in males the coelomopore opens as a simple slit or is raised on a
papilla.

The paired, hermaphroditic gonads of neomenioid aplacophorans each
contain oogenic and spermatogenic tissues. Ovaries usually comprise the
medial walls and testes the lateral walls of the tubular gonads (see
Fig. 3D). Lummel (1930) noted that gonads of *Hypomenia nierstraszi*
were longitudinally differentiated with the female elements located
anteriorly and the male elements posteriorly. In contrast, the anterior
portions of the gonadal tubules of *Micromenia fodiens* contain only
spermatozoan forming elements, while the remainder contain both
oogonia and spermatogonia (Schwabl, 1955).

A specimen of *Chaetoderma nitidulum* described by Leloup (1950)
as hermaphroditic was reexamined by Salvini-Plawen (1972b) and found
to be purely male. Salvini-Plawen suggested that Leloup mistook the
digestive gland for an ovary.

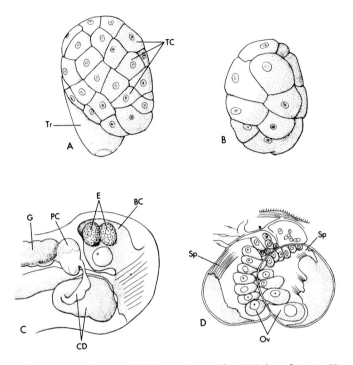

Fig. 3. (A and B) Embryos of *Halomenia gravida;* (C) brooding in *H. gravida;* (D) diagrammatic representation of a gonadal cross section of *Proneomenia sopita.* BC, Brood chamber; CD, coelomoduct; E, embryos; G, gonads; Ov, oocytes; PC, pericardium; Sp, spermatids; TC, test cells; Tr, trunk. (A, B, C, after Heath, 1918; D, after Pruvot, 1891.)

1.3.2 Anatomy of the Reproductive System

The gonads of aplacophorans are tubular organs occupying the medial dorsal portion of the body above the gut throughout much of the length of the animal. Gonads are reported to be paired in all neomenioids except, perhaps, *Nematomenia banyulensis norvegica,* which Odhner (1921) described as having a single gonad. Chaetodermatoids are reported by some reviewers (e.g., Fischer-Piette and Franc, 1960) to have a single, middorsal gonad. However, Heath (1905) described paired gonads in *Limifossor talpoides,* and Heath (1911) and Scheltema (1973) noted that the gonad in juveniles of *Chaetoderma* exists as paired tubes.

Whether the gonad is singular or paired, two gonopericardial ducts arise from its posterior end and run posteriorly to open into the anterior end of the pericardial cavity (Fig. 2), with but two exceptions. In

Phyllomenia austrina gonoducts open directly into the pallial cavity and lack connection with the pericardium (Salvini-Plawen, 1970b). Schwabl (1963) reported a single gonopericardial duct in *Prochaetoderma californicum*.

Terminology for parts of the reproductive tract of the Aplacophora remains in a confused state. According to Goodrich (1895), we are seeing in these animals retention of the most primitive molluscan condition; paired coeloms give rise to bilaterally symmetrical, serially connected (from anterior to posterior) gonads, pericardium, and kidneys. The bilateral cavities are fused at the pericardium into a single medial chamber. The "kidneys" forming, as they are assumed to do, from mesoderm, are not nephridial but are enlarged coelomoducts. No certain renal function has been demonstrated for the ducts connecting the pericardium to the mantle cavity in aplacophorans, and diverticula of the ducts which serve for storage of endogenous and exogenous sperm indicate a clear role in reproduction. The swollen glandular terminal portions of the ducts in the neomenioids have been assumed to be shell glands, and this name is well entrenched in the literature. The same term has been applied to the female chaetodermatoid mantle epithelium. I shall mostly follow Scheltema (1973) and utilize terms as follows: ducts from the gonad to the pericardium are *gonopericardial ducts*; ducts from the pericardium to the pallial cavity are *coelomoducts*. The openings of the coelomoducts into the mantle cavity are thus coelomopores.

The paired coelomoducts arise ventrally and laterally from the pericardium. Scheltema (1973) has detailed the configuration of the pericardium of *Chaetoderma* and illustrated extensive paired, ventrolateral extensions of the pericardium from which the coelomoducts arise on the ventral, inner walls. The coelomoducts usually consist of two segments: (1) an anteriorly directed tubule, which is usually simple in structure but may be glandular and in neomenioids may bear small, saccular seminal vesicles; and (2) a posteriorly directed segment, which is glandular and in neomenioids is partly or entirely swollen by high, columnar secretory cells, the so-called shell gland. In many chaetodermatoids as well, the posterior segment of the coelomoduct is enlarged. In neomenioids, seminal receptacles, from one to dozens, open into the coelomoduct at the sharp angle where its anteriorly directed segment joins the posteriorly directed portion (Fig. 2A and B). Seminal vesicles and seminal receptacles are absent in chaetodermatoids.

Hoffman (1949), and after him Salvini-Plawen (1972a), suggested that because of histological similarity homologous parts of the reproductive system in neomenioids and chaetodermatoids are, respectively, the shell gland and the open glandular gutter or funnel into which the coelomoducts open. However, Salvini-Plawen (1972a) and Scheltema (1973) have noted that the glandular epithelium around the coelomopore of chaetodermatoids occurs *only* in females. Furthermore, in both sexes of apparently all chaetodermatoids the enlarged, posteriorly directed segment of the enclosed coelomoduct appears structurally perfectly homologous to the same region of the neomenioid coelomoduct system. The histology of parts of the coelomoduct in the two aplacophoran groups is too plastic (e.g., see Heath, 1911) to be utilized, as did Hoffman (1949) and Salvini-Plawen (1972a), in a rigid interpretation of homology. The glandular, ciliated funnel in the mantle cavity of chaetodermatoids should be referred to by Scheltema's name, "the columnar glandular epithelium of the female mantle cavity."

The function of the glands of the coelomoducts is not clear. Pruvot (1890) maintained that eggs were naked above, but invested with a thin shell below, the glandular region in *Nematomenia banyulensis*; hence shell gland. Heath (1918), however, noted a thin "cuticle" surrounding zygotes and embryos in the brood chamber of *Halomenia gravida* which was also present around ovarian oocytes. Heath thus defined the layer as a "vitelline membrane," as opposed to a "chorion." This being the case, no "shell" is added to zygotes passing through the shell gland of *H. gravida*. Obviously, the precise function of the coelomoduct glands has yet to be elucidated. Heath's (1911, pp. 54, 71) evidence for excretory function, based on histological and cytological investigations, should not be totally discounted.

The coelomoducts open individually into the mantle cavity in all Chaetodermatoidea and some Neomenioidea (Fig. 2A and C). In other neomenioids, however, the terminal portions of the ducts are fused and thus open together into the mantle cavity via a common coelomopore (Fig. 2B).

Variation in the specific structure of the coelomoducts and other organs probably related to reproductive functions and found in the mantle cavity is great in neomenioids. An exhaustive review of these structures is not within the framework of the present summary. So-called copulatory spicules and associated sacs and musculature are found in the mantle cavities of many different neomenioids (Fig. 1D) (Fischer-Piette and Franc, 1960; Hyman, 1967). Special saccular diverticula of the mantle cavity are found in a few neomenioids; Heath (1911) referred to these sacs as "cloacal coecums," although he had no notion as to their function. Heath noted that secretory activity of the coecal epithelia appeared to correlate with increased reproductive activity (gonadal ripeness, etc.). The brood chamber of *Halomenia gravida* is illustrated in Fig. 3C.

1.3.3 Origin of the Germ Cells

Owing to the fact that there are no complete ontogenies, there is no direct evidence on the source of germ cells in any aplacophoran.

1.3.4 Cytodifferentiation of the Gametes

Spermatogenesis has been partially described for *Chaetoderma nitidulum, Neomenia carinata,* and *Nematomenia* sp. by Franzén (1955) and for *Micromenia fodiens* by Schwabl (1955). The processes of spermiogenesis in the Aplacophora involve no unique features. Spermatozoa of chaetodermatoids have a bulbous head, and Retzius (1904) drew a sperm of *C. nitidulum* with a long, curving anterior process similar to that found in chiton spermatozoa (see Section 2.3.5.2). Spermatozoa of neomenioids are slender, elongate, and pointed (Salvini-Plawen, 1972a). The midpiece of *Chaetoderma* spermatozoa is very short and consists of a small number of spherical mitochondria; the neomenioid midpiece is elongate. Franzén has noted that these two spermatozoan types correlate with the probable modes of fertilization practiced by the species possessing them: the short-headed *Chaetoderma* type is primitive and functions in distance swimming for external fertilization; the slender *Neomenia* type of sperm is considered more advanced and is related to copulation and internal fertilization.

Oogenesis has apparently not been extensively investigated in any aplacophoran. Many of the older descriptive anatomies contain figures of the gonad with developing eggs (Pruvot, 1891; Wirén, 1892a,b; Heath, 1911) (Fig. 3D). To this day, the illustrations of gamete formation provided by Wirén (1892a,b) remain the best for the Aplacophora. Wirén's drawings indicate that follicle cells are intimately associated with growing oocytes of *Chaetoderma nitidulum*. Interestingly, Wirén mistakenly assumed neomenioids to be dioecious. Schwabl (1955) provided an extended written description of oogenesis in *Micromenia fodiens*; oocytes are apparently proliferated only in the midregion of the gonad and grow larger as they are moved posteriorly.

1.3.5 Gametogenic Cycles within Populations

Baba (1951) reported that the breeding season of *Epimenia verrucosa* is from June to October at Amakusa, Japan. This is the only aplacophoran species for which a restricted breeding season has been noted. For

TABLE I

APLACOPHORAN REPRODUCTIVE PERIODS

Order, genus, and species	Ripe or reproductive season	Place	Evidence	Reference
Chaetodermatoidea				
Chaetoderma erudita	Summer	Alaska	Spawned sperm	Heath (1911)
Chaetoderma nitidulum	July–August	Sweden	Mature sperm	Franzén (1955)
Neomenioidea				
Epimenia verrucosa	June– October	Amakusa, Japan	Breeding	Baba (1951)
Nematomenia banyulensis	October	Mediterranean, France	Spawned	Pruvot (1890)
Nematomenia sp.	July–August	Sweden	Mature sperm	Franzén (1955)
Neomenia carinata	November– February	England	Spawned in laboratory	Thompson (1960)
Neomenia carinata	August	Sweden	Mature sperm	Franzén (1955)
Strophomenia lacazei	June	Mediterranean, France	Ripe	Pruvot (1899)
Stylomenia salvatori	June	Mediterranean, France	Ripe	Pruvot (1899)

others we have only scattered bits of information indicating when preserved animals were collected together with some notation on their reproductive states.

Thompson (1960) reported that a specimen of *Neomenia carinata* collected in November deposited developing eggs late the following February. If, as Thompson assumed, the animal did not fertilize its own eggs, copulation must have occurred before the animal was collected in November. Taken together, these facts give an extended midwinter reproductive season for *N. carinata* in the Irish Sea. However, there are no data to substantiate an assumption that breeding does not occur continuously. The same is true of other aplacophorans for which developmental information is available.

Table I lists species for which some knowledge of sexual state and season exists.

1.3.6 Factors Influencing Gametogenesis

There is no information on this topic.

1.3.7 Reproductive Behavior and Spawning

Without actual observations, the assumption exists that members of the Chaetodermatoidea spawn their gametes free into seawater, where fertilization occurs. This assumption is based on the following: (1) the sexes are separate, and females lack seminal receptacles or other sperm-storing organs such as are prominent in neomenioids; and (2) according to Franzén (1955), the "primitive" type of spermatozoa of *Chaetoderma nitidulum* is typically found in molluscs where fertilization occurs externally.

In the Neomenioidea much complex reproductive anatomy, including muscularized spicules and seminal receptacles, is best understood by the assumption that copulation and internal fertilization are universal. Heath (1918) interpreted animals in his collections as having been preserved *in copula*. Findings by Heath (1911, 1918), Thiele (1913), Baba (1940), and Salvini-Plawen (personal communication) of complete or partial brooding of embryos and larvae in five neomenioid species further substantiate the notion. The eggs produced by *Nematomenia banyulensis* and studied by Pruvot (1890) were already fertilized; this was also true of eggs which Baba (1940) and Thompson (1960) found being laid, respectively, by *Epimenia verrucosa* and *Neomenia carinata*.

Species of *Pruvotina* studied by Thiele (1913) and Salvini-Plawen (personal communication) brood their embryos completely through to juvenile stages. *Nematomenia banyulensis*, and apparently *Rhopalomenia aglaopheniae*, spawn their fertilized eggs into the surrounding sea where development, via a more-or-less well-defined larval stage, ensues (Pruvot, 1890, 1892). This is probably also the pattern in *N. carinata* (Thompson, 1960). *Halomenia gravida* broods its embryos throughout much, if not all, of their development (Heath, 1918), and *E. verrucosa* apparently broods its embryos through the earlier period of development (Baba, 1940, 1951).

Baba described two methods of spawning in *E. verrucosa*. In the first (Baba, 1938, 1940, 1951), fertilized but uncleaved eggs, individually encapsulated, passed from the mantle cavity in two mucous masses, each containing 65–90 eggs. Embryos so produced did not survive through metamorphosis, suggesting that such spawning was premature. The second method (Baba, 1940) consisted of brooding the fertilized eggs through early development and releasing them as trochophores or metamorphosed juveniles among which survival was good.

Thompson (1960) found on several occasions that an aquarium specimen of *N. carinata* spawned during the night (as Baba, 1940, found for *E. verrucosa*) and produced two piles of eggs per spawning. The eggs, in advanced cleavage stages, were individually encapsulated and ". . . free of each other and the substratum."

1.4 Development

Very little is known of developmental patterns in the Aplacophora. Brief to extended descriptions are available for parts of the embryology of eight neomenioid species; nothing has been reported on the embryology of the Chaetodermatoidea.

1.4.1 Fertilization

Franzén (1955), on the basis of sperm morphology, suggested that fertilization is external in chaetodermatoids. In contrast, all available evidence indicates internal fertilization in neomenioids: copulation has probably been seen (Heath, 1918); numerous sperm-storing organs are present (see Section 1.3.2); and fertilized eggs have been seen in the process of being laid (Pruvot, 1890, 1892; Baba, 1940; Thompson, 1960) or found brooded in the mantle cavity (Heath, 1911; Thiele, 1913; Salvini-Plawen, personal communication). As judged from observations that zygotes lying in the mantle cavity (Heath, 1918) or just spawned free (Baba, 1940) are undergoing maturation divisions, it may be assumed that the oocytes pass out of the coelomoducts very soon after they are inseminated.

Maturation of neomenioid oocytes occurs after fertilization (10–15 min after egg deposition in *Epimenia verrucosa*, according to Baba, 1951), and the production of two polar bodies from the surface of the zygotes indicates that they are primary oocytes when fertilized. Two polar bodies were also seen by Pruvot (1890) next to zygotes of *Nematomenia banyulensis*, while Heath (1918) pointed out that zygotes and embryos of *Halomenia gravida* possessed but one.

connect the folds to the dorsal aorta as in the ovary of some species. The spermatogenic cells form layers on the folds of varying thickness and continuity, depending on the stage of gametogenesis. The innermost spermatogonial layer is overlaid with a spermatocyte layer, then a spermatid layer, and finally an epithelium. The cells of the covering epithelium, probably equivalent to the follicle cells in the ovary, change from a cuboidal to a thin, flat shape as the spermatogenic cells proliferate. Sperms accumulate in the lumen of the testis.

A pair of gonoducts arises from near the dorsoposterior end of the gonad. Each gonoduct extends along the anterior surface of the pericardium and then curves ventrolaterally to open into the pallial groove, at a level one to nine gills anterior to the excretory pore, near the boundary between the sixth and seventh shell valves. Risbec (1946) noted an anomalous arrangement in a specimen of *P. (Aerilamma) primordia* which had both gonoducts opening on the left side and none on the right.

The female gonoducts, or oviducts, begin as simple ducts lined with a ciliated epithelium, but after a short distance they are lined with elongated secretory cells. Moreover, in some species, e.g., *Craspedochiton* (=*Loboplax*) *violaceus* (Plate, 1901), glandular diverticulums arise from the proximal portions of the oviduct; these are referred to as "slime sacs" by Hyman (1967). The glandular portion of the oviducts probably produces much of the mucus that envelops the eggs when they are spawned (see Section 2.3.8). Lawrence *et al.* (1965) showed that the size of the oviducts (shell glands) varied seasonally in *Cryptochiton stelleri* and was greatest when the gonads were largest, just before spawning.

The male gonoducts, or sperm ducts, are simple ducts lined with a ciliated epithelium continuous with that of the dorsal surface of the testis without secretory cells (Matthews, 1956), but they often contain yellowish-brown globules (Hyman, 1967). Sometimes the ducts are enlarged, either proximally or terminally, as "seminal vesicles," although they have not been reported to store sperms.

2.3.4 Origin of the Gonads and Germ Cells

The gonads appear rather late in development, after settling, either as a single rod of cells (Hammarsten and Runnström, 1924, 1925) or as a pair of solid rods of cells (Higley and Heath, 1912) extending anteriorly from the pericardium (Fig. 3). When paired, the rods soon fuse into one structure. Gonoducts extend laterally to the pallial groove without any connection to the pericardium or kidneys. Presumably a coelomic space develops within the gonad primordium and the inner

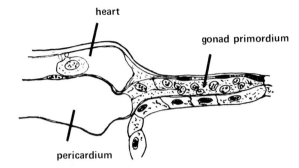

FIG. 3. Drawing of a section through the unpaired genital primordium in a juvenile of *Acanthochitona discrepans* (Brown). (From Hammarsten and Runnström, 1925.)

lining of cells forms the germinal epithelium, but this is yet to be established.

2.3.5 Gametogenesis

2.3.5.1 OOGENESIS

Light microscopic analyses of oogenesis in chitons were reported by Haller (1882), Garnault (1888), Pelseneer (1898), Schweikart (1940a,b), Lyngnes (1924), Knorre (1925), Stephenson (1934), and Matthews (1956). Histochemical and electron microscopic analyses include those of Gabe and Prenant (1949), Cowden (1961), Donato (1963), Nimitz and Giese (1964), Davenport and Davenport (1965), Selwood (1968, 1970), Anderson (1969), Richter and Götting (1974), and Richter (1976).

In *C. septentriones*, Selwood (1968) has described primordial germ cells that differentiate into oogonia within the germinal epithelium. According to Selwood, each oogonium undergoes four mitotic divisions, and of the resulting cells, those adjacent to the blood sinuses of the ovarian folds become primary oocytes, while the overlying cells become follicle cells. Anderson (1969) also identified oogonia ("stage I cells") along the ovarian wall of *M. muscosa*.

The oocytes can be divided into previtellogenic, vitellogenic, and full-grown phases (Fig. 4). Nuclear activity is most intense during the previtellogenic phase when ribosomal RNA is synthesized and accumulated in the cytoplasm, which becomes very basophilic. Basic proteins associated with cytoplasmic RNA are particularly prominent in previtellogenic oocytes (Davenport and Davenport, 1965). Cowden (1961)

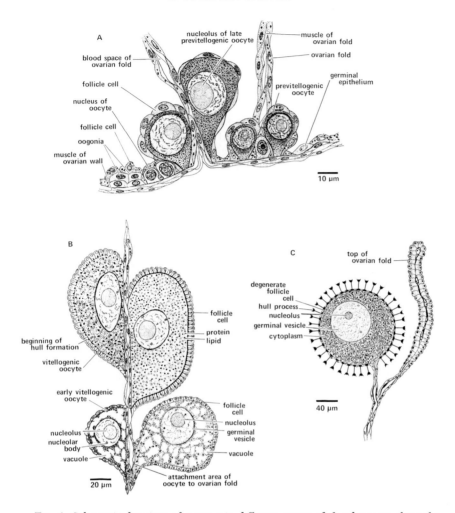

FIG. 4. Schematic drawings of oocytes in different stages of development along the ovarian folds of *Chiton septentriones* (Ashby). (A) Previtellogenic oocytes near the base of the ovarian folds. (B) Vitellogenic oocytes along the middle portion of an ovarian fold. (C) Full-grown oocyte near the top of an ovarian fold. Note different scale bars. (From Selwood, 1968.)

showed that, with completion of the previtellogenic stage, RNA synthesis ceased and did not begin again until after the trochophore larva developed (however, see Section 2.3.7.7). In a cytometric analysis of growing oocytes of *Chiton olivaceus*, Donato (1963) showed that changes in the volume ratios of the nucleus, nucleolus, and cytoplasm were similar to those in other animals. Eddy (1975) illustrated two forms of the

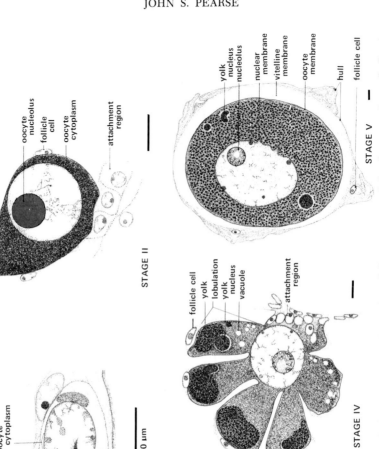

Fig. 5. Schematic drawings of oocytes of *Lepidochitona cinereus* (Linné) in different stages of development. Stage I: Previtellogenic oocytes embedded in the germinal epithelium. Stage II. Previtellogenic oocyte protruding from the germinal epithelium. Stage III; Previtellogenic oocyte with yolk nucleus and vacuoles. Stage IV: Vitellogenic oocyte showing characteristic lobulations. Stage V: Full-grown oocyte free in the ovarian lumen and surrounded by the hull. All scale bars 10 μm; note the different sizes. (From Richter and Götting, 1974).

"nuage," clumps of electron-dense granules found in germ cells of many animals, in an oocyte of *K. tunicata.*

Richter and Götting (1974) and Richter (1976) have divided the previtellogenic phase into three stages (Fig. 5). In stage I, the single nucleolus is formed and nucleolar material passes through the nuclear envelope. In stage II, the nucleolus becomes vacuolated and yolk nuclei (Balbiani bodies), consisting of ribosomes surrounded by mitochondria, form in the cytoplasm. The oocyte is usually pear-shaped in stage III, with an elongated stalk attached to the ovarian wall, and large systems of endoplasmic reticulums, Golgi complexes, and vacuoles are produced in the cytoplasm. These vacuoles, called "areolae" by Gabe and Prenant (1949), mark the beginning of the vitellogenic phase.

The vitellogenic oocyte is often characterized by lobulations formed by infoldings of the cell membrane (e.g., Garnault, 1888; Matthews, 1956; Nimitz and Giese, 1964; Anderson, 1969; Barnes, 1972; Richter and Götting, 1974; Richter, 1976) (Figs. 5 and 6). However, other workers did not note these lobulations in the species they studied (e.g., Stephenson, 1934; Johns, 1960; Selwood, 1968). The dense masses of yolk nuclei disperse, and piles of annulate lamellas contact the cytoplasmic vacuoles, which become filled with mucopolysaccharides and acid lipid material. The source of the yolk appears to be mainly *de novo* synthesis (Richter and Götting, 1974), possibly by mitochondrion transformations (Richter, 1976). However, Nimitz and Giese (1964) noted a decrease in lipid and glycogenlike material in the follicle cells during vitellogenesis, and they attributed this reduction to a transfer of material. Moreover, Selwood (1968) suggested that material was transferred from the blood sinuses of the ovarian folds to growing oocytes through the attachment areas. Other organelles in growing oocytes include striated rootlets in connection with microtubules, and cortical granules in the peripheral ooplasm (Anderson, 1969; Richter, 1976).

Most full-grown oocytes range in size from about 150 to 300 μm (Table II) and are filled with yolk, which is usually pigmented dark green. They are surrounded by a hull ("chorion") of complex origin and organization, often with numerous folds and processes (Fig. 7), as described especially by Ihering (1878), Sabatier (1886), Schweikart (1904a,b), and Lyngnes (1924), and reviewed by Hoffmann (1929–1930). Risbec (1946) grouped the eggs of chitons into three categories according to the arrangement of the hull. In the first category, the hull is smooth and spherical (e.g., *Lepidopleurus asellus*, Christiansen, 1954); in the second it is made up of a number of cupules (e.g., *C. porosus*, Brewin, 1942; *P. primordia*, Risbec, 1946; *L. cinereus*, Lyngnes, 1924;

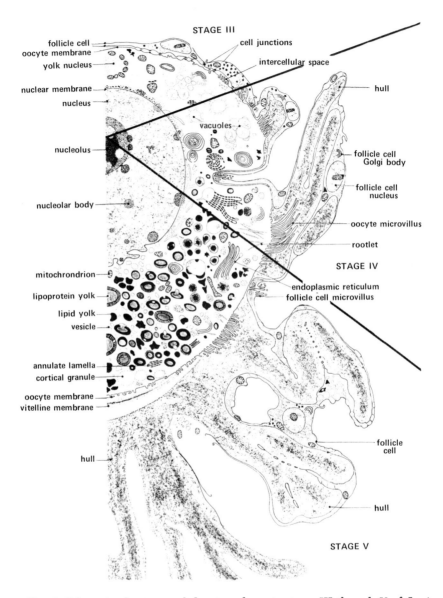

STAGE III

follicle cell
oocyte membrane
yolk nucleus

cell junctions

intercellular space

nuclear membrane

hull

nucleus

vacuoles

nucleolus

follicle cell
Golgi body

follicle cell
nucleus

nucleolar body

oocyte microvillus

rootlet

STAGE IV

mitochrondrion

endoplasmic reticulum
follicle cell microvillus

lipoprotein yolk

lipid yolk
vesicle

annulate lamella
cortical granule

oocyte membrane
vitelline membrane

follicle
cell

hull

hull

STAGE V

FIG. 6. Schematic ultrastructural drawing of oocyte stages III through V of *Lepidochitona cinereus* (Linné) showing accumulation of the yolk and development of the hull. The cytoplasmic organelles are disproportionately enlarged. (From Richter, 1976.)

TABLE II

Sizes Reported for Full-Grown Oocytes of Different Species of Chitons

Species	Diameter (μm)	Reference
Hanleyidae		
Hemiarthrum setulosum Dall [a]	600–800	Simpson (1977)
Ischnochitonidae		
Ischnochiton albus (Linné)	295	Lynges (1924)
Ischnochiton acomphus Hull and Risbec	196	Hull and Risbec (1930–1931)
Ischnochiton radians Carpenter	180	Thorpe (1962)
Lepidozona mertensii (Middendorff)	200	Thorpe (1962)
Stenoplax heathiana Berry	400	Heath (1899)
Lepidochitona cinereus (Linné) [a]	110–150	Richter and Götting (1974)
Tonicella lineata (Wood)	215	J. S. Pearse (unpublished data)
Tonicella marmoreus (Fabricius)	210	Lynges (1924)
Cyanoplax dentiens (Gould) [a]	200–250	J. S. Pearse and R. D. Lindberg (unpublished data)
Cyanoplax hartwegii (Carpenter)	170	J. S. Pearse (unpublished data)
Schizoplacidae		
Schizoplax brandtii (Middendorff) [a]	370–420	Kussakin (1960)
Callistoplacidae		
Middendorffia polii (Philippi) [a]	~200	Kowalevsky (1883)
Chaetopleuridae		
Chaetopleura apiculata (Say)	180–190	Grave (1932)
Mopaliidae		
Mopalia ciliata (Sowerby)	200–225	Barnawell (1951); Thorpe (1962)
Mopalia hindsii (Reeve)	225–240	Barnawell (1951)
Mopalia muscosa (Gould)	175–250	Barnawell (1951); Anderson (1969)
Mopalia imporcata (Carpenter)	180	Thorpe (1962)
Mopalia lignosa (Gould)	220–240	Barnawell (1951); Thorpe (1962)
Mopalia lowei Pilsbry	200	Thorpe (1962)
Mopalia porifera Pilsbry	170	Thorpe (1962)
Plaxiphora aurata (Spalowsky)	270	Simpson (1977)
Plaxiphora primordia (Hull)	240	Risbec (1946)
Katharina tunicata (Wood)	~175	Nimitz and Giese (1964)
Chitonidae		
Chiton granoradiatus Leloup [b]	195–240	Murti and Nagabhushanam (1968)
Chiton marmoratus Gmelin	200	Lewis (1960)
Chiton pelliserpentis Quoy and Gaimard	~120	Johns (1960)

TABLE II (*Continued*)

Species	Diameter (μm)	Reference
Chiton septentriones (Ashby) [c]	~150	Selwood (1968)
Chiton tuberculatus Linné	240	Lewis (1960)
Onithochiton lyelli (Sowerby)	150–175	Pearse (1978)
Acanthopleura gemmata (Blainville)	125	Stephenson (1934)
Acanthopleura granulata (Gmelin)	320	Lewis (1960)
Acanthopleura haddoni Winckworth	150–175	Pearse (1978)
Acanthochitonidae		
Acanthochitona curiosus (Iredale and Hull)	350	Risbec (1946)
Cryptoconchus porosus (Burrow)	400–600	Brewin (1942)
Cryptochiton stelleri (Middendorff)	250	Okuda (1947)

[a] Embryos reported to be brooded in the pallial groove; see Table VI.
[b] = *Chiton burmanus* Pilsbry *fide* Bullock (1972).
[c] = *Chiton pelliserpentis* Quoy and Gaimard *fide* Johns (1960) and Bullock (1972).

Matthews, 1956; Richter and Götting, 1974; *M. muscosa*, Fig. 7); and in the third it possesses numerous projections or spines (e.g., *Chaetopleura apiculata*, Grave, 1932; *Acanthopleura gemmata*, Stephenson, 1934; *Ischnochiton acomphus*, Hull and Risbec, 1930–1931; *Acanthochitona curiosus*, Risbec, 1946; *Schizoplax brandtii*, Kussakin, 1960; *Cyanoplax hartwegii*, Fig. 7). However, there is much variation in hull morphology among different species of chitons, and a thorough analysis of the various types remains wanting. The elaborate hull may function to hold the eggs to the substrate or together in clumps after they are spawned (Kowalevsky, 1883; Hull and Risbec, 1930–1931; Risbec, 1946; Kussakin, 1960; Lewis, 1960), or to hold them in the mucus that surrounds them (Heath, 1899).

Matthews (1956) distinguished two "membranes" in the egg hull of *L. cinereus* based on different responses to chemical treatments; both appeared to be mainly proteins with disulfide linkages, but the outer layer also was sensitive to acids and formanide, suggesting electrovalent linkages. Selwood (1970) described three layers in the hull of the eggs of *C. septentriones*: an inner mucopolysaccharide layer, an intermediate protein layer, and an outer lipid layer. Anderson (1969) believed that the hull of the eggs of *M. muscosa* was secreted by the follicle cells, while Richter (1976) proposed that all layers of the hull of eggs of *L. cinereus* might be secreted by microvilli of the oocyte (see Fig. 6) and that, in addition, a vitelline envelope is secreted by the oocyte just

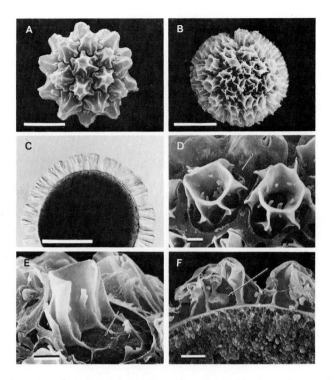

FIG. 7. Scanning electron micrographs and a Nomarski photomicrograph (C) of full-grown oocytes taken from the ovaries of chitons to show features of the hull. (A) *Cyanoplax hartwegii* (Carpenter). (B–F) *Mopalia muscosa* (Gould). (A–C) Whole full-grown oocytes without follicle cells. (D) Two hull cupules without follicle cells adjacent to a cupule covered with its follicle cell (arrow). (E) Partially broken cupule showing a portion of the hull in direct contact with the surface of the oocyte (arrow). (F) Portion of a broken oocyte showing yolk (below), hull, and covering of follicle cells with the nucleus of a follicle cell suspended in a hull cupule (arrow). Specimens from Pacific Grove, California. Scale bar: (A–C) 100 μm, (D–F) 10 μm.

before spawning. If the latter process is the case, the follicle cells serve mainly to shape the hull. In any case, the follicle cells degenerate near the completion of vitellogenesis and are sloughed off from the full-grown oocytes, which then are free in the ovarian lumen with their characteristic hulls.

Nimitz and Giese (1964) has found that oogenesis requires 2 years in *K. tunicata*, with the oocytes reaching 30–50 μm during the first year and about 175 μm by the end of the second year, when they are spawned. Other species reach sexual maturity within a year (see Table IV), and oogenesis must be completed in less time. Oocytes up to about 50 μm are present at all times in adult specimens of *Acantho-pleura haddoni*, and some of these grow to a full size of 150–175 μm over a period of 3 months, in the spring (Pearse, 1978; see also Fig. 9). In species with monthly or semimonthly spawning rhythms (see Table III and Section 2.3.8) the vitellogenic growth phase may be completed within a period of a month or less.

2.3.5.2 SPERMATOGENESIS

Events of spermatogenesis have been followed with light microscopy (e.g., Matthews, 1956; Johns, 1960; Nimitz, 1964; Nimitz and Giese, 1964; Barnes, 1972), but we have yet to benefit from a detailed electron microscopic analysis. In *C. pelliserpentis*, Johns (1960) identified large cells arranged in loose groups along the testicular folds as stem cells or primordial germ cells because they seemed to give rise to pear-shaped (7 × 20 μm) primary spermatogonia containing large nuclei (6.5–8 μm in diameter). The spermatogonia multiplied, thickening the surface of the folds; they first formed pockets and then continuous bands of primary spermatocytes with lightly staining nuclei. Secondary sper-

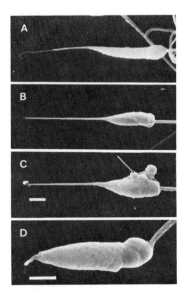

FIG. 8. Scanning electron micrographs of sperms taken from the testes of chitons to show the long anterior filament. (A) *Stenoplax heathiana* Berry. (B) *Ceratozona squalida* (C. B. Adams). (C) *Mopalia muscosa* (Gould); the plasmalemma has torn, revealing the spherical mitochrondria (arrow). (D) *Chaetopleura gemma* Dall. Specimens of *C. squalida* from Ft. Pierce, Florida; those of the other species from Pacific Grove, California. Scale bars: 1 μm; (A–C) are all at the same magnification.

matocytes, with densely staining nuclei (1.5–2 μm in diameter), resulted from division of the primary spermatocytes and differentiated into small spermatids (1 μm in diameter) with little cytoplasm. These overlaid the spermatocytes just under the epithelium lining the testicular lumen. Spermiogenesis was completed as the spermatids elongated into 1×3 μm triangular heads and developed a tail well over 10 μm long. The sperms migrated between the epithelial cells and accumulated within the testicular lumen.

Sperms of chitons are unusually pointed, and in many species they have a long, anterior filament (Ihering, 1878; Retzius, 1904, 1905; Franzén, 1955) (Fig. 8) similar to that found in some aplacophoran sperms (see Section 1.3.4). Rothschild and Tyler (1955) and Thompson (1973) referred to this anterior filament as an acrosomal filament or an acrosome, respectively. However, the sperms did not display an acrosome reaction during egg water tests by Thompson (1973). Moreover, no acrosome has been found in electron micrographs by A. Franzén (unpublished data) or by R. M. Woollacott and Pearse (unpublished data). Rather, the anterior filament appears to contain only an extension of the nucleus. There are four to six mitochondria and, in most species examined, these are placed eccentrically near the posterior end of the sperm head.

2.3.6 Reproductive Cycles

Most seasonal studies on chiton reproduction have followed changes in gonad size (gonad indexes) and/or in the microscopic appearance of fresh gonad smears. Because the gonad contains mainly gametogenic cells and gametes, and few accessory cells, changes in gonadal size are good indications of changes in reproductive stages. Gonads increase in size as they accumulate gametes and decrease, often precipitously, during spawning (Fig. 9). Himmelman (1976) showed that, in smaller animals, gonad size ws not proportional to animal size, so simple gonadal indexes (gonad weight times 100 divided by animal weight) can be misleading. However, workers usually have used large animals in which gonad size was proportional to animal size (e.g., Giese and Hart, 1967; Himmelman, 1975, 1978; Pearse, 1978). Histological analyses of chiton gonads over a period of a year or more have been relatively few (e.g., Johns, 1960; Nimitz, 1964; Barnes, 1972; Pearse, 1978). These studies all indicate that the reproductive condition of the gonads can be divided into four fairly distinct stages: (1) proliferation of gametogenic

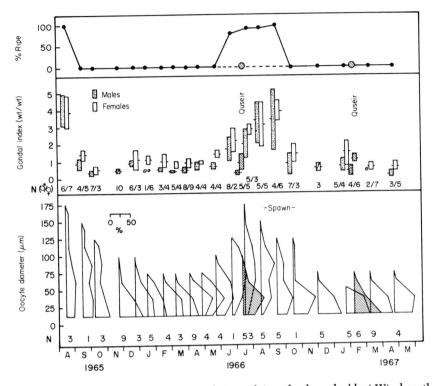

FIG. 9. Reproductive changes in populations of *Acanthopleura haddoni* Winckworth in the Gulf of Suez and northern Red Sea ("Quseir," hatched). Top: Percentages of animals containing numerous gametes. Middle: Means and standard deviations of the gonadal indexes. Bottom: Averaged oocyte size distributions. (From Pearse, 1978.)

cells, (2) vitellogenesis or accumulation of sperms, (3) ripe, filled with gametes, and (4) spawned, nearly empty of gametes.

Well-defined annual reproductive cycles with discrete spawning periods have been found in populations of most species of chitons that have been examined (Table III). Most chitons of the Pacific coast of North America spawn mainly during the coldest part of the year, in the late winter and spring. Of the 12 species examined to date, only *Mopalia muscosa*, which spawns throughout the year, and *M. hindsii*, which begins spawning in the fall in central California, deviate much from this pattern. Five species in Puerto Rico were examined by Glynn (1970), and all spawned mainly in the fall, during the warmest part of the year. Records from other parts of the world are too few to show any consistent pattern. However, it is clear that reproductive rhythms,

ERRATA

Giese/Pearse: *Reproduction of Marine Invertebrates,* Vol. V

Pages 47-48 as printed in this book are in error. See amended version below:

TABLE III

SPAWNING TIMES REPORTED FOR SPECIES OF CHITONS,
ARRANGED BY LATITUDE FROM NORTH TO SOUTH

Species and location	< Months of Year >	Reference
	J F M A M J J A S O N D	
Lepidopleurus asellus (Spengler)		
Oslo, Norway	————	Christiansen (1954)
Tonicella marmoreus (Fabr.)		
Oslo, Norway	– – – – – –	Lynges (1924)
Lepidochitona cinereus (Linné)		
Oslo, Norway	———	Thorson (1946); Christiansen (1954)
Helgoland, Germany	————	Hoffmann (1931)
List, Germany	————————	Richter and Götting (1974)
Netherlands	————————	Benthem Jutting (1936)
Easthaven, Scotland	——	Baxter and Jones (1978)
Whitstable, England	<u>Lunar</u> – –	Matthews (1956)
Arcachon, France	—– ··	Magne (1937)
Acanthochitona discrepans (Brown)		
Arcachon, France	—– ··	Magne (1937)
Adriatic Sea	– – – – – – – – – – – –	Leloup and Volz (1938)
Lepidopleurus cajetanus (Poli)		
Adriatic Sea	– – – – – – – – – – – –	Leloup and Volz (1938)
Middendorffia polii (Philippi)		
Adriatic Sea	——— – – – – – – ———	Leloup and Volz (1938)
Chiton olivaceus (Spengler)		
Adriatic Sea	– – – – – – – – – – – –	Leloup and Volz (1938)
Chaetopleura apiculata (Say)		
Cape Cod, Massachusetts	<u>Lunar</u>	Grave (1922)
Schizoplax brandtii (Middendorff)		
Kurile Islands	– – – – – – – – —	Kussakin (1960)
Tonicella insignis (Reeve)		
British Columbia	—————– – – –	Himmelman (1975, 1976, 1979)

TABLE III (*Continued*)

Species and location	< Months of Year >	Reference
	J F M A M J J A S O N D	
Tonicella lineata (Wood)		
British Columbia		Himmelman (1975, 1976, 1979)
Oregon		Barnes (1972)
Katharina tunicata (Wood)		
British Columbia		Himmelman (1976, 1978)
Central California		Giese (1969)
Mopalia hindsii (Reeve)		
British Columbia		Himmelman (1976)
Central California		Giese *et al.* (1959)
Mopalia ciliata (Sowerby)		
British Columbia		Himmelman (1976)
Mopalia laevior (Pilsbry)		
British Columbia		Himmelman (1976)
Mopalia lignosa (Gould)		
British Columbia		Himmelman (1976)
Mopalia muscosa (Gould)		
British Columbia		Himmelman (1976)
Central California		Barnawell (1951)
Southern California		Boolootian (1964); Monroe and Boolootian (1965)
Cryptochiton stelleri (Middendorff)		
Central California		Tucker and Giese (1962)
Hokkaido, Japan		Okuda (1947)
Cyanoplax dentiens (Gould)		
Central California		Heath (1907)
Nuttallina thomasi Pilsbry		
Central California		Heath (1905)
Stenoplax heathiana Berry		
Central California		Heath (1899)
Chiton tuberculatus Linné		
Bermuda		Arey and Crozier (1919)
Puerto Rico		Glynn (1970)
Barbados		Lewis (1960)
Chiton marmoratus Gmelin		
Puerto Rico		Glynn (1970)
Barbados		Lewis (1960)
Chiton squamosus Linné		
Puerto Rico		Glynn (1970)

TABLE III (*Continued*)

Species and location	J	F	M	A	M	J	J	A	S	O	N	D	Reference
Tonica schrammi (Shuttleworth)													
Puerto Rico								———					Glynn (1970)
Acanthopleura granulata (Gmelin)													
Puerto Rico	- - - - - -							_ _ _ _Lunar_ _ _			Glynn (1970)
Barbados										—————			Lewis (1960)
Panamá (Atlantic)	—————————Lunar—————————												Glynn (1970)
Acanthochitona hemphilli (Pilsbry.)													
Panamá (Atlantic)	———————————————————————												Glynn (1970)
Acanthopleura haddoni Winckworth													
Gulf of Suez							- - - —— - - -						Pearse (1978)
Onithochiton lyelli (Sowerby)													
Gulf of Suez	- - - - - - - - - Lunar? - - - - - - - - -												Pearse (1978)
Chiton granoradiatus Leloup [a]													
Eastern India			- - - - - - ——				_ _?_ _					Murti and Nagabhushanam (1968)
Acanthopleura gemmata (Blainville)													
Great Barrier Reef			—————				_ _ _Lunar_ _ _				Stephenson (1934)
Chiton septentriones (Ashby) [b]													
Southeast Australia					- - - ———							Selwood (1968)
Chiton pelliserpentis Quoy and Gaimard													
Southeast New Zealand	_ _ _Lunar_ _ _												Johns (1960)
Cryptoconchus porosus (Burrow)													
Southeast New Zealand						_ _Lunar_ _							Brewin (1942)
Plaxiphora aurata (Spalowsky)													
Macquarie Island			—— - - - - -						- - - - -		Simpson (1977)
Hemiarthrum setulosum Dall													
Macquarie Island	- - - - - ——— - - -												Simpson (1977)

[a] = *Chiton burmanus* Pilsbry *fide* Bullock (1972).
[b] = *Chiton pelliserpentis* Quoy and Gaimard *fide* Johns (1960) and Bullock (1972).

as determined from spawning times, may vary in the same species in
different localities as well as in the same locality in different years.
Glynn (1970), for example, has reported that *A. granulata* spawns
during the late summer and fall in Puerto Rico while it spawns
throughout the year on the Atlantic coast of Panama. Similarly, I found
that populations of *A. haddoni* in the northern Red Sea were not in
reproductive synchrony with those in the northern Gulf of Suez
(Pearse, 1978) (Fig. 9). Johns (1960) and Bullock (1972) argued that
C. septentriones of southeast Australia is synonymous with *C. pelliserpentis* of New Zealand, and these two forms spawn at different times.
Gonadal index records over a 10-year period in central California for
K. tunicata show considerable variation from year to year (Fig. 10).
The reasons for such variations are unclear, but Himmelman (1975,
1976, 1978) presented evidence suggesting that differences in the time
of the spring plankton bloom, which stimulates spawning in this species,
could lead to different spawning times (see Section 2.3.8).

A number of species of chitons display a monthly or semimonthly
spawning pattern within their annual rhythm (monthly: *A. gemmata*,

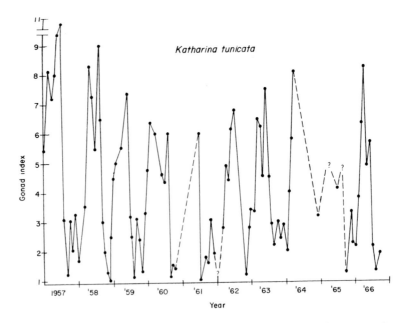

Fig. 10. Changes in mean gonadal indexes over a 10-year period in a population
of *Katharina tunicata* (Wood) in central California. Precipitous drops indicate periods
of spawning. (From Giese, 1969.)

C. apiculata, and *L. cinereus;* semimonthly: *A. granulata, C. pelliser-pentis, C. porosus,* and perhaps *Onithochiton lyelli;* see Table III for references). Histological analyses of gonads taken from chitons sampled sequentially during the month have not been done. However, these monthly or semimonthly rhythms may be mainly spawning rhythms superimposed on annual gametogenic rhythms, and they are discussed in more detail in Section 2.3.8. On the other hand, at least a portion of the gametogenic cycle may be completed during each monthly or semimonthly period. Monthly gametogenic cycles have been described, for example, for the sea urchin *Centrostephanus coronatus* (Pearse, 1972).

2.3.7 Factors Regulating Gametogenesis

2.3.7.1 Temperature

Although seasonal changes in sea temperatures appear to be important in regulating gametogenesis in many marine invertebrates (see review in Giese and Pearse, 1974), there is little information on their importance in chitons. In most chitons of the northeast Pacific gametogenesis occurs mainly during the late summer and fall, when sea temperatures are highest, and spawning occurs in the late winter and spring when they are lowest (see Table III), suggesting that higher sea temperatures may be important for initiating gametogenesis. Himmelman (1978) argues that gonadal growth in *K. tunicata* is most closely correlated with the decline in sea temperatures in the autumn, and in central California, when temperatures begin to decline later than in British Columbia, gonadal growth begins later. Moreover, Glynn (1970) has suggested that the absence of a well-defined seasonal reproductive cycle in Panamanian populations of *A. granulata* and its presence in Puerto Rican populations, is related to temperature fluctuation differences in the two areas.

2.3.7.2 Light

Because many chitons are intertidal, it might be expected that they use photoperiod as a clue for maintaining reproductive seasonality. Moreover, many chitons have photoreceptors (aesthetes), and some have well-developed eyes with lenses located in the shell valves (Boyle, 1969). However, most evidence is contrary to the notion that reproduction is synchronized by changes in photoperiod. Where followed for several years, the timing of changes in gonad sizes varies considerably from year to year (e.g., Giese, 1969; Himmelman, 1975, 1976, 1978, 1979) (see Fig. 10), suggesting that consistent influences such as photoperiod are of little or no importance in synchronizing reproduction.

There is evidence, however, that some chitons that spawn at a particular time of day or with a semimonthly or monthly rhythm are synchronized by light, including moonlight (see Section 2.3.8).

2.3.7.3 TIDES

Although there is evidence that spawning in some chitons is synchronized by tidal rhythms (see Section 2.3.8), there is no evidence that tidal rhythms influence gametogenesis.

2.3.7.4 PARASITES

A number of parasites have been reported in chitons (Johns, 1960; Hyman, 1967; Bullock and Boss, 1971) and, when the chitons are heavily infected, their reproductive activities might be adversely affected. However, I have found no references to specific effects of parasites on reproduction. Matthews (1956) mentioned that the blood vessels in the ovarian walls of L. cinereus sometimes contained numerous spores of the sporozoan parasite Minchinia chitonis (Lankester), but she did not indicate any effect of these parasites on reproduction. Baxter and Jones (1978) found that infection by these parasites led to enhanced growth in older individuals of L. cinereus, and infected animals reached much larger sizes than uninfected animals. The infected larger animals might have had larger gonads and higher fecundities than the uninfected smaller animals, or the enhanced somatic growth could have been at the expense of gonadal growth. These interesting alternative possibilities merit further study.

2.3.7.5 AGE

There are few data on the influence of age on reproduction, and most of them concern the age at which sexual maturity is reached (Table IV). These data suggest that chitons reach sexual maturity within 1–2 years. Most data also indicate that chitons live only a few years (see review in Hyman, 1967). Glynn (1970) showed that C. tuberculatus lived about 2 years, rather than 8 years as estimated by Crozier (1918a,b). Annual growth lines resulting from seasonal growth patterns were found in the shell valves of L. cinereus by Baxter and Jones (1978), and these lines showed that the animals rarely lived more than 5 years.

There is no evidence that chitons become reproductively senescent with age, although males of some species may live longer than females (see Section 2.3.2).

TABLE IV

AGE AND SIZE REPORTED FOR CHITONS AT FIRST SEXUAL MATURITY,
ARRANGED IN ORDER OF INCREASING SIZE

Species	Length (mm)	Age (yr)	Reference
Lepidochitona cinereus (Linné)	5–10	1	Matthews (1956); Baxter and Jones (1978)
Lepidopleurus asellus (Spengler)	6	1–2	Christiansen (1954)
Schizoplax brandtii (Middendorff)	6	∼ 2	Kussakin (1960)
Cyanoplax dentiens (Gould)	8–10	1	Heath (1905)
Nuttallina thomasi Pilsbry	8–10	1	Heath (1905)
Chaetopleura apiculata (Say)	10–12	1	Grave (1932)
Chiton tuberculatus Linné	32–41	1	Glynn (1970)
Katharina tunicata (Wood)	33–36	2	Heath (1905)
Acanthopleura granulata (Gmelin)	34–38	1	Glynn (1970)
Cryptochiton stelleri (Middendorff)	120–175	2	Heath (1905)

2.3.7.6 FOOD

Chitons generally are omnivorous grazers that feed on encrusting plants and animals. There is no direct evidence that fluctuations in food abundance or quality influence gametogenesis. Monroe and Boolootian (1965) suggested that a drop in gonadal size in a population of *M. muscosa* was due to disruption of feeding following the dumping of sand and rock in the area. Himmelman and Carefoot (1975) found that the feeding rate of individuals of *K. tunicata* increased during the period when the gonads were growing, from the fall through the spring. However, the algal caloric value also varied seasonally and was highest in the fall and lowest in the spring so that the energy intake was nearly constant in the animals throughout the year.

There is indirect evidence that nutrient stores may be related to reproduction. The size of the digestive gland in *K. tunicata* varies inversely with the size of the gonad (Lawrence *et al.*, 1965; see also Giese and Pearse, 1974, Fig. 11), although this relationship is not consistent from year to year (Giese and Hart, 1967). Such an inverse relationship suggests that nutrients are accumulated in the digestive gland and then transferred to the gonad during gametogenesis; perhaps some critical level of stores in the digestive gland is required before gametogenesis begins. This suggestion is supported by the finding of Nimitz and Giese (1964) that starvation experiments initiated at the beginning of gonad growth had little effect on the course of gametogenesis, even after 5 months. Moreover, the measurement of total lipids, carbohydrates, and proteins suggests that considerable buildup of nutrient materials occurs in the gonads (gametes) during gonad growth, probably at the expense of stores in other body components (see Section 2.3.7.7).

2.3.7.7 BIOCHEMICAL COMPOSITION

Giese and his associates followed the biochemical composition of chitons with particular reference to seasonal fluctuations that might be related to reproduction. Three species of chitons in central California were studied: *C. stelleri* by Tucker and Giese (1962), Lawrence *et al.* (1965), and Vasu and Giese (1966); *K. tunicata* by Giese *et al.* (1959), Giese and Araki (1962), Nimitz and Giese (1964), Lawrence *et al.* (1965), Giese and Hart (1967), and Lawrence and Giese (1969); and *M. hindsii* by Giese *et al.* (1959) and Giese and Araki (1962). Giese reviewed much of this work, along with new data, in 1966 and 1969 (Giese, 1966, 1969).

In the studies cited above, the body was divided into discrete "components" (i.e., shell, mantle, foot, gonads, digestive gland, gut, body fluid), and these were analyzed with respect to their size (wet weight) and chemical composition (protein, nonprotein nitrogen, lipid, carbohydrate, nucleic acid, ash, water). Component size and chemical composition were found to be relatively stable throughout the year except for the gonads and to a lesser extent the digestive glands, which tended to vary inversely in size with the gonads (e.g., Lawrence *et al.*, 1965).

Giese and his associates expressed the chemical composition of the various body components in terms of percentage of the total dry weight of the component, or "levels." When so expressed, the lipid levels of the gonads of *K. tunicata* and *M. hindsii* were found to be relatively constant throughout the year, with about 20% in the ovaries and 10% in the testes, as determined by chloroform–methanol extraction methods. Tucker and Giese (1962), using the Soxhlet method (which extracts neutral lipids but not structural lipids and phospholipids) also found relatively constant lipid levels in the testes of *C. stelleri*, with about 5–10% throughout the year, but lipid levels fluctuated from about 25–30% in ripe ovaries to about 10% in spent ovaries. Protein and carbohydrate levels in the gonads of all three species fluctuated inversely and seasonally, with high protein levels (30–60%) and low carbohydrate levels (3–6%) when the gonads were largest, just before spawning, and low protein levels (10–30%) and high carbohydrate levels (20–45%) when the gonads were smallest. Because percentage values are relative, these fluctuations are difficult to interpret; it is not readily apparent, for example, whether a decrease in the level of one substance reflects a decrease in the amount of that substance in the gonad or dilution by the increase of another substance. Moreover, because the gonads fluctuate in size, percentage values provide little insight into absolute changes during the year in the amounts of different materials present. To avoid these difficulties, the fluctuations in the absolute amounts of the major biochemical constituents in the ovary or testis of a 40-gm individual of *K. tunicata* are shown in Fig. 11, as calculated from the percentage values given by Giese and Hart (1967). Gonadal indexes in animals above 20 gm wet weight vary little with total animal size (Himmelman, 1976); consequently, the absolute values given in Fig. 11 are proportional to animal size (e.g., 20-gm animals would have about half the amounts of materials in their gonads as those calculated for the 40-gm animals).

The total dry weight of the gonads in *K. tunicata* increased between January and April, the main period of gamete accumulation, and dropped in April and May during spawning (Fig. 11). These changes,

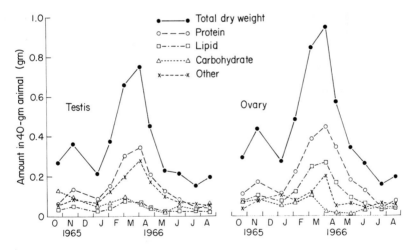

Fig. 11. Changes in the total dry weight and the amount of protein, lipid, carbohydrate, and other material in the gonads of 40-gm individuals of *Katharina tunicata* (Wood) in central California. Values calculated from data provided by Giese and Hart (1967) on gonadal indexes (sexes combined) and levels (percentages) of water, protein, lipid, and carbohydrate in the testes and ovaries. "Other" represents the difference between the total dry weight and the sum of the calculated protein, lipid, and carbohydrate weights.

calculated from Giese's data collected in 1965 and 1966, indicate one
rather abrupt spawning period. However, as shown in Fig. 10, there is
considerable year-to-year variation in this population, and in some
years spawning occurred two or three times. Giese and Hart (1967)
found no differences in the gonadal indexes of males and females, but
they did find that the testes contained more water than the ovaries
(means of 72 versus 67% wet weight). Consequently, the ovaries
generally were larger than the testes in terms of total dry weight.
From Fig. 11, it can be estimated that a 40-gm animal sheds about
600 mg dry weight in gametes during the year. With about 70% of the
gonads being water, this value is equivalent to about 2 gm total wet
weight, or about 5% of the animal's total wet weight. This estimate
compares favorably to Tucker and Giese's (1962) finding that individuals
of C. stelleri lost about 5–15% of their wet weight during spawning.
Because about 20% of the weight of K. tunicata is inorganic shell and
another 25% of the weight is body fluid (Giese, 1969), the total organic
wet weight of a 40-gm animal is about 22 gm, and the annual output
of gametes is about 9% of the total organic wet weight. This value may
be compared with Glynn's (1970) estimate, derived from different
techniques, that individuals of A. granulata and C. tuberculatus annually
release gametes equivalent to about 5–10% and 34% of their total ash-free
dry weight, respectively.

The increase in the total dry weight of the gonads in K. tunicata
was due mainly to an increase in the amount of protein (Fig. 11). In
contrast, the total amount of carbohydrate, which remained relatively
low throughout the year, declined in both the testes and ovaries during
gonad growth. Histochemical studies by Nimitz and Giese (1964) also
indicated that glycogen content in the gonads decreased during gametogenesis. Moreover, carbohydrate, as estimated by both histochemistry
(Nimitz and Giese, 1964) and chemical determinations (Giese and
Hart, 1967), decreased in other organs (foot, mantle, digestive gland)
during gametogenesis, suggesting its use for gamete production.

The amount of lipid in the ovaries of individuals of K. tunicata increased as the size of the ovary increased, presumably because lipid,
as well as protein, accumulated as energy stores in the eggs. The
amount of lipid in the testes also increased during the initial increase
in the size of the testes. Nimitz and Giese (1964) showed histo-
chemically that neutral lipid virtually disappeared from the gonadal
walls during gametogenesis in both males and females, and in females
neutral lipid accumulated in the eggs. Lawrence and Giese (1969)
found much higher proportions of neutral lipid in ripe ovaries of
K. tunicata than in ripe testes, and the level of neutral lipid declined

in the ovaries during spawning. Moreover, they found that the proportion of phospholipids increased as the size of the testes increased, perhaps as a result of synthesis of sperm cell membranes, as well as accumulated energy stores. These data suggest that lipid is an important synthetic material during spermatogenesis, as well as an energy store in the accumulated eggs and sperms.

In contrast with the data obtained for *K. tunicata*, as well as for *C. stelleri* and *M. hindsii*, Lawrence (1970) did not find any sex differences in either total or neutral lipid levels in *A. granulata* and *C. tuberculatus* sampled in Costa Rica. In both males and females of both species, lipid constituted about 13–14% of the dry weight, about half of which was neutral lipid. These levels are comparable to those in the testes of the California species. Lawrence (1970) was able to examine these tropical chitons only once during July, when their gonads were probably growing (Glynn, 1970), and seasonal variations may exist. Nevertheless, from his data it appears that the nutritional basis of these tropical species may differ from that of the temperate species studied.

A substantial portion of the total dry weight of the gonads (and other organs) of individuals of *K. tunicata* was not accounted for by determinations of the protein, lipid, and carbohydrate. This difference between the total dry weight of the gonad and the sum of the absolute amounts of protein, lipid, and carbohydrate is plotted as "other" in Fig. 11. In part, this material was inorganic ash, which had mean levels of about 5 and 10% of the dry weight of the ovaries and testes, respectively (Giese, 1969). Another portion of this material was nucleic acid; the mean level of nonprotein nitrogen in ripe and spent testes was 4.9 and 3.8%, respectively, while in ripe and spent ovaries it was 1.7 and 2.2% (Giese and Araki, 1962). The amount of DNA undoubtedly increased as sperms accumulated during growth of the testes. Similarly, the amount of RNA undoubtedly increased as oocytes accumulated in the ovaries, although from histochemical analyses, RNA accumulates mainly during the early phases of oogenesis, and not just before spawning as suggested in Fig. 11 (see Section 2.3.5.1).

2.3.7.8 ENDOCRINES

Martoja (1967) presented histochemical evidence for the presence of neurosecretory cells in the buccal ganglia of three species of chitons, *L. cinereus*, *C. olivaceus*, and *Acanthochitona discrepans*. Because of their association with the digestive system and apparent lack of change in relation to the gonads, she discounted their role in reproduction. However, Vicente (1970) and Vicente and Gasquet (1970), using histo-

chemistry and electron microscopy, found neurosecretory cells throughout the nervous systems of *C. olivaceus* and *A. discrepans,* and these
species seemed to have a cycle of neurosecretory products in correlation
with gonad changes. The cells showed high secretory activity and apparently released products directly into the blood. Perhaps they also
stimulated associated "juxtacommissural" cells to release products
into the blood. Such a system appeared to Vicente and Gasquet (1970)
to be similar to the neurosecretory system found in opisthobranchs, but
not as well organized, and they felt that experimental studies would
be valuable. In one such study, Watanabe and Cox (1975) unsuccess-
fully tried to induce spawning in *Mopalia lignosa* and *M. muscosa* by
injecting individuals with nerve extracts. Clearly, research in this
promising area has hardly begun.

2.3.8 Reproductive Behavior and Spawning

There is no evidence of sexual pairing during spawning in chitons.
However, Heath (1907) mentioned that small numbers of the hermaphroditic, brooding chiton *C. dentiens* were grouped together during the
breeding season. Okuda (1947) found large numbers of *C. stelleri* in the
Japanese intertidal during the spring and suggested that they were
"ascending from deep water for spawning." Such behavior was not noted
in this species in California by Tucker and Giese (1962), but D. R. Lindberg (unpublished observations) found *C. stelleri* in the intertidal on
Southeast Farallon Island, off California, only in late February and
early March, during the spawning season.
    Individuals of several species of chitons have been reported to
become active just before spawning in the laboratory (Metcalf, 1892;
Brewin, 1942; Barnes, 1972). Clark (1855), Brewin (1942), Nagabhushanam and Murti (1969), and Barnes (1972) noted that animals tended
to position themselves vertically in aquariums just before spawning.
Immediately before spawning the girdle arches upward to form a
spout at the posterior end, or on both sides near the posterior end,
and then the gametes are expelled in one or two streams (Fig. 12).
Thorpe (1962) saw variable behavior in the laboratory during the
spawning of several species of chitons with different animals remaining
stationary, moving intermittently, or moving continuously as the gametes
were expelled. He described egg laying as a two-stage process with
the eggs accumulating in the pallial groove before being expelled.
Chitons that brood their eggs merely retain them in the pallial groove
until the larvae or juveniles are released (see Section 2.4.4).

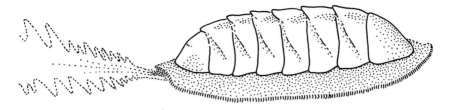

FIG. 12. Drawing of a spawning male specimen of *Lepidochitona cinereus* (Linné). (From Matthews, 1956.)

Spawning may last for more than an hour (e.g., Metcalf, 1892; Heath, 1905; Christiansen, 1954), and Thorpe (1962) found that, once started, spawning was difficult to interrupt in the laboratory. Four males of *M. lignosa* were seen by Watanabe and Cox (1975) to release sperms in spurts lasting 3–5 min at 5- to 15-min intervals, while three females released a steady stream of eggs. In *L. cinereus*, Matthews (1956) found that each spawning outburst lasted about 2.5 hr during which time 1500–2000 eggs were released. After such a spawning outburst many eggs remained in the ovary, so spawning probably could be repeated. Tucker and Giese (1962) reported that one female of *C. stelleri*, weighing 582 gm, spawned for more than a week in the laboratory, losing 30 gm in weight during that period. A male, weighing 461 gm, lost 72 gm during the course of spawning over 1 week. Individuals may spawn several times during the spawning period; Glynn (1970), for example, mentioned that one individual of *C. tuberculatus* spawned 6 times in the laboratory over a 17-day period.

Mucus, presumably produced in the oviducts, often surrounds the expelled eggs and holds them together as small clumps of a few eggs each (e.g., *K. tunicata*, Heath, 1905; *Mopalia ciliata*, Thorpe, 1962; *Tonicella lineata*, Barnes, 1972), as masses with several hundred eggs (e.g., *P. primordia*, Risbec, 1946; *L. asellus*, Christiansen, 1954), as coiled egg strings with 100,000–200,000 eggs (e.g., *S. heathiana*, Heath, 1899), or as long egg strings reaching nearly 1.5 m in length which must contain millions of eggs (e.g., *C. stelleri*, Okuda, 1947). In some species apparently little or no mucus coats the eggs. Hull and Risbec (1930–1931), for example, described the spawning of separate eggs by an individual of *I. acomphus* and claimed that the eggs stuck to objects with their hooked spines. Similarly, Kowalevsky (1883), Kussakin (1960), and Lewis (1960) reported that in the species they examined the eggs stuck together by the interlocking of spines on the hull of the eggs. Nagabhushanam and Murti (1969), moreover, saw females of *Chiton granoradiatus* (=*C. burmanus* Pilsbry *fide* Bullock, 1972) spawn

3000–4000 eggs at a time, and all the eggs remained loose and separate. Allan (1959) mentioned that a female of *Chiton* (*Amaurochiton*) *glaucus* freely spawned in captivity, covering itself and the bottom of the tank with about 4 million powdery eggs.

Matthews (1956) found that individuls of *L. cinereus* from different populations differed in egg laying in both the field and the laboratory. The eggs released by animals from southeast England (Whitstable), were encased in gelatinous strings which did not break up even after strong agitation of the water. In contrast, animals from several localities along the southwest English coast released eggs one by one with only a thin loosely bound covering of mucus. Similar differences have been noted by other authors working with *L. cinereus*. Clark (1855) observed an animal from Norfolk, England, expelling eggs and a stream of "flaky white matter like a fleecy cloud" which entangled the eggs and prevented them from being washed away. At Arcachon, France, the eggs were reported to be contained in a soft "nidosome" (Magne, 1937), on the west coast of Sweden they were laid in clusters of from 7 to 16 upon small stones (Lovén, 1855), while at Helgoland they were released separately (Hoffmann, 1931). According to Matthews (1956) these differences appear to be genetic and may reflect the effect of different selective regimes on the dispersal characteristics of the eggs and larvae.

When present, the mucus surrounding the egg masses usually does not last long. Okuda (1947) reported that the egg strings of *C. stelleri* were "soon" broken into fragments by waves; Heath (1899) found that the large egg masses of *S. heathiana* lasted only several days; and Thorpe (1962) said that the translucent material enclosing the eggs of *M. ciliata* dispersed within 12–15 hr. On the other hand, Matthews (1956) found that when the eggs of *L. cinereus* were encased in egg strings, they remain encased until hatching.

Kowalevsky (1879) reported that males spawned before females in the several species of chitons he examined. Metcalf (1892, 1893) also observed this with several species of chitons, as did Heath (1899, 1905). According to Heath (1905), when kept isolated in tide pools males but not females spawned, while when together both sexes spawned. He suggested that the males release substances in their seminal fluid that stimulate nearby females to spawn. Other species in which spawning by males was reported to precede, and perhaps stimulate, spawning by females include *L. cinereus* (Hoffmann, 1931), *C. stelleri* (Okuda, 1947), *L. asellus* (Christiansen, 1954), and *C. granoradiatus* (= *C. burmanus* Pilsbry *fide* Bullock, 1972) (Murti and Nagabhushanam, 1968; Nagabhushanam and Murti, 1969). Magne (1937) stimulated females of

L. cinereus and *A. discrepans* to spawn by adding suspensions of sperms from respective males; sperm suspensions of *A. discrepans* even provoked females of *L. cinereus* to spawn, although they were incapable of fertilizing the eggs. Himmelman (1975) also found that he could stimulate ripe individuals of *Tonicella insignis* and *T. lineata* to spawn with the addition of sperm suspensions. However, he also could stimulate them to spawn with plankton suspensions (see below), and Barnes (1972) saw females of *T. lineata* spawning in the laboratory in isolation from males. Grave (1932) reported for *Chaetopleura apiculata* that "when the sexes are placed together, the males are the first to respond and sperm may spread widely through the water before eggs begin to appear," but he also noted that "individuals spawn readily when isolated." Other authors have also noted females to spawn without being stimulated by males (Brewin, 1942; Matthews, 1956; Johns, 1960; Thorpe, 1962; Glynn, 1970). On the other hand, Crozier (1922) claimed that with *C. tuberculatus* "the presence of one or more neighboring females serves in some way to activate the discharge of sperm by males, the spermatic substances secondarily inducing the liberation of eggs." This claim has not been substantiated for any chiton. Nevertheless, the possibility of males stimulating females to spawn (or vice versa) seems to have selective advantage, by increasing synchronized spawning and fertilization, and deserves more critical examination.

Heath (1905) observed individuals of *K. tunicata, Lepidozona cooperi, L. mertensii,* and *M. muscosa* spawning in quiet tide pools and suggested that a period of quiet water was necessary for these species to spawn. Grave (1922) also found that quiet water seemed necessary for spawning of *C. apiculata* in the laboratory, and spawning was inhibited by agitated water. Okuda (1947) mentioned that *C. stelleri* spawned on a "calm day when low tides occur in the early morning," and Stephenson (1934) claimed that *A. gemmata* only spawned when there was a calm sea and mild weather. It is not clear from these observations, however, how quiet water is related to spawning, and spawning simply may be difficult to observe in agitated water. Moreover, in a series of experiments, Matthews (1956) found that individuals of *L. cinereus* spawned when the water was agitated by an electric stirrer for 2–3 hr, but not when kept still.

Grave (1922, 1932, 1937) successfully obtained large numbers of spawned gametes from individuals of *C. apiculata* by placing ripe animals in bowls with running seawater in the morning and then washing them free of sediment and placing them in bowls half-filled with water in the evening. When so treated, most of the animals spawned in the evening between 2000 and 2300. The spawning showed a monthly rhythm (see below) and was most intense on the second evening after the animals were collected. Christiansen (1954) and Cowden (1961) used Grave's techniques to obtain gametes from individuals of L. *asellus* and C. *tuberculatus*, respectively, with the animals also spawning in the evening. Spawning by chitons in the evening has been noted for other species as well (e.g., Magne, 1937; Johns, 1960; Thorpe, 1962; Glynn, 1970; Barnes, 1972). On the other hand, Brewin (1942) and Nagabhush-anam and Murti (1969) noted that most of the individuals of the species they observed spawning in the laboratory did so in the late morning, between 1000 and 1200. All these observations suggest that light may act as a spawning clue and that experimental light manipulations might be most rewarding. Such experiments may be difficult, however. Barnes (1972), for example, obtained spawned eggs and sperms from freshly collected individuals of T. *lineata* which were held in aquariums, but he could not stimulate animals to spawn by providing quiet water conditions, by alternating temperature or light, or by providing electrical shock. Similarly, Watanabe and Cox (1975) obtained spawned gametes from individuals of M. *lignosa* and M. *muscosa* held in stagnant bowls of slightly warmed seawater, but were unable to induce spawning consistently either by this method or by mechanical, electrical, or chemical (potassium chloride, nerve extracts) stimulation.

Himmelman (1975, 1976, 1979) showed with laboratory and field observations that individuals of T. *insignis* and T. *lineata* could be stimulated to spawn by substances in the spring plankton bloom. Individuals of these species did not spawn in the laboratory when those in the field spawned, but they did spawn when returned to the field or when freshly collected plankton was provided to them in the laboratory. Increases in the number of diatoms of one species, *Skeletonema costatum*, corresponded most closely to the spawning of these species. The larvae of chitons, including those of T. *insignis* and T. *lineata*, do not feed on plankton, nor do the juveniles of these species (Barnes, 1972). The spawning stimulus provided by the plankton therefore probably serves solely to synchronize spawning and to signal forthcoming summer conditions favorable for the juveniles. Three other species of chitons, K. *tunicata*, *Mopalia laevior*, and M. *ciliata*, also spawned during the time of the spring plankton bloom, and Himmelman (1976, 1978) suggested that they similarly may be stimulated to spawn by substances released by the plankton. The time of spawning of M. *hindsii*, M. *lignosa*, and M. *muscosa*, however, did not seem to be related to the plankton blooms.

Several species of chitons spawn with monthly or semimonthly rhythms (see Table III). Although it is not clear whether these rhythms are gametogenic or spawning rhythms, or both, it seems best to treat them as spawning rhythms, because most data are on spawning and because the monthly or semimonthly rhythms are superimposed on annual reproductive rhythms. Grave (1922), for example, reported that individuals of *C. apiculata* from Cape Cod spawned in the laboratory mainly near the time of the third lunar quarter during the summer; Stephenson (1934) observed that individuals of *A. gemmata* from the Great Barrier Reef spawned in bowls mainly on the night of the full moon during the spring and summer; and Matthews (1956) found that individuals of *L. cinereus* from Whitstable, England, spawned in the laboratory and field mainly between the time of the full moon and third lunar quarter in May. Barnes (1972) also presented data suggesting that *T. lineata* in Puget Sound spawned near the time of the full moon during the spring, but Himmelman (1976) argued that the variable spawning times he found made such synchronization unlikely for this species. Individuals of *O. lyelli* in the Gulf of Suez spawn mainly in the summer, and heavy spawning occurred near the time of an extreme spring tide on the new moon (Pearse, 1978).

Semimonthly spawning in the summer was found for individuals of *C. pelliserpentis* in New Zealand by Johns (1960; also in Knox, 1963). Spawning in the field seemed to correspond more closely to the rhythm of the spring tides than the moonlight rhythm, which was out of phase with the tides. Moreover, Johns (1960) saw spawning in the laboratory during evenings near the time of the high water level of the spring tides (Fig. 13), and he suggested that the spawning rhythm was related mainly to a tidal rather than a moonlight rhythm. Semimonthly spawning patterns with spawning in the laboratory near the phases of both the full and new moon, and the spring tides in the field, also have been reported for *C. porosus* in New Zealand (Brewin, 1942) and *A. granulata* in the Caribbean (Glynn, 1970). However, Brewin (1942) noted that individuals of *C. porosus* spawned in the middle of the day with little relation to the daily tidal cycle, and she did not believe that the spawning rhythm could be causally related to the tidal cycle. Similarly, Glynn (1970) found that individuals of *A. granulata* spawned most frequently in the early evening and not according to the daily tidal cycle, and he argued that the semimonthly spawning cycle correlated better with the moonlight cycle than the tidal cycle.

Spawning in relation to the daily tidal cycle also was proposed by Thorpe (1962), who saw laboratory spawning of over 30 individuals of *M. ciliata* collected at different times of the year. These spawnings were nearly always near the time of the lower high water level of the daily

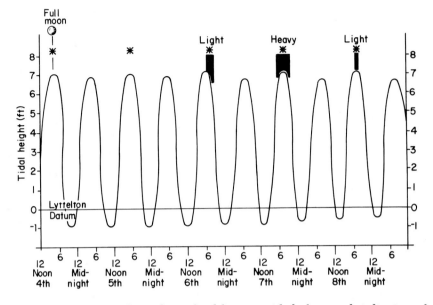

Fig. 13. Spawning observed in the laboratory (dark bars and indication of whether "heavy" or "light") and probable spawning times in the field (asterisks) of *Chiton pelliserpentis* Quoy and Gaimard in relation to the daily tidal cycle following the full moon of April 1958 in southern New Zealand. (From Johns, 1960.)

mixed tidal cycle of central California. However, most of these times were in the middle of the night, and the animals may have been responding to light clues. Moreover, these times were within 3–9 hours after the lower low water level when the animals were collected, and ripe animals may simply have spawned after several hours of stress. Similarly, D. Bonar (in Miller, 1977) found that the chitons he worked on tended to spawn near the time of the afternoon high tide several hours after they were collected on the low tide. Thorpe (1962) further noted that specimens of an undescribed species of *Mopalia* spawned mainly in the early morning, 11–12 hr after they were collected, and near the time of the higher low water level.

2.4 Development

Development has been described in detail for *Middendorffia polii* (=*M. caprearum* and *Nuttallina cinerea*, see Kass, 1957) by Kowalevsky (1879, 1882, 1883), for *S. heathiana* (as *Ischnochiton magdalenensis*) by

Heath (1899), and for *A. discrepans* by Hammarsten and Runnström (1924, 1925), and is well reviewed by Hoffmann (1929–1930) and Hyman (1967), as well as more briefly by Wada (1968). Descriptions are also available for the development of *L. cinereus* by Lovén (1855), Hoffmann (1931), and Matthews (1956); *Chiton olivaceus* by Kowalevsky (1883); *C. squamosus* (probably =*C. tuberculatus*, R. C. Bullock, unpublished data), and *C. marmoratus* by Metcalf (1892, 1893); *I. acomphus* by Hull and Risbec (1930–1931); *C. apiculata* by Grave (1932); *P. primordia* by Risbec (1946); *C. stelleri* by Okuda (1947); *L. asellus* by Christiansen (1954); *C. porosus* by Brewin (1958); *M. ciliata* by Thorpe (1962); *C. granoradiatus* (=*C. burmanus, fide* Bullock, 1972) by Nagabhushanam and Murti (1968); *T. lineata* by Barnes (1972); and *M. lignosa* and *M. muscosa* by Watanabe and Cox (1975). All these species have free-swimming larvae and, although individuals of *M. polii* and *L. cinereus* (rarely) brood their eggs, the embryos are released when they hatch as trochophores (see Section 2.4.4). Simpson (1977) briefly outlined the developmental stages of the embryos he found brooded by *Hemiarthrum setulosum*. The development of all chitons studied is remarkably similar and varies among species mainly in temporal details (Table V). Developmental rates may vary considerably even within the same species, depending on environmental conditions, especially temperature (e.g., Magne, 1937; Matthews, 1956; Costello and Henley, 1971; Barnes, 1972). There has been, however, virtually no experimental work on the development of chitons.

2.4.1 Fertilization

Fertilization occurs after the eggs are spawned, probably often while they are within the pallial groove. During the spawning of *C. tuberculatus*, for example, Crozier (1922) saw a "thick gelatinous stream" of sperms pass under the girdle of a female, and then numerous eggs emerged surrounded by the sperms (see also Section 2.3.8). Crozier (1922) also observed sperms to "agglutinate" when passing over the ctenidia of *C. tuberculatus*, but this phenomenon apparently has no bearing on fertilization (Southwick, 1939). The events at fertilization, however, have not been described in detail for any species of chiton. It is unclear how the sperms penetrate the egg hull, especially since they apparently lack an acrosome (see Section 2.3.5.2). How the eggs of the ovoviviparous *Callistochiton viviparus* are fertilized remains particularly enigmatic.

Miller (1977) reported that chitons have chemotactic sperms, as also found in hydroids and ascidians. Sperms of a variety of chitons, in-

TABLE V

DEVELOPMENTAL TIMES REPORTED FOR SPECIES OF CHITONS

Species	Time to reach developmental stages (hr)							Time in plankton (days)	Reference
	First cleavage	Gastrula	Proto-troch	Hatching	Larval eyes	Shell valves	Settlement		
Lepidopleuridae									
Lepidopleurus asellus (Spengler)	2	~12-14	~18-20	20-21	~114	~114	240-480	9-19	Christiansen (1954)
Ischnochitonidae									
Ischnochiton acomphus Hull and Risbec				~72	<72	<72	~72	0	Hull and Risbec (1930-1931)
Stenoplax heathiana Berry	2.2			168		96-120	168-171	<0.1	Heath (1899)
Lepidochitona cinereus (Linné)	1-2	4-5	8-12	12-96	74-100	74-100	80-110	0-4	Hoffmann (1931); Matthews (1956)
Tonicella lineata (Wood)	2	18	23-24	48	110	120-150	160	4.5	Barnes (1972)
Chaetopleuridae									
Chaetopleura apiculata (Say)	1.5-1.7	~13		25-36	<50	49-78	96-240	2.5-9	Grave (1932); Costello and Henley (1971)
Mopaliidae									
Mopalia ciliata (Sowerby)	1-1.5	10-12	18-24	36-42	~96-120	~96-120	120-192	3-6	Thorpe (1962)

Mopalia lignosa (Gould)	1–1.5	~10–11	~15	19	~48–60	156	130	4.5	Watanabe and Cox (1975)
Mopalia muscosa (Gould)	1–1.5	~10–11	~16	20	60	320	220–300	8–11	Watanabe and Cox (1975)
Plaxiphora primordia (Hull)	1.5		48	72	<72	72–76	~72	0	Risbec (1946)
Chitonidae									
Chiton granoradiatus Leloup [a]	0.2		7	8–8.5	68–70	93	116	3.5	Nagabhushanam and Murti (1968)
Acanthochitonidae									
Cryptoconchus porosus (Burrow)	3	22–28	33–48	60–72	>120	>120	>120	>3	Brewin (1958)
Cryptochiton stelleri (Middendorff)	2		18–20	70–120	<96	<96	82–140	0.5–3	Okuda (1947)

[a] = *Chiton burmanus* Pilsbry *fide* Bullock (1972).

cluding *C. stelleri, K. tunicata, M. ciliata, M. lignosa, M. muscosa,* and
T. lineata, were attracted to both seawater extracts (egg water) and
alcoholic extracts of eggs without agglutinating. Presumably this behavior aids sperms in locating eggs. However, unlike those of hydroids
and ascidians, chiton sperms showed a complete absence of species
specificity. Moreover, D. Bonar (personal communication) found that
hybrid larvae could be formed from gametes of *K. tunicata, M. muscosa,*
and *T. lineata,* although these hybrids have not been raised to maturity.

Anderson (1969) found cortical granulelike bodies in the eggs of
C. apiculata, as did Richter (1976) in eggs of *L. cinereus* (see Fig. 6),
but it is not known if the contents of these granules are released at
fertilization. However, a space forms between the egg hull and the
cell membrane during fertilization. This space was attributed to shrinkage of egg cytoplasm by Southwick (1939) and Barnes (1972).

Spawned eggs may remain viable for 24–40 hr after release (Graves,
1932). The germinal vesicle breaks down when the eggs are spawned.
Heath (1899) killed females of *S. heathiana* when they were spawning
and believed he could identify the first meiotic spindle in the eggs in the
ovary. Costello and Henley (1971) report that the egg of *C. apiculata*
is usually developing the first meiotic spindle when it is shed. Meiosis
is not completed, however, until after fertilization, and the resulting
polar bodies lodge between the egg membrane and the hull, marking,
the location of the animal pole.

2.4.2 Embryonic Development

Cleavage is holoblastic, nearly equal, spiral, and determinant. The
fate of each blastomere during the early development of *S. heathiana*
was followed in exquisite detail by Heath (1899). Heath commented that
the highly pigmented cells are particularly favorable material for
following cell lineages. The first three quartets of micromeres produce
the ectoderm and its derivatives (apical tuft and larval eyes, nervous
system, stomodeum, radula, foot glands, shell valves); the 4d cell
forms from one macromere, the endomesodemal cell, which produces
most or all of the mesodermal derivatives (muscles, heart, kidneys,
gonad); and the remaining fourth quartet of micromeres and macromeres form the digestive tube (pharynx, esophagus, stomach, intestine,
and their glandular diverticulums).

The macromeres invaginate during gastrulation followed by the endomesodermal cell (Figs. 14 and 15), which lodges near the blastopore

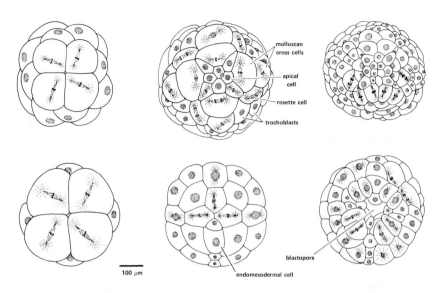

FIG. 14. Drawings of early cleavage in *Stenoplax heathiana* Berry showing development of the molluscan cross, the appearance of the endomesodermal cell (the 4d cell), and the beginning of gastrulation. Upper row: Anterior (animal) view. Lower row: Posterior (vegetal) view. (From Heath, 1899.)

and divides in two to form a pair of posteriolateral cells. Matthews (1956) also presented evidence of gastrulation by delamination. Endomesenchyme cells proliferate, partly filling the blastocoel. Although Kowalevsky (1883) figured the early formation of paired coelomic pouches in *M. polii*, neither Hammarsten and Runnström (1925) nor Matthews (1956) could detect the formation of coelomic pouches until after metamorphsis in *A. discrepans* or *L. cinereus*, respectively. The position of the blastopore shifts from the posterior pole anteriorly along the midventral surface, as a consequence of rapid proliferation of the posteriodorsal ectodermal cells, until it is located on the midventral surface (Fig. 15). It sinks inward in front of the stomodeal invagination and nearly closes.

At the anterior pole of the embryo, the micromeres form the tetraradiate molluscan cross alternating with four clusters of trochoblasts and capped with the apical and rosette cells, which form the annelidan cross (Fig. 14). The apical cells become the sensory cells bearing the apical tuft, while the trochoblasts form the prototroch, consisting of two rows of ciliated cells. The prototroch divides the embryo into pretrochal and posttrochal regions; the former with the apical tuft is

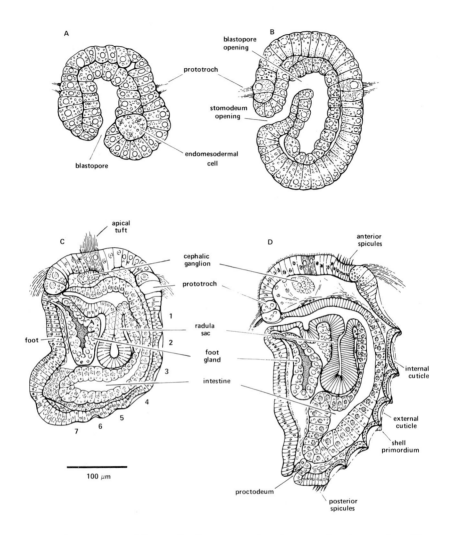

FIG. 15. Drawings of longitudinal sections of embryos and a larva of *Middendorffia polii* (Philippi) showing development of the main organ systems. (A) Late gastrula with prototroch and beginning of the ventral displacement of the blastopore. (B) Ventral displacement of the blastopore and stomodeal invagination complete. (C) Embryo near the time of hatching with apical tuft, cephalic ganglion, radula sac, foot gland, ventral foot, and dorsal spaces marking the position of the shell valves (numbered 1 to 7). (D) Larva near the time of settlement after loss of the apical tuft. (From Kowalevsky, 1883.)

mainly larval, while the latter will form the definitive adult. With the formation of the prototroch, the embryo moves around within the egg hull and is nearly ready to hatch.

2.4.3 Larval Development, Settlement, and Metamorphosis

The embryo hatches when the egg hull splits open by unknown means. The embryo pushes through the opening, using rapid body flexions until it can swim free. Most chiton embryos hatch as trochophores (or "pseudotrochophores"; see Salvini-Plawen, 1973), before the appearance of the foot, shell valves, or larval eyes, but in some species (e.g., *C. stel-*

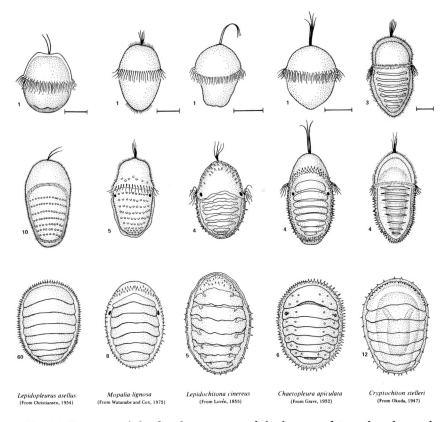

Lepidopleurus asellus
(From Christiansen, 1954)

Mopalia lignosa
(From Watanabe and Cox, 1975)

Lepidochitona cinereus
(From Lovén, 1855)

Chaetopleura apiculata
(From Grave, 1932)

Cryptochiton stelleri
(From Okuda, 1947)

FIG. 16. Drawings of the dorsal appearance of the larvae and juveniles of several species of chitons. Upper row: Larvae at hatching. Middle row: Larvae at settling. Bottom row: Juveniles shortly after settling. Numbers give age in days after fertilization. Scale bars: 100 μm.

leri, S. heathiana) these structures are well developed by the time of hatching (Fig. 16 and Table V). The embryos of some populations of *L. cinereus* hatch within a day after fertilization as trochophores which swim for 0.5–4 days before settling, while others, in eggs encased in gelatinous egg strings (see Section 2.3.8), hatch about 3 days after fertilization as creeping larvae nearly ready to settle (Matthews, 1956).

Pigmented cells form the two larval eyes located laterally just behind the prototroch. Their photosensitivity or function has not been demonstrated, but larvae are sensitive to light, being photopositive before settling and photonegative during and after settling (e.g., Okuda, 1947). Kowalevsky (1883) believed he could demonstrate nerve connections to the eyes which Matthews (1956) also found. Heath (1904) could not find such connections, however. The eyes gradually become indistinct after settlement, when they are covered by the expanding mantle and first or second shell valve, but they may be retained in the lining of the pallial groove well into adult life (Heath, 1904; Matthews, 1956).

As the larva develops, the dorsal and ventral surfaces become distinct (Fig. 15). The ventral surface forms the foot with tall cells bearing short cilia. Near the posterior lip of the stomodeum the larval foot gland forms as a simple invagination and, in some species, this apparently multiplies to form numerous mucus secretory glands in the adult. Dorsally a thin cuticle forms, and transverse rows of clear cells mark the division of the shell valves. The clear cells probably secrete the cuticle, and small cells between the rows form the shell valves (Matthews, 1956). In most species, the first seven shell valves form nearly simultaneously as irregularly shaped calcareous deposits, and according to Heath (1899) and Okuda (1947) the tegmentum forms before the articulamentum. Details of calcification await further study. The first valve forms at the level of the prototroch, either as the first to form (e.g., Thorpe, 1962), simultaneously with the first seven valves (e.g., Lovén, 1855), or somewhat after the second to seventh shell valves form (e.g., Watanabe and Cox, 1975). The eighth shell valve may form nearly at the same time as the first seven valves (e.g., *C. olivaceus*, Kowalevsky, 1883), 1–2 days later when the larva settles (e.g., Hull and Risbec, 1930–1931; Risbec, 1946; Matthews, 1956; Thorpe, 1962), or much later, e.g., 3 days after settlement in *C. stelleri* (Okuda, 1947; see, however, Fig. 16), 5 days after settlement in *T. lineata* (Barnes, 1972), and 5 weeks after settlement in *M. lignosa* (Watanabe and Cox, 1975). Hyman (1967) speculated that the initial seven plates may be a "phylogenetic reminiscence," since the oldest known fossil chiton had only seven shell valves.

With their dorsal shell valve primordiums, ventral foot, prototroch, and apical tuft, chiton larvae have a very characteristic appearance, and there is little variation among those that have been described

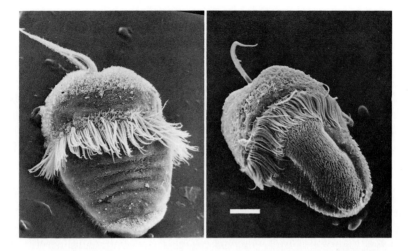

FIG. 17. Scanning electron micrographs of *Katharina tunicata* (Wood). Left: Dorsal view of a 3-day-old larva showing indentations separating the future shell valves. Right: Ventral view of an 8-day-old larva showing a well-developed foot. Scale bar: 50 μm; both at same magnification. (Originals by F.-S. Chia and R. M. O'Clair, University of Alberta.)

(Figs. 16 and 17). They are mainly, if not wholly, nonfeeding larvae serving in short-range dispersal. Grave (1932) mentioned that the larvae of *C. apiculata* developed more rapidly if supplied with phytoplankton, but he provided no details, and it is not clear whether his larvae actually fed. In any case, the time chiton larvae spend swimming is relatively short, usually ranging from a few hours to a few days (Table V).

Settlement is gradual, and in the laboratory the larvae may spend time dropping to the bottom of the dish and creeping, and then swimming up in the water again (Christiansen, 1954). Although most species that have been raised appear to settle indiscriminately, there is some evidence of substrate selection. Heath (1899) noted that larvae of *S. heathiana* seemed to prefer to settle on mussel shells and on the green alga. *Ulva* sp., as Watanabe and Cox (1975) noted for *M. lignosa*. Barnes and Gonor (1973) demonstrated that larvae of *T. lineata* had a strong preference for settling on encrusting coralline algae, *Lithophyllum* sp. and *Lithothamnion* sp., the major adult food. The larvae settled on tiles soaked in water extracts of the corallines, as well as the corallines themselves, but not on other subsrates such as clean rocks, tiles, or pieces of foliose brown or red algae. The substance inducing settlement lost its effect when boiled. D. Bonar (personal communication) confirmed Barnes and Gonor's observations on larvae settling on a wide variety of coralline algae and, in addition, on the brown alga *Hedophyllum sessile*. Settlement and metamorphosis occurred even when the anterior third or posterior portions of the larva were surgically removed. Without suitable substrates, larvae delayed settling for nearly a month. However, Bonar found that different batches of larvae varied so much in their settling behavior that precise analysis of the factors leading to settlement was very difficult.

The prototroch is lost at settlement, and its disappearance marks the beginning of juvenile life. Metamorphosis is not well delineated and mainly involves growth; Thorson (1946) and Christiansen (1954) argued that there is no metamorphosis in chitons. The larval pretrochal region gradually diminishes as the region anterior to the first shell valve forms the anterior portion of the mantle and overgrows the mouth, and within a short time after settlement the animal has the appearance of a tiny chiton (Fig. 16). Baxter and Jones (1978) showed that individuals of *L. cinereus* maintain a constant geometric shape during post-settlement growth (after reaching about 1–2 mm in length) and that there is an isometric relationship of length to height and breadth.

2.4.4 Brooding and Ovoviviparity

A relatively large number of chitons brood their embryos in the pallial (mantle) groove (Table VI). Approximately equal numbers of embryos are held on each side of the foot, and these often nearly fill all available space in the pallial groove (Fig. 18). In addition to

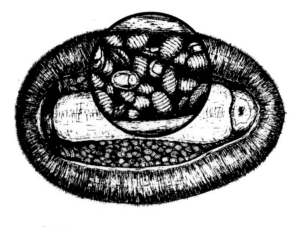

|——— 5 mm ———|

FIG. 18. Drawing of a specimen of *Ischnochiton (Ovatoplax) mayi* (Pilsbry) showing juveniles brooded in the pallial groove. (Original by H. Elizabeth Turner, Tasmanian Museum and Art Gallery.)

these brooding species, *Callistochiton viviparus* apparently is ovovivi-parous, based on Plate's (1899) finding of about 15 embryos, some with 7 shell valves, in the ovary of a single specimen.

Species of chitons that brood are found among about half of the families of chitons, the Mopaliidae and Acanthochitonidae being the only large families without known brooding species. The brooding habit probably evolved independently many times, as most of the brooding species do not seem to be closely related, but there are some apparent exceptions. Bullock (1972), for example, considered *Chiton barnesii* to be closely related to *Chiton nigrovirescens* and placed them in the same genus, *Radsia*. He suggested that brooding in these two species afforded "evidence that the translocation of a single individual could lead to the successful colonization of a new area." Perhaps *C. nigrovirescens* was derived from a single brooding female of *C. barnesii* which was transported from South America to South Africa on floating debris.

Nearly all species of chitons known to brood are found in cold temperate or subpolar seas. There are many more brooding species in the Southern Hemisphere than in the Northern Hemisphere, and Tasmania, in particular, seems notable for the number of brooding species found there (Turner, 1978). The predominance of brooding species in the high southern latitudes may be related to the large proportion of water to land there, and to the west wind drift that transports water west to east around Antarctica and away from most land masses. Most chitons have rather short larval dispersal phases (see Table V), and their larvae might be susceptible to being swept away from near-shore areas suitable for settlement. However, some species of brooding chitons have wide distributions spanning large bodies of water (e.g., *S. brandtii, Ischnochiton constanti, I. imitator,* and espe-cially *H. setulosum*), which indicates a considerable capability for dispersal, probably on floating plant debris (Simpson, 1977).

Most brooding chitons are small, and they rarely exceed 20 mm maximum length (Table VI). Brooding species of other taxa also are often small, and Menge (1975) has argued that because of internal space limitations small animals are unable to produce enough eggs for any of the resulting offspring to have much chance of surviving a pelagic larval stage. Chia (1974) has proposed that small species are energy-limited and, by brooding their offspring, individuals of such species do not have to commit as much energy to reproduction as do those with pelagic larvae. In contrast to the many thousands to millions of eggs spawned by species with pelagic larvae, most brooding species of chitons produce only a couple of hundred embryos or fewer at any one time. Moreover, the number of brooded embryos is closely related to animal size (Kus-

TABLE VI
Chitons Reported to Retain Their Embryos [a]

Species	Locality	Approximate maximal adult length (mm)	Approximate number of embryos brooded	Approximate embryo diameter (mm)	Stage of embryo release	Reference
Hanleyidae						
Hemiarthrum setulosum Dall	Subantarctic Islands and Antarctic Peninsula	15	10–40	0.8	Juvenile	Martens and Pfeffer (1886); Simpson (1977 and unpublished data)
Ischnochitonidae						
Ischnochiton constanti (Vélain)	South Africa and Tristan da Cunha	~20	60–70	0.2		Dell (1962, 1964); J. D. Taylor (unpublished data)
Ischnochiton (Lophyropsis) imitator (Smith)	South American and Antarctic Peninsulas	20	200+	0.4		Plate (1899); Leloup (1956)
Ischnochiton (Heterozona) subviridis (Iredale and May)	Tasmania	50	300	0.4	Juvenile	Iredale and Hull (1923); Turner (1978)
Ischnochiton (Ovatoplax) mayi (Pilsbry)	Tasmania	20	40–70	0.8	Juvenile	Turner (1978)
Lepidochitona cinereus (Linné) [b]	Europe (Helgoland, Germany)	20		0.1–0.2		Knorre (1925)
Cyanoplax dentiens (Gould)	California	20	200	0.4–0.5	Juvenile	Heath (1905); Pearse and Lindberg (unpublished data)
"Chiton" torri Suter	Stewart Island, New Zealand	~10	16	0.3	Juvenile	R. C. Bullock (unpublished data)

Schizoplacidae						
Schizoplax brandtii (Middendorff)	Kurile Island to British Columbia	25	52–1112	0.4	Juvenile	Kussakin (1960); Smith (1966)
Callochitonidae						
Callochiton (= *Eudoxoplax*) *inornata* (Tenison-Woods) [c]	Tasmania	100	20	0.2		Turner (1978)
Callochiton (= *Paricoplax*) *crocinus* (Reeve)	Southern Australia and Tasmania	50	100+	0.3		Turner (1978)
Callistoplacidae						
Callistochiton asthenes (Berry)	Southern California and Guadalupe Island	10	3–18	0.2–0.3	Juvenile	Smith (1966); Ferreira (1979); D. R. Lindberg (unpublished data)
Callistochiton leei Ferreira	Guadalupe Island	8.5	2–2 +	0.5	Juvenile	Ferreira (1979)
Callistochiton (= *Callistassecla*) *maulei* (Iredale and May) [d]	Tasmania	25	3	0.3	Juvenile	Turner (1978)
Callistochiton viviparus Plate [e]	Chile	~ 10	~ 15	0.3	Juvenile	Plate (1899)
Nuttallina thomasi Pilsbry	California	15	~ 80		Juvenile	Heath (1905); Smith (1966)
Middendorffia polii (Philippi)	Southern Europe	20		0.2	Trochophore	Kowalevsky (1883)
Chitonidae						
Chiton barnesii Gray	Chile	35				Plate (1899)
Chiton nigrovirescens (Blainville)	South Africa	25	~ 50	0.7–0.8	Juvenile	Dell (1964); Smith (1966)

[a] Yakovleva (1952) lists *Ischnochiton*, *Lepidochitona*, *Tonicella*, and *Chiton* as genera containing species that brood; I have been unable to document brooding in any species of *Tonicella*.

[b] Egg masses with embryos found in the pallial groove of one preserved specimen; normally not a brooder.

[c] Eggs found in the pallial groove of one preserved adult; may not be a brooder.

[d] Juveniles found on the dorsal surface of two preserved adults; may not be a brooder.

[e] Juveniles found in the ovary of one specimen; ovoviviparous.

sakin, 1960; R. D. Simpson, unpublished data). The eggs of brooding species are about the same size as those of species producing free-swimming larvae (cf. Tables II and VI), so brooding species expend much less energy for reproduction than spawning species. While such low fecundities may be related to small size and either internal space or energy limitations, individuals of some species of brooding chitons attain quite large sizes (e.g., *I. subviridis*), and the evolution of brooding, at least in these species, must be attributed to factors unrelated to size. Similarly, most individuals of some species known to produce free-swimming larvae (e.g., *L. asellus*, *L. cinereus*, *C. apiculata*, *T. lineata*) are as small as those of most brooding species.

Dell (1962) pointed out that species of chitons can be arranged in a sequence to show more and more protection of the young. Species that do not brood their eggs often stick them together in small masses, and these masses initially pass through the pallial groove. In *L. cinereus*, the egg masses may be retained within the pallial groove, as described for a single specimen by Knorre (1925) or, more usually, expelled at spawning (Clark, 1855; Lovén, 1855; Hoffmann, 1931; Magne, 1937; Matthews, 1956; Richter and Götting, 1974). Thorson (1946) suggested that Knorre's report be considered a case of the eggs being "accidentally" retained in the pallial cavity rather than true brooding. However, this case may be viewed as a good example of how brooding could become established under a suitable selective regime, and Matthews (1956) reported considerable variation in how the eggs were spawned among different populations of this species (see Section 2.3.8). Kowalevsky (1883) found that brooding was obligatory in *M. polii* and that the embryos died or developed abnormally when not brooded by the parent. The young of *M. polii* were released as trochophores with a prototroch, but with developing shell valves; they swam for a few hours to a few days in culture dishes before settling. Most brooding species, and the single known ovoviviparous species, apparently retain the embryos until tiny juvenile chitons, with seven or eight fully formed shell valves, are released.

Acknowledgments

This review was written mainly while I was a visiting faculty member of the Museum of Comparative Zoology, Harvard University, where I was able to use the excellent library and scanning electron microscope facilities and receive help and encouragement particularly from K. J. Boss, R. D. Turner, and R. M. Woollacott. I also made extensive use of the superb library at the Hopkins Marine Station of Stanford University. Early drafts benefited enormously from critical readings by R. C. Bullock, A. C. Giese, J. M. Lawrence, D. R. Lindberg, V. B. Pearse, H.-P. Richter, R. D. Turner, and R. M. Woollacott. I am also indebted to R. C. Bullock, D. R. Lindberg, R. D. Simpson, H. E. Turner, and J. D. Taylor for providing unpublished information used in Table I, III, and VI, and to H.-P. Richter for providing Figs. 1, 5, and 6; A. H. Hines for collecting specimens, and K. Chu, L. Fleischman, E. Seling, and R. M. Woollacott for aid in preparing Figs. 7 and 8; L. Meszoly for redrawing Figs. 11, 12, and 14–16; F.-S. Chia for Fig. 17; H. E. Turner for Fig. 18; and R. Buchsbaum for help in the final preparation of all the figures.

References

Allan, J. (1959). "Australian Shells," 2nd Ed. Georgian House, Melbourne.

Anderson, E. (1969). Oocyte-follicle cell differentiation in two species of amphineurans (Mollusca), Mopalia mucosa and Chaetopleura apiculata. J. Morphol. 129, 89–126.

Arey, L. B., and Crozier, W. J. (1919). The sensory responses of Chiton. J. Exp. Zool. 29, 157–260.

Barnawell, E. B. (1951). The biology of the genus Mopalia in San Francisco Bay. M.Sc. Thesis, Univ. of California, Berkeley.

Barnes, J. R. (1972). Ecology and reproductive biology of Tonicella lineata (Wood, 1815) (Mollusca-Polyplacophora). Ph.D. Thesis, Oregon State Univ., Corvallis.

Barnes, J. R., and Gonor, J. J. (1973). The larval settling response of the lined chiton Tonicella lineata. Mar. Biol. 20, 259–264.

Baxter, J. M., and Jones, A. M. (1978). Growth and population structure of Lepidochitona cinereus (Mollusca: Polyplacophora) infected with Minchinia chitonis (Protozoa: Sporozoa) at Easthaven, Scotland. Mar. Biol. 46, 305–313.

Benthem Jutting, T. van (1936). Amphineura. Fauna Nederland 8, 75–87.

Boolootian, R. A. (1964). On growth, feeding and reproduction in the chiton Mopalia muscosa of Santa Monica Bay. Helgol. Wiss. Meeresunters. 11, 186–199.

Boyle, R. P. (1969). Fine structure of the eyes of Onithochiton neglectus (Mollusca: Polyplacophora). Z. Zellforsch. Mikrosk. Anat. 102, 313–332.

Brewin, B. I. (1942). The breeding habits of Cryptoconchus porosus (Burrow). Trans. Proc. R. Soc. N.Z. 72, 186–190.

Brewin, B. I. (1958). Note on the development of Cryptoconchus porosus (Burrow). Trans. R. Soc. N.Z. 85, 715–716.

Bullock, R. C. (1972). The systematics and distribution of the Chitoninae of the world (Mollusca: Polyplacophora). Ph.D. Thesis, Harvard Univ., Cambridge, Massachusetts.

Bullock, R. C., and Boss, K. J. (1971). Lithophaga aristata in the shellplate of chitons (Mollusca). Breviora No. 369, 1–7.

Charnov, E. L., and Bull, J. (1977). When is sex environmentally determined? Nature (London) 266, 828–830.

Chia, F.-S. (1974). Classification and adaptive significance of developmental patterns in marine invertebrates. Thalassia Jugosl. 10, 121–130.

Christiansen, M. E. (1954). The life history of *Lepidopleurus asellus* (Spengler) (Placophora). *Nytt Mag. Zool.* **2**, 52–72.

Clark, W. (1855). On the phaenomena of the reproduction of the chitons. *Ann. Mag. Nat. Hist.* **16**, 446–449.

Costello, D. P., and Henley, C. (1971). "Methods for Obtaining and Handling Marine Eggs and Embryos," 2nd Ed. Mar. Biol. Lab., Woods Hole, Massachusetts.

Cowden, R. R. (1961). A cytochemical investigation of oogenesis and development to the swimming larval stage in the chiton, *Chiton tuberculatum* L. *Biol. Bull. (Woods Hole, Mass.)* **120**, 313–325.

Crozier, W. J. (1918a). Growth and duration of life of *Chiton tuberculatus*. *Proc. Natl. Acad. Sci. U.S.A.* **4**, 322–325.

Crozier, W. J. (1918b). Growth of *Chiton tuberculatus* in different environments. *Proc. Natl. Acad. Sci. U.S.A.* **4**, 325–328.

Crozier, W. J. (1920). Sex-correlated coloration in *Chiton tuberculatus*. *Am. Nat.* **54**, 84–88.

Crozier, W. J. (1922). An observation on the "cluster-formation" of the sperms of chiton. *Am. Nat.* **56**, 478–480.

Davenport, R., and Davenport, J. C. (1965). Cytoplasmic basic proteins in the oocytes of three species of molluscs. *Exp. Cell Res.* **39**, 74–80.

Dell, R. K. (1962). Stages in the development of viviparity in the Amphineura. *Nature (London)* **195**, 512–513.

Dell, R. K. (1964). Antarctic and Subantarctic Mollusca: Amphineura, Scaphopoda and Bivalvia. *Discovery Rep.* **33**, 93–250.

Dolph, C. I., and Humphrey, D. G. (1970). Chromosomes of the chiton, *Katherina tunicata*. *Trans. Am. Microsc. Soc.* **89**, 229–232.

Donato, A. (1963). Esame degli ovociti in accrescimento di *Chiton olivaceus* Spengl. (Moll. Anf. Plac.). *Atti Soc. Peloritana, Sci. Fis., Mat. Nat.* **9**, 255–272.

Eddy, E. M. (1975). Germ plasm and the differentiation of the germ cell line. *Int. Rev. Cytol.* **43**, 229–280.

Ferreira, A. J. (1979). The genus *Callistochiton* Dall, 1879 (Mollusca: Polyplacophora) in the eastern Pacific, with the description of a new species. *Veliger* **21**, 444–466.

Fischer-Piette, E., and Franc, A. (1960). Classe des Polyplacophores. *In* "Traité de Zoologie" (P.-P. Grassé, ed.), Vol. 5, pp. 1701–1785. Masson, Paris.

Franzén, Ä. (1955). Comparative morphological investigations into the spermiogenesis among Mollusca. *Zool. Bidr. Uppsala* **30**, 399–456.

Gabe, M., and Prenant, M. (1949). Contribution à l'histologie de l'ovogenèse chez les polyplacophores. *Cellule* **53**, 99–116.

Garnault, P. (1888). Recherches sur la structure et le développement de l'oeuf et de son follicule chez les chitonides. *Arch. Zool. Exp. Gen.* **6**, 83–116.

Giese, A. C. (1966). Lipids in the economy of marine invertebrates. *Physiol. Rev.* **46**, 244–298.

Giese, A. C. (1969). A new approach to the biochemical composition of the mollusc body. *Oceanogr. Mar. Biol. Ann. Rev.* **7**, 175–229.

Giese, A. C., and Araki, G. (1962). Chemical changes with reproductive activity of the chitons, *Katherina tunicata* and *Mopalia hindsii*. *J. Exp. Zool.* **151**, 259–267.

Giese, A. C., and Hart, M. A. (1967). Seasonal changes in component indices and chemical composition in *Katharina tunicata*. *J. Exp. Mar. Biol. Ecol.* **1**, 34–46.

Giese, A. C., and Pearse, J. S. (1974). Introduction: General principles. *In* "Reproduction of Marine Invertebrates" (A. C. Giese and J. S. Pearse, eds.), Vol. 1, pp. 1–49. Academic Press, New York.

Giese, A. C., Tucker, J. S., and Boolootian, R. A. (1959). Annual reproductive cycles of the chitons, *Katherina tunicata* and *Mopalia hindsii*. *Biol. Bull. (Woods Hole, Mass.)* **117**, 81–88.

Glynn, P. W. (1970). On the ecology of the Caribbean chitons *Acanthopleura granulata* Gmelin and *Chiton tuberculatus* Linné: Density, mortality, feeding, reproduction, and growth. *Smithsonian Contrib. Zool.* No. 66, 1–21.

Grave, B. H. (1922). An analysis of the spawning habits and spawning stimuli of *Chaetopleura apiculata* (Say). *Biol. Bull. (Woods Hole, Mass.)* **42**, 234–256.

Grave, B. H. (1932). Embryology and life history of *Chaetopleura apiculata*. *J. Morphol.* **54**, 153–160.

Grave, B. H. (1937). *Chaetopleura apiculata*. *In* "Culture Methods for Invertebrate Animals" (P. S. Galtsoff, F. E. Lutz, P. S. Welch, and J. G. Needham, eds.), pp. 519–520. Cornell Univ. Press (Comstock), Ithaca, New York. (Reprinted by Dover, New York, 1959.)

Haller, B. (1882). Die Organisation der Chitonen der Adria. I. *Arb. Zool. Inst. Univ. Wien, Zool. Stn. Triest* **4**, 1–74.

Hammarsten, O. D., and Runnström, J. (1924). Ein Beitrag zur Kenntniss der Ontogenie der Placophoren. *Ark. Zool.* **16**(19), 1–5.

Hammarsten, O. D., and Runnström, J. (1925). Zur Embryologie von *Acanthochiton discrepans* Brown. *Zool. Jahrb., Abt. Anat. Ontog. Tiere* **47**, 261–318.

Heath, H. (1899). The development of Ischnochiton. *Zool. Jahrb., Abt. Anat. Ontog. Tiere* **12**, 567–656.

Heath, H. (1904). The larval eye of chitons. *Proc. Acad. Nat. Sci. Philadelphia* **56**, 257–259.

Heath, H. (1905). The breeding habits of chitons of the California coast. *Zool. Anz.* **29**, 390–393.

Heath, H. (1907). The gonad of certain species of chitons. *Zool. Anz.* **32**, 10–12.

Higley, R. M., and Heath, H. (1912). The development of the gonad and·gonoducts in two species of chitons. *Biol. Bull. (Woods Hole, Mass.)* **22**, 95–97.

Himmelman, J. H. (1975). Phytoplankton as a stimulus for spawning in three marine invertebrates. *J. Exp. Mar. Biol. Ecol.* **20**, 199–214.

Himmelman, J. H. (1976). Factors regulating the reproductive cycles of some west coast invertebrates. Ph.D. Thesis, Univ. of British Columbia, Vancouver.

Himmelman, J. H. (1978). The reproductive cycle of *Katharina tunicata* Wood and its controlling factors. *J. Exp. Mar. Biol. Ecol.* **31**, 27–41.

Himmelman, J. H. (1979). Factors regulating the reproductive cycles of two Northeast Pacific chitons, *Tonicella lineata* and *T. insignis*. *Mar. Biol.* **50**, 215–225.

Himmelman, J. H., and Carefoot, T. H. (1975). Seasonal changes in calorific value of three Pacific Coast seaweeds, and their significance to some marine herbivores. *J. Exp. Mar. Biol. Ecol.* **18**, 139–151.

Hoffmann, H. (1929–1930). Polyplacophora. *In* "H. G. Bronn's Klassen und Ordnungen des Tier-Reichs," Vol. 3, Part 1, pp. 135–453. Akad. Verlag, Berlin.

Hoffmann, H. (1931). Beiträge zur Kenntnis der Chitoanen. I. Über die Fortpflanzung und Entwicklung von *Trachydermon cinereus* L. *Z. Morphol. Oekol. Tiere* **20**, 719–732.

Hull, A. F. B., and Risbec, J. (1930–1931). The loricates of the New Caledonian region (class Mollusca-Order Loricata). *Aust. Zool.* **6**, 277–286, 372–386.

82   JOHN S. PEARSE

Hyman, L. H. (1967). "The Invertebrates." Vol VI: Mollusca I," McGraw-Hill, New York.

Ihering, H. von (1878). Beiträge zur Kenntniss der Anatomie von Chiton. *Morphol. Jahrb.* **4**, 128–146.

Iredale, T., and Hull, A. F. B. (1923). A monograph of the Australian loricates (Phylum Mollusca-Order Loricata). I. Systematics and structure. *Aust. Zool.* **3**, 186–194.

Johns, P. M. (1960). *Chiton pelliserpentis* (Mollusca; Amphineura), a study in the taxonomy of a species in relation to its breeding biology and ecology. M.Sc. Thesis, Univ. of Canterbury, Christchurch, New Zealand.

Kass, P. (1957). Notes on Loricata. *Basteria* **21**, 83–87.

Knorre, H. von (1925). Die Schale und die Rückensinnesorgane von *Trachydermon (Chiton) cinereus* L. und die ceylonischen Chitonen der Sammlung Plate. *Jena. Z. Naturwiss.* **61**, 469–632.

Knox, G. A. (1963). Problems of speciation in intertidal animals with special reference to New Zealand shores. *In* "Speciation in the Sea" (J. P. Harding and N. Tebble, eds.), pp. 7–29. Systematics Assoc., London.

Kowalevsky, A. (1879). Ueber der Entwickelung der Chitonen. *Zool. Anz.* **2**, 469–473.

Kowalevsky, A. (1882). Weitere Sudien über die Entwickelung der Chitonen. *Zool. Anz.* **5**, 307–310.

Kowalevsky, A. (1883). Embryogénie du Chiton Polii (Philippi). *Ann. Mus. Hist. Nat. Marseille, Zool.* **1**(5), 1–46.

Kussakin, O. G. (1960). Biological peculiarities of the Far Eastern chiton *Schizoplax brandtii* (Middendorff). *Zool. Zh.* **39**, 1145–1150. (In Russ.; Engl. sum.)

Lawrence, A. L., Lawrence, J. M., and Giese, A. C. (1965). Cyclic variations in the digestive gland and glandular oviduct of chitons (Mollusca). *Science* **147**, 508–510.

Lawrence, J. M. (1970). The lipid composition of the organs of two species of tropical chitons. *Caribb. J. Sci.* **10**, 1–3.

Lawrence, J. M., and Giese, A. C. (1969). Changes in the lipid composition of the chiton, *Katharina tunicata*, with the reproductive and nutritional state. *Physiol. Zool.* **42**, 353–360.

Leloup, E. (1956). Polyplacophora. *Rep. Lund Univ. Chile Exped. 1948–1949* **27**, 1–93.

Leloup, E., and Volz, P. (1938). Die Chitonen (Polyplacophoren) der Adria. *Thalassia* **2**(10), 1–63.

Lewis, J. B. (1960). The fauna of rocky shores of Barbados, West Indies. *Can. J. Zool.* **38**, 391–435.

Lovén, S. (1855). Om utvecklingen hos slägtet Chiton. *Oefversigt K. Vetensk.-Akad. Foerh.* **12**, 169–173. [Also *Ann. Mag. Nat. Hist.* **17**, 413–416 (1856) and *Arch. Naturgesch.* **22**, 206–210 (1856).]

Lyngnes, R. (1924). Zur Kenntnis der Eihüllen der Chitonen. *Skr. Vidensk. Selsk. (Oslo)* **20**, 1–19.

Magne, A. (1937). Recherches sur l'activité sexuelle et la croissance de deux espèces d'Amphineures du bassin d'Arcachon *Ischnochiton cinereus* L. et *Acantochites discrepans* Brown. *Bull. Stn. Biol. Arcachon* **34**, 147–159.

Martens, E., and Pfeffer, G. (1886). Die Mollusken von Sud-Georgien. *J. Wiss. Anst. Hamburg* **3**, 65–135.

Martoja, M. (1967). Sur l'existence de cellules neurosécrétrices chez quelques Mollusques Polyplacophores. *C. R. Acad. Sci.* **264**, 1461–1463.

Matthews, F. G. C. (1956). The breeding behavior, embryology and larval ecology of *Lepidochitona cinereus* (L). Ph.D. Thesis, Univ. of London.

Menge, B. A. (1975). Brood or broadcast? The adaptive significance of different reproductive strategies in the two intertidal sea stars *Leptasterias hexactis* and *Pisaster ochraceus*. *Mar. Biol.* **31**, 87–100.

Metcalf, M. M. (1892). Preliminary notes upon the embryology of Chiton. *Johns Hopkins Univ., Circ.* **11**, 79–80.

Metcalf, M. M. (1893). Contributions to the embryology of Chiton. *Johns Hopkins Univ., Stud. Biol. Lab.* **5**, 249–267.

Miller, R. L. (1977). Chemotactic behavior of the sperm of chitons (Mollusca: Polyplacophora). *J. Exp. Zool.* **202**, 203–212.

Monroe, H. C., and Boolootian, R. A. (1965). Reproductive biology in the chiton *Mopalia muscosa*. *Bull. South. Calif. Acad. Sci.* **64**, 223–228.

Murti, K. G., and Nagabhushanam, R. (1968). Observations of the spawning and reproductive cycle of *Chiton granoradiatus* Leloup. *Broteria, Ser. Cienc. Nat.* **37**, 223–233.

Nagabhushanam, R., and Murti, K. G. (1968). Studies on the development of *Chiton granoradiatus* Leloup. *Broteria, Ser. Cienc. Nat.* **37**, 209–214.

Nagabhushanam, R., and Murti, K. G. (1969). Physiology of reproduction in *Chiton granoradiatus* Leloup. *Proc. Symp. Mollusca, Cochin, 1968, Mar. Biol. Assoc. India, Symp. Ser.* No. 3, Part 2, 580–584.

Nimitz, M. A. (1964). The histology and histochemistry of the chiton *Katherina* in relation to the reproductive cycle. Ph.D. Thesis, Stanford Univ., Stanford, California.

Nimitz, M. A., and Giese, A. C. (1964). Histochemical changes correlated with reproductive activity and nutrition in the chiton *Katherina tunicata*. *Q. J. Microsc. Sci.* **105**, 481–495.

Nishikawa, S., and Ishida, S. (1969). Some aspects on chromosomes of Japanese chitons. *J. Shimonoseki Univ. Fish.* **17**, 179–183. (In Jpn.; Engl. sum.)

Okuda, S. (1947). Notes on the post-larval development of the giant chiton, *Cryptochiton stelleri* (Middendorff). *J. Fac. Sci., Hokkaido Univ., Ser. 6* **9**, 267–275.

Pearse, J. S. (1972). A monthly reproductive rhythm in the diadematid sea urchin *Centrostephanus coronatus* (Verrill). *J. Exp. Mar. Biol. Ecol.* **8**, 167–186.

Pearse, J. S. (1978). Reproductive periodicities of Indo-Pacific invertebrates in the Gulf of Suez. IV. The chitons *Acanthopleura haddoni* Winckworth and *Onithochiton lyelli* (Sowerby), and the abalone *Haliotis pustulata* Reeve. *Bull. Mar. Sci.* **28**, 92–101.

Pelseneer, P. (1898). Recherches morphologiques et phylogénétiques sur les Mollusques archaïques. *Mem. Courr. Acad. R. Belg.* **57**, 1–113.

Pelseneer, P. (1926). La proportion relative des sexes chez les animaux et particulièrement chez les Mollusques. *Acad. R. Belg., Cl. Sci., Mem., Collect.* **8**, 1–258.

Plate, L. (1897). Die Anatomie und Phylogenie der Chitonen. *Zool. Jahrb., Suppl.* **4**, 1–243.

Plate, L. (1899). Die Anatomie und Phylogenie der Chitonen. *Zool. Jahrb., Suppl.* **5**, 15–216.

Plate, L. (1901). Die Anatomie und Phylogenie der Chitonen. *Zool. Jahrb., Suppl.* **5**, 281–600.

Retzius, G. (1904). Zur Kenntnis der Spermien der Evertebraten I. *Biol. Untersuch.* 11, 1–32.

Retzius, G. (1905). Zur Kenntnis der Spermien der Evertebraten II. *Biol. Untersuch.* 12, 79–102.

Richter, H.-P. (1976). Feinstrukturelle Untersuchungen zur Oogenese der Käferschnecke *Lepidochitona cinereus* (Mollusca, Polyplacophora). *Helgol. Wiss. Meeresunters.* 28, 250–303.

Richter, H.-P., and Götting, K. J. (1974). Oogenese und Sexualität der Käferschnecke *Lepidochitona cinereus* (Mollusca, Polyplacophora). *Helgol. Wiss. Meeresunters.* 26, 42–62.

Risbec, J. (1946). Études anatomiques sur les amphineures de la Nouvelle-Calédonie. *J. Soc. Ocean.* 2, 129–190.

Rothschild, Lord, and Tyler, A. (1955). Acrosomal filaments in spermatozoa. *Exp. Cell Res., Suppl.* 3, 304–311.

Sabatier, A. (1886). Quelques observations sur la constitution de l'oeuf et de ses enveloppes chez les chitonides. *Trav. Lab. Zool., Fac. Sci. Montpellier Stn. Zool. Cette.* 5, 101–110.

Salvini-Plawen, L. von (1973). Zur Klärung des "Trochophora"–Begriffes. *Experientia* 29, 1434–1435.

Schweikart, A. (1904a). Die Bildung der Eihüllen und ihrer Anhänge bei den Chitonen. *Zool. Anz.* 27, 636–648.

Schweikart, A. (1904b). Beiträge zur Morphologie und Genese der Eihüllen der Cephalopoden und Chitonen. *Zool. Jahrb., Suppl.* 6, 353–406.

Selwood, L. (née Bedford) (1968). Interrelationships between developing oocytes and ovarian tissues in the chiton *Sypharochiton septentriones* Ashby (Mollusca, Polyplacophora). *J. Morphol.* 125, 71–103.

Selwood, L. (née Bedford) (1970). The role of the follicle cells during oogenesis in the chiton *Sypharochiton septentriones* Ashby (Polyplacophora, Mollusca). *Z. Zellforsch. Mikrosk. Anat.* 104, 178–192.

Simpson, R. D. (1977). The reproduction of some littoral molluscs from Macquarie Island (Sub-Antarctic). *Mar. Biol.* 44, 125–142.

Simroth, H. (1892–1894). Amphineura und Scaphopoda. In "Klassen und Ordnungen des Thier-Reichs" (H. G. Bronn, ed.), Vol. 3, Part I, pp. 128–355. Winterische, Leipzig.

Smith, A. G. (1960). Amphineura. In "Treatise on Invertebrate Paleontology" (R. C. Moore, ed.), Part I, pp. 41–76. Univ. of Kansas Press and Geol. Soc. Am., New York.

Smith, A. G. (1966). The larval development of chitons (Amphineura). *Proc. Calif. Acad. Sci.* 32(15), 433–446.

Southwick, W. E. (1939). The "agglutination" phenomenon with spermatozoa of *Chiton tuberculatus. Biol. Bull. (Woods Hole, Mass.)* 77, 157–165.

Stephenson, A. (1934). The breeding of reef animals. Part II. Invertebrates other than corals. *Great Barrier Reef Exped. 1928–1929, Sci. Rep.* 3(9), 247–272.

Thiele, J. (1906). Ueber die Chitonen der Deutschen Tiefsee-Expedition. *Wiss. Ergeb. Dtsch, Tiefsee-Exped. "Valdivia" 1898–1899.* 9, 327–336.

Thompson, T. E. (1973). Euthyneuran and other molluscan spermatozoa. *Malacologia* 14, 167–206.

Thorpe, S. R. (1962). A preliminary report on spawning and related phenomena in California chitons. *Veliger* 4, 202–210.

Thorson, G. (1946). Reproduction and larval development of Danish marine bottom invertebrates. *Medd. Dan. Fisk. Havunders., Ser. Plankton* **4**, 1–523.

Tucker, J. S., and Giese, A. C. (1962). Reproductive cycle of *Cryptochiton stelleri* (Middendorff). *J. Exp. Zool.* **150**, 33–43.

Turner, E. (1978). Brooding of chitons in Tasmania. *J. Malacol. Soc. Aust.* **4**, 43–47.

Vasu, B. S., and Giese, A. C. (1966). Variations in the body fluid nitrogenous constituents of *Cryptochiton stelleri* (Mollusca) in relation to nutrition and reproduction. *Comp. Biochem. Physiol.* **19**, 737–744.

Vicente, N. (1970). Observations sur l'ultrastructure d'un organe juxtacommissural dans le système nerveux du chiton (Mollusque Polyplacophore). *C. R. Seances Soc. Biol. Marseille* **164**, 601–605.

Vicente, N., and Gasquet, M. (1970). Étude du système nerveux et de la neurosécrétion chez quelques Mollusques Polyplacophores. *Tethys* **2**, 515–546.

Wada, S. K. (1968). Amphineura, Gastropoda, Scaphopoda, Pelecypoda. *In* "Invertebrate Embryology" (M. Kumé and K. Dan, eds.; transl. by J. C. Dan), pp. 485–525. NOLIT Publ., Belgrade.

Wagner, R. (1839). Observations on the generative system of some of the lower animals. *Proc. Zool. Soc. London* **7**, 177–178.

Watanabe, J. M., and Cox, L. R. (1975). Spawning behavior and larval development in *Mopalia lignosa* and *Mopalia muscosa* (Mollusca: Polyplacophora) in central California. *Veliger* **18**, Suppl., 18–27.

Yakovleva, A. M. (1952). Shell-bearing mollusks (Loricata) of the seas of the U.S.S.R. *In* "Keys to the Fauna of the U.S.S.R.," (A. A. Strelkov, ed.), No. 45, pp. 1–127. Zool. Inst. Acad. Sci. U.S.S.R., Moscow–Leningrad. (Engl. transl., Smithsonian–NSF Israel Program Sci. Transl., Jerusalem, 1965.)

Chapter 3

MONOPLACOPHORA

3.1 Introduction 87
3.2 Asexual Reproduction 88
3.3 Sexual Reproduction 88
    3.3.1 Sexual Dimorphism 88
    3.3.2 Anatomy and Histology of the Reproductive System 88
    3.3.3 Gametogenesis 90
    3.3.4 Mode of Gamete Discharge 91
3.4 Postembryonic Development 92
    References 93

3.1 Introduction

Living monoplacophorans are little modified abyssal representatives of a molluscan class known only as fossils from the Cambrian to the Devonian until the description of *Neopilina galatheae* from a collection at abyssal depths (3570 m) of the Pacific Ocean off Costa Rica (Lemche, 1957). Subsequently this and five other species were found at depths ranging from 2730 to 6280 m (Rosewater, 1970; Menzies *et al.*, 1973), but these were all very similar and were all placed in the genus *Neopilina*; however, see McLean, 1979). These few Paleozoic relics are of importance beyond their numbers not only because they are the only extant representatives of this class, but because their indications of serially repeated organs have been interpreted as remnants of metamerism. If this is accepted, they represent a molluscan body condition not far removed from that of a metameric ancestor closely linked to similar ancestors of the annelids and arthropods and are consequently more primitive than any other living molluscs. Their reproductive biology may be expected to show the original or primitive molluscan mode and, with allowance made for their present ecology, serve as a standard for judging the degree of specialization found in other molluscs. Because of the abyssal

habitat of living monoplacophorans, it has been difficult to secure
material for study, and what is known about reproduction is primarily
inferred from the descriptions of the anatomy and histology of
N. galatheae by Lemche and Wingstrand (1959). Despite the subsequent collection of other material of this species and the discovery of
new species, no new information has been published on *N. galatheae*,
and only cursory descriptions of external features are available for the
new species:

3.2 Asexual Reproduction

Except for a few brackish and freshwater prosobranchs (see Volume
4, Section 1.2), no molluscs are known to reproduce asexually. There
is no direct evidence of asexual reproduction in monoplacophorans,
and it seems improbable in this group.

3.3 Sexual Reproduction

3.3.1 Sexual Dimorphism

Fortuitously, of the two specimens of *N. galatheae* sectioned, one
was a mature male and the other a mature female. No external sex
characteristics are known and the anatomy of males and females is
identical. There are no indications of hermaphroditism or protandry.

3.3.2 Anatomy and Histology of the Reproductive System

The extensively lobed gonads are located lateroventrally in the
posterior half of the body, contained within the visceral hemocoel
and lying on its floor below the coiled intestine and digestive gland.
The lobes of the gonads are free in the blood of the visceral hemocoel
and are not otherwise vascularized, internally or externally. Two pairs
of gonads are described, but only in the male specimen could the gonad
definitely be seen to be two pairs rather than one. The gonads are little
more than simple lobed sacs with a very thin external lining of
connective tissue, gametogenic cells lining the interior wall, and mature

gametes in the lumen. Each gonad is joined to one of the third and fourth pairs of nephridia by short, thin-walled gonoducts lined with a low, ciliated epithelium. The two pairs of nephridia that receive the gonoducts are anatomically and histologically identical to the other nephridial pairs: thin-walled, lobed sacs lined with a low, vacuolate epithelium. In the two mature specimens examined by Lemche and Wingstrand, mature eggs and a larger quantity of mature sperms were found in the nephridia into which the gonoducts emptied. These nephridia may therefore serve as storage organs for gametes despite their apparent lack of modification. Mature gametes can be liberated only through the nephridial ducts and renal pores. The short nephridial ducts are lined with columnar epithelium in all nephridial pairs, and all renal pores are simple openings into the pallial groove at the base of a gill. The renal pores are not elevated on papillae, and there are no copulatory structures. There is no anatomical provision at any point for the elaboration of secretions around the gametes, and they are apparently spawned freely through the nephridial sac via its duct and pore.

Lemche and Wingstrand point to the dorsal position of the gonad in polyplacophorans and other molluscs and the ventral position in *Neopilina* as evidence that the gonads are not homologous throughout the molluscs. They indicate that the dorsally placed gonads in poly-placophorans correspond in position to a pair of large flattened, saccate structures in *Neopilina*, which they identify as dorsal body coeloms. From one but not the other of the two sectioned specimens, these authors describe strings of cells which they interpret as nephrostomic ducts, connecting the second, third, and fourth pairs of nephridia, hence also connecting the gonads with the dorsal coeloms. Nephrostomic funnels were not observed but surmised. Nephridial pairs five and six were found joined to the pericardial coeloms by similar connections, but again no ductal lumina were observed, and the authors later said that such connections were only probable (Lemche and Wingstrand, 1960, p. 1813).

Both the interpretation of the dorsal structures as paired coeloms and the interpretation of remnant coelomic metamerism rest largely on these putative nephrostomic ducts and the resulting anatomical con-fluence of dorsal coeloms, nephridia, and gonads. However, the material was admittedly unsuited for such detailed interpretation, and the apex of both sectioned specimens of *N. galatheae* had been damaged in the region of the dorsal coeloms (Lemche and Wingstrand, 1959). Lemche and Wingstrand (1959, p. 59 footnote, 1960, p. 1820) and Lemche (1962) indicated that sections of an undamaged specimen of *Neopilina* (*Vema*) *ewingii* Clark and Menzies, 1959 showed distinct broad connections between the anterior parts of the dorsal coeloms and the sugar glands of the pharynx, to which they are histologically similar. These connections clearly indicate that the original material was misinterpreted because of damage and poor histological condition. These large, dorsally extended sugar glands, like the large, coiled intestine, are possibly adaptations related to feeding upon large quantities of sediment, the nutritive source most commonly utilized by benthic animals in the deep sea (Menzies *et al.*, 1973). The difficulties of accommodating *Neopilina* in the molluscan pattern (Clark, 1964; Morton and Yonge, 1964), particularly with reference to the homology of the gonads, largely disappear along with the "dorsal coeloms". A more plausible assessment of their relation to the other molluscs is that given by Stasek (1972), who distinguishes between the pseudometameric serial repetition of some structures in *Neopilina* and other molluscs and the regular metameric repetition of organ sets in the segmented bodies of annelids and arthropods. The claim of nephrostomes or connections between the sugar glands and nephridia and from them to the gonads, resting as it does on poor histological evidence and lacking any rational functional interpretation, must be rejected as not definitely demonstrated and moreover extremely unlikely. Without connections between the nephridia and the dorsal coelom, the difference in position of the gonads in *Neopilina*, the polyplacophorans, and the other extant molluscan classes appears to be a trivial reason for rejecting gonad homology throughout the phylum.

In summary, the anatomy of the reproductive system of *Neopilina* is very simple and resembles that of lower prosobranch gastropods and protobranch bivalves with the gonads discharging through the nephridia. The available material has been overinterpreted; it is not even definitely established whether the gonads consist of one or of two pairs. The interpretive diagrams of the urogenital and coelomic systems provided by the original describers should not be accepted as definitive.

3.3.3 Gametogenesis

As in many other molluscs such as oysters (Galtsoff, 1964; see also Sections 5.3.5.2 and 6.3), not all oocytes mature synchronously within an individual. In the female specimen of *N. galatheae*, large, free oocytes in the germinal vesicle stage were abundant in the lumina of the ovaries and nephridia, while immature oocytes of various sizes were still present in the ovarian wall. A continuous lining of small un-differentiated cells 10–15 μm high was found between scattered oocytes; they may represent a continuous undifferentiated germinal epithelium. As vitellogenesis proceeds and the oocytes enlarge, they become domed, with a broad base on the ovarian wall. Larger oocytes become somewhat clavate, with a small area of attachment to the wall. The accumulation of many free mature oocytes in the lumen indicates that gamete discharge by the individual female is phasic. Mature eggs have a thin, granule-free cortical layer, with the remainder of the cytoplasm densely packed with basophilic yolk granules. Many are oblong in shape, apparently not as a result of compression since the long axes are not aligned. Mature oocytes are approximately 200 μm in diameter, with an eccentrically placed nucleus 60 μm in diameter. There are no indications of nurse or follicle cells in the ovary, and the mature oocytes in both the ovaries and nephridia are without any chorion or other extracellular covering such as is found in those of polyplacophorans (Section 2.3.5.1).

In the testes, spermatogenic cells form a thick internal layer, with the usual regular progression of stages of spermatogenesis and spermiogenesis from the base to the lumen and with mature sperms oriented so that the tails extend into the lumen. The nuclei of spermatogonia or early spermatocytes next to the testis wall are 2–3 μm in diameter, and fully mature sperms have simple, rounded heads 1 μm in diameter. The gonoducts and nephridia contain only mature sperms.

3.3.4 Mode of Gamete Discharge

The predominant mode of reproduction in benthic animals of the deep sea and polar environments includes large, yolky eggs, nonpelagic or nonplanktotrophic larvae, and often some form of brood protection (Menzies *et al.*, 1973). However, some deep sea invertebrates have small eggs and presumably pelagic, planktotrophic larvae. The extant monoplacophorans appear to belong to the free-spawning group with free larvae, judging from indirect evidence on *N. galatheae*. There are no copulatory or glandular accessory reproductive structures or other modifications of the adult reproductive system which could be related to encapsulation or egg mass formation. The eggs are simple, naked, and of a moderate size and yolkiness comparable to those of many free-spawning prosobranchs and bivalves, while the sperms are of a simple type. These features and the presence of free mature gametes in the nephridia all indicate that *Neopilina* spawns free gametes directly into the surrounding water and that fertilization and subsequent embryonic development are external, free in the water. Nothing further can be surmised about the nature of early embryonic development.

3.4 Postembryonic Development

In *N. galatheae*, later development must include a larval stage with a symmetrical, exogastrically coiled larval shell, since a 0.15-mm-diameter protoconch of this type was found still attached to the apex of one of the adult specimens of *N. galatheae* by Lemche and Wingstrand (1959). An abrupt metamorphosis to the adult form must take place, since the coiled protoconch lies on the apex of the adult shell on its right side and, coiling abruptly, changes to the open, cap-shaped adult shell form. This metamorphosis and coiling change is strikingly reminiscent of the late development and metamorphosis of limpet-form prosobranch gastropods with the exception of the endogastric coiling resulting from torsion during early gastropod development. A small, empty shell with a noncoiled, bulbous protoconch was described and figured by Menzies (1968) as that of a monoplacophoran, but this identification was not definitively established. Since the dimensions of the mature egg and the larval protoconch are approximately the same in *N. galatheae*, larval development, whether it be pelagic or demersal, cannot include significant larval growth, and the larvae of *N. galatheae* should be expected to be a primarily lecithotrophic dispersal stage, possibly of short duration. Abyssal benthic population density is low, and free larvae may be retained by benthic abyssal invertebrates as an adaptation utilizing currents for dispersal.

Acknowledgments

Preparation of this review was supported in part by a National Oceanic and Atmospheric Administration (maintained by the U.S. Department of Commerce) Institutional Sea Grant to Oregon State University. Despite critical comments in this review, I freely acknowledge that writing it was possible only because Henning Lemche and K. G. Wingstrand recognized the unique and important nature of the Galathea material and were extraordinarily successful in extracting all possible information from it despite severe limitations.

References

Clark, R. B. (1964). "Dynamics in Metazoan Evolution. The Origin of the Coelome and Segments." Oxford Univ. Press (Clarendon), London and New York.

Galtsoff, P. S. (1964). The American oyster *Crassostrea virginica* Gmelin. *U.S. Fish Wildl. Serv., Fish. Bull.* **64**, 1–480.

Lemche, H. (1957). A new living deep sea mollusc of the Cambro-Devonian class Monoplacophora. *Nature (London)* **179**, 413–416.

Lemche, H. (1962). The primitive mollusc that emerged from our oceans. *Scientia (Milan)* **56**, 1–8.

Lemche, H., and Wingstrand, K. G. (1959). The anatomy of *Neopilina galatheae* Lemche, 1957 (Mollusca, Tryblidiacea). *Galathea Rep.* **3**, 9–72.

Lemche, H., and Wingstrand, K. G. (1960). Classe des Monoplacophores. *In* "Traité de Zoologie" (P.-P. Grassé, ed.), Vol. 5, Part 2, pp. 1787–1821. Masson, Paris.

McLean, J. H. (1979). A new monoplacophoran limpet from the continental shelf off southern California. *Contrib. Sci. Natur. Hist. Mus. Los Angeles County* **307**, 1–19.

Menzies, R. J. (1968). New species of *Neopilina* of the Cambro-Devonian class Monoplacophora from the Milne–Edwards Deep of the Peru–Chile Trench, R/V Anton Bruun. *Proc. Symp. Mollusca, Cochin, 1968, Mar. Biol. Assoc. India, Symp. Ser.* 3, Part 1, 1–19.

Menzies, R. J., George, R. Y., and Rowe, G. T. (1973). "Abyssal Environment and Ecology of the World Oceans." Wiley, New York.

Morton, J. E., and Yonge, C. M. (1964). Classification and structure of the Mollusca. *In* "Physiology of the Mollusca" (K. M. Wilbur and C. M. Yonge, eds.), Vol. 1, pp. 1–58. Academic Press, New York.

Rosewater, J. (1970). Monoplacophora in the South Atlantic Ocean. *Science* **167**, 1485–1486.

Stasek, C. R. (1972). The molluscan framework. *In* "Chemical Zoology. Vol. II: Mollusca" (M. Florkin and B. T. Scheer, eds.), pp. 1–44. Academic Press, New York.

Chapter 4

SCAPHOPODA

M. McFadien-Carter

4.1 Introduction 95
4.2 Asexual Reproduction 96
4.3 Sexual Reproduction 96
    4.3.1 Anatomy of the Reproductive Organs 96
    4.3.2 Spermatogenesis 98
    4.3.3 Oogenesis 100
    4.3.4 Spawning 101
4.4 Development 104
    4.4.1 Fertilization and Cleavage 104
    4.4.2 Localization of Morphogenetic Factors 105
    4.4.3 Blastula and Gastrula 106
    4.4.4 Trochophore and Veliger Stages 106
    References 110

4.1 Introduction

Scaphopods are benthic molluscs found in both shallow and deep waters of all the major seas. They are burrowing organisms, commonly found with the anterior tip down. Whether or not the posterior tip is exposed varies with the species (Gainey, 1972). Water circulates (Yonge, 1937) and gametes are expelled (Dinamani, 1964) through the posterior aperture.

Anatomical and reproductive studies have, with few exceptions, centered on different species of the genus *Dentalium*. It is because of such emphasis in past studies that most of this chapter is devoted to reproduction in this genus. Unfortunately, available information on the Scaphopoda contains many minor discrepancies. Through my own observations and a comparative survey of the literature, I feel that these contradictions are due in part to species variations and in part to the lack of time given to some of the studies. Also, recent improvements in techniques and modern understanding of reproductive cycles in general have modified the conclusions of some otherwise excellent nineteenth-century works. Two of the most thorough of these works on the reproduction and embryology of *Dentalium* are Lacaze-Duthiers' (1856–1857) account of gametogenesis, fertilization, and development through 35 days after fertilization and Kowalevsky's (1883) study of specimens to 6 days after fertilization. Although he could not maintain the juveniles for prolonged study, Kowalevsky advanced his predecessor's work by utilizing histological techniques.

4.2 Asexual Reproduction

Asexual reproduction is unknown among scaphopods.

4.3 Sexual Reproduction

Scaphopods are dioecious, but according to Reverberi (1971) hermaphroditic individuals of *Dentalium* are found occasionally. Gametes are ejected through the right nephridial aperture. Fertilization is external, and larvae have been observed to settle within 2–6 days after fertilization depending on the species or on external conditions (Kowalevsky, 1883; Verdonk, 1968a).

4.3.1 Anatomy of the Reproductive Organs

The origin of the gonad in this class has, to my knowledge, not yet been studied. In both sexes, it is a single, unpaired organ divided into lobes. It lies within the hemocoel, posterior to the digestive diverticula. The anterior end of this long organ leads to the right excretory organ (Figs. 1 and 2).

As spawning approaches, the sexes may be identified externally in translucent-shelled species such as *Dentalium antillarum* Orbigny and *Cadulus fusiformis* Pilsbry and Sharp. Depending on the species, the ovary appears reddish or brownish gray. Under a dissecting microscope eggs are yellowish or brownish gray, biscuit-shaped objects packed tightly like stacked coins. The testis is a milky white tube (Verdonk, 1968a) and consists of a clear membrane lined with clear or whitish,

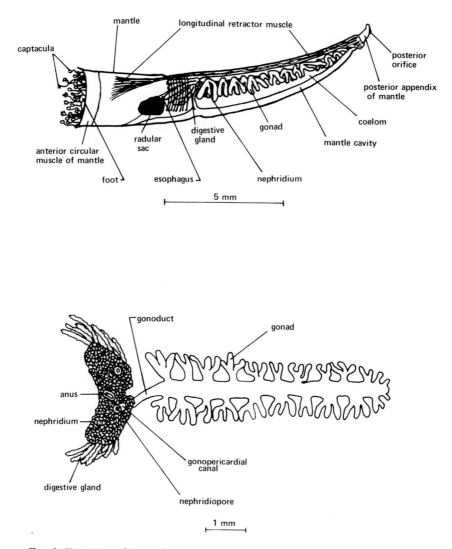

FIG. 1. Top: Lateral view through the mantle of a male specimen of *D. neohexagonum* with the shell removed showing the gonad in relation to other structures. Bottom: Ventral view of the gonad of *D. neohexagonum;* representation similar to that given by Lacaze-Duthiers (1856–1857) for *D. vulgare.*

small, coarsely granular cells (Lacaze-Duthiers, 1856–1857). The ovary, likewise, is covered by a transparent membrane. However, internally it appears yellowish and the parenchyma consists of larger, more finely granular cells than in the testis.

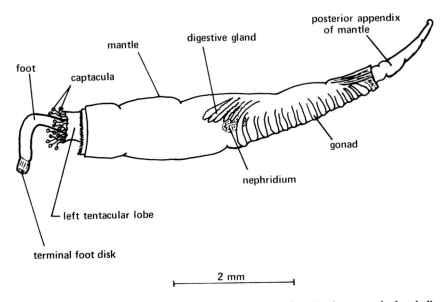

FIG. 2. Lateral view through mantle of a specimen of *C. fusiformis* with the shell removed showing the gonad in relation to other structures.

I have observed that at least in *Dentalium neohexagonum* Sharp and Pilsbry the structure of the posterior portion is distinctly sex-related. In the male, the posterior circular muscle is wider, the posterior appendix of the mantle is more blunt, and the musculature at the posterior edge of the posterior orifice of the mantle is thicker than in the female (Fig. 3). The female posterior appendix is longer and more tubelike than that of the male. Such differences may relate to gamete deposition. The male may require powerful musculature to eject sperms into the surrounding water, while the female requires a longer, tubelike posterior for oviposition on sediments.

4.3.2 Spermatogenesis

To my knowledge, spermatogenesis has not yet been studied in the Scaphopoda. This is the result of various difficulties. The major problem is that most specimens are already mature when collected. Lacaze-Duthiers (1856–1857) failed to observe the complete process of spermatogenesis. However, his limited observations on the testis led him to conclude that spermatogenesis in *Dentalium* was similar to the process in bivalves (Fig. 4). Mature sperms are primitive and have a pointed head. A constriction in the anterior of the body makes the posterior of the head the larger of the two portions (Lacaze-Duthiers, 1856–1857).

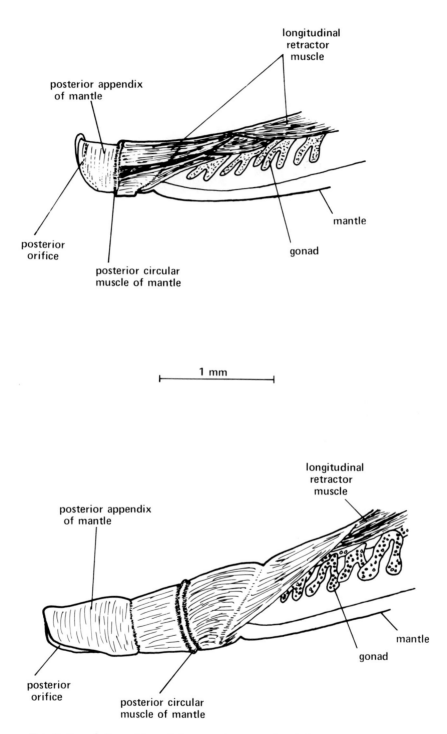

Fɪɢ. 3. Sexual dimorphism of the posterior appendix of the mantle in *D. neohexa-gonum.* Top: Male. Bottom: Female.

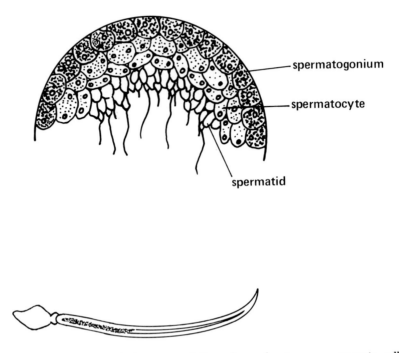

spermatogonium

spermatocyte

spermatid

Fig. 4. Top: Section of the testis of *D. vulgare* showing spermatogenic cells. (After Lacaze-Duthiers, 1856–1857.) Bottom: Drawing of a sperm of *D. occidentale*. Maximum width of head, 2 μm.

My observations on *D. neohexagonum* and *Dentalium occidentale* Stimpson 1851 confirm this description (Fig. 4).

4.3.3 Oogenesis

Nucleolar activity during oogenesis is much as it is in other molluscs (Raven, 1958). That is, the single nucleolus grows to 50 to 200 times its original volume. Nucleolar buds form cyclically to the end of oogenesis and then dissolve into the karyolymph (Arvy, 1950; Raven, 1958; Reverberi, 1971).

Yolk formation in the Mollusca has been described by Raven (1958), who stated that during oogenesis the cytoplasm in the primordial oocyte became heterogeneous and that numerous clear areolae appeared. Arvy (1950) observed this process in *Dentalium entale* Deshayes and noted that yolk granules were formed in or near these areolae. Also, the close connection between the mitochondria and the areolae led

Raven (1958) to suggest that the mitochondria may contribute to yolk granule formation.

The cytochemistry of oogenesis has been studied in *Dentalium* with respect to DNA and RNA. In most molluscs chromosomes are Feulgen-positive in early stages. This DNA reaction of the chromatin disappears during oocyte growth "until the tetrads become visible preceding maturation" (Raven, 1958). Timmermans *et al.* (1970) studied Feulgen reactions in eggs of *Dentalium dentale* Linné that had been (1) artificially released and (2) spontaneously released. In both cases a Feulgen-positive area was found in the cortex of the vegetal pole of the eggs immediately after breakdown of the germinal vesicle. They suggested several possibilities for the origin of these Feulgen-positive granules, including (1) origin attributable to the germinal vesicle, and (2) origin as a result of unmasking of the DNA as the nucleoplasm and cytoplasm mix during maturation. There have been fewer studies on the RNA content in *Dentalium* eggs. The cytoplasm of early oocytes is rich in RNA, but the RNA content decreases with growth, and Arvy (1950) observed a completely negative RNA reaction in mature eggs.

Development of eggs in scaphopods appears to occur with varying degrees of synchrony depending upon the species. While Rokop (1977) observed synchronous development of eggs in the scaphopod *Cadulus californicus* Pilsbry and Sharp, Lacaze-Duthiers (1856–1857) noted that eggs of *Dentalium vulgare* (DaCosta) did not develop synchronously but at widely varying rates. Figure 5 shows the stages of egg development as de Lacaze-Duthiers (1856–1857) observed them. Initially there is the uniformly immature stage (Fig. 5A). As oogenesis proceeds (Fig. 5B–E), the rate of egg development within a specimen of *Dentalium* varies, and both immature and full-grown oocytes may be seen in a specimen which is not yet ready for oviposition.

As oocytes mature they detach and accumulate in the oviduct, where they remain until the ovary is fully packed with mature, yellowish gametes. At this time the ovary may appear yellow to reddish brown and stippled in texture. The full-grown yellowish oocyte has a germinal vesicle which usually disappears when the eggs are shed (Lacaze-Duthiers, 1856–1857; Verdonk, 1968a) (Fig. 5F and G).

4.3.4 Spawning

According to Lacaze-Duthiers (1856–1857), when the gonad is filled with eggs, they are liberated singly. However, Dinamani (1964) described oviposition differently. According to Dinamani the female

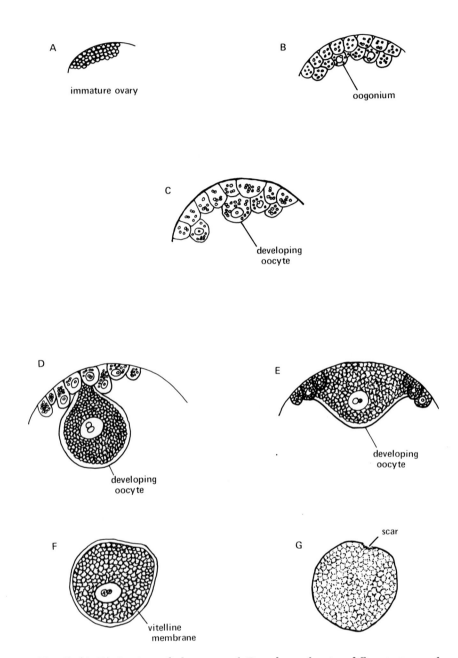

Fig. 5. (A–E) Sections of the ovary of *D. vulgare* showing different stages of oogenesis. (After Lacaze-Duthiers, 1856–1957.) (F) Full-grown oocyte of *D. neohexagonum;* diameter, 75 μm. (G) Spawned egg of *D. neohexagonum* after disappearance of the germinal vesicle. Diameter, 75 μm.

Dentalium conspicuum Melvill ejects a cloud from the posterior aperture. This cloud is deposited on the mud surface as a ribbon 6–7 cm long and consists of small groups of 4–5 eggs which total to about 1000 eggs in the entire ribbon.

When the eggs of both *D. antillarum* and *D. dentale* are deposited, they are ellipsoidal, with the animal and vegetal poles at the ends of the short axis (Verdonk, 1968a; Timmermans *et al.*, 1970). In the eggs of *D. dentale*, the disintegrating nucleolus is still visible, and six or eight alternating dark and light bands are noticeable at the vegetal pole, while the chromosomes are condensed near the animal pole. At 4–6 min after deposition the eggs are spherical (Timmermans *et al.*, 1970). In some cases the vitelline membrane remains with *Dentalium* eggs at deposition, but it is shed rapidly (Raven, 1958). The large, centrally located germinal vesicle is usually lacking in spontaneously shed eggs (Timmermans *et al.*, 1970). The "micropyle" noted by Lacaze-Duthiers (1856–1857) is merely a scar and not an entry point for sperms (Raven, 1958) (Fig. 5G).

I have found no recent papers describing detailed aspects of male spawning or what factor(s) initiate the process. Lacaze-Duthiers (1856–1857) observed, however, that sperms were ejected from the posterior aperture in bursts of milky clouds. Later Verdonk (1968a) noted that, when sperms of *D. antillarum* were artificially released by breaking the shell, they remained immobile for 20 min. Subsequently the sperms became active and fertilized the eggs.

Rokop (1974, 1977) has done the only work of which I am aware relating seasonality and spawning in scaphopods. He worked with *C. californicus*, a scaphopod found at depths of from 48 to 2340 m (Keen, 1971) and concentrated on seasonality studies for deep sea populations. He observed that ovarian length and oocyte sizes of deep sea *C. californicus* increased between October and July and then decreased between July and the subsequent October. The diameter of testicular follicles followed the same increase and decrease sequence in synchrony with the ovaries. However, oocyte sizes were different from one year to the next, indicating that the reproductive cycle, although synchronous and seasonal, was not completely predictable. Rokop (1977) discussed several possible advantages of reproductive seasonality in deep sea populations of *C. californicus* and proposed that very likely (1) it was advantageous to the species to spawn simultaneously with shallow water populations of the same species and/or (2) it increased the likelihood of fertilization because of the scaphopods' limited mobility and paucity in numbers.

4.4 Development

4.4.1 Fertilization and Cleavage

The beginning of maturation in *Dentalium* is spontaneous after spawning and does not require fertilization. It is between deposition and metaphase of the first reduction division that fertilization occurs. Subsequent to first metaphase only fertilized eggs continue to mature.

Both animal and vegetal pole plasm are present in recently deposited eggs of *Dentalium*. Initially the animal pole plasm is a disk of yolk-free protoplasm which increases slightly after deposition. After fertilization it grows rapidly, eventually covering the animal hemisphere when the polar bodies have developed. The vegetal pole plasm, which is continuous with an ectoplasmic layer, is initially dense and lacking in yolk granules.

Polyspermy in scaphopods has been demonstrated for *D. antillarum* by Verdonk's (1968a) studies. Wilson (1904a,b), however, observed that polyspermy only occurred rarely for *D. entalis* Linné. Tyler and Scheer (1937) confirmed Wilson's observations on *D. entalis* but showed that polyspermy could be induced in this species' eggs by transferring fertilized eggs to acidified seawater and reexposing them to sperms. One must conclude that polyspermy is variable, depending on the species, and may be subject to environmental conditions.

The two polar bodies form 15–35 min after fertilization (Verdonk, 1968a). Wilson (1904a) noted the association of polar body extrusion and ameboid movement in *Dentalium*. Ameboid movements were strongest at the vegetal pole. Raven (1958) relates this phenomenon to similar activity in other molluscan classes and notes that it precedes a transition to polar lobe formation at the vegetal pole. Formation of the first polar lobe occurs 55 min after fertilization in *Dentalium* (Verdonk, 1968a), and spiral cleavage begins with the development of this polar lobe. Verdonk (1968a) has succinctly described initial cleavages:

> While the lobe rounds off, a cleavage furrow is formed at the animal side of the egg and a little later also at the vegetative side. When the egg reaches the trefoil stage no connexion between the lobe and the blastomeres can be observed. After 4–5 min., a connexion between the polar lobe and one of the blastomeres appears and subsequently the lobe fuses with the CD blastomere.
>
> About 30 min after the beginning of first cleavage a second polar lobe is formed which is well rounded with respect to the four blastomeres.
>
> Finally, at third cleavage, a polar lobe is present, which, however, does not separate completely from the D-blastomere. A first quartette of micromeres is

split off at the animal side of the egg. In subsequent cleavages, which follow each other rapidly a polar lobe is not formed.

The cortex is divided among blastomeres with cleavage patterns being determined by cortical factors. Transection of the polar lobe indicates that cytoplasmic activity is independent of the nucleus. Such transection results in cortex changes, hence determination of the position and direction of cleavage spindles (Wilson, 1904a,b; Raven, 1958).

4.4.2 Localization of Morphogenetic Factors

Much of the research on scaphopod embryology has centered on determination of the factors influencing morphological characteristics and is reviewed by Reverberi (1971). Wilson (1904a,b), in his classical studies on germinal localization in *Dentalium*, removed the polar lobes to determine localization of the factor(s) controlling apical tuft formation. The apical tuft appears in the late gastrula and later becomes the velum. He concluded that germinal prelocalization was definitely indicated in *Dentalium*. Raven (1958) questioned Wilson's conclusions, because in many organisms centrifugation of uncleaved eggs produces many normal embryos. Raven felt that centrifugation did not displace "organ forming substances." He suggested that the "substances are (at least in part) pre-formed but they are not prelocalized."

However, Wilson's conclusion has been confirmed. Verdonk (1968a) studied the effects of removal of the first polar lobe and of centrifugation, using apical tuft formation as an index for results. Like Wilson, he showed that apical tuft formation was principally controlled by the polar lobe. He also noted that apical tuft formation was affected by centrifugation and finally concluded that "the morphogenetic factors of the first polar lobe are restricted to its cortex." Subsequently he noted that bilaterality and dorsoventrality were predetermined in the uncleaved egg (Verdonk, 1968b). Geilenkirchen et al. (1970) varied the percentage and the site of removal of the polar lobe at different times during the development of *D. dentale* and demonstrated that the apical tuft factor was initially in the animal side of the first polar lobe. In other attempts to determine the exact factors that determine the development of certain morphological factors Timmermans et al. (1970) and Verdonk et al. (1971) observed that removal of the Feulgen-positive granules from the cortex of the vegetal half of the uncleaved egg did not influence apical tuft formation. They concluded that the DNA particles were related to posttrochal or adult features.

Verdonk et al. (1971) performed quantitative experiments to determine the influence of removal of the vegetal half of the first polar lobe on apical tuft formation. They found that removal of up to 70% by volume of the vegetal half of the first polar lobe reduced posttrochal development but did not influence the apical tuft. However, when more than 70% was removed, the tuft was not formed. The authors thus concluded that the "polar lobe area is quantitatively determined in the uncleaved egg."

The various studies on localization of morphogenetic factors suggest the importance of both the cortex and DNA particles in the uncleaved egg for the normal development of *Dentalium*. Too, the studies support Wilson's (1904a,b) conclusion that certain directive substances are prelocalized in the *Dentalium* egg.

4.4.3 Blastula and Gastrula

The blastula is composed half of small and half of large cells. This cell size difference serves to distinguish the animal and vegetal poles. The latter consists of the larger cells and very quickly becomes flattened preceding invagination and formation of the archenteron (Kowalevsky, 1883) (Fig. 6A). The process of gastrulation has been observed as early as 4 hr after initial cleavage (Verdonk, 1968a). Although Wilson (1904b) has shown that part of the vegetal portion of the egg is essential to normal gastrulation, this part is neither the first nor the second polar lobe. Loss of polar lobes, however, prevents mesoderm development (Crampton, 1896).

The invaginated gastrula has a wide, transverse axis. Large blastomeres from the external surface detach, entering the blastocoel where in late stages they arrange irregularly but bilaterally near the blastopore and give rise to the mesoderm (Kowalevsky, 1883) (Fig. 6B).

The gastrula becomes free-swimming with a ciliated tuft and three groups of ciliated cells at the midpoint of the body (Fig. 6C). The blastopore in the early gastrula opposes the cephalic pole (Fig. 6B), later shifting toward the ventral surface (Kowalevsky, 1883) (Fig. 6C). As the preoral portion assumes a conical shape, the posterior portion lengthens and the cells around the blastopore become the stomodeum. The latter eventually becomes the buccal mass.

4.4.4 Trochophore and Veliger Stages

The larva begins to take the shape of a trochophore, although it lacks the excretory organ and neither the anus nor the cerebral ganglion plate have appeared. According to different studies the trochophore

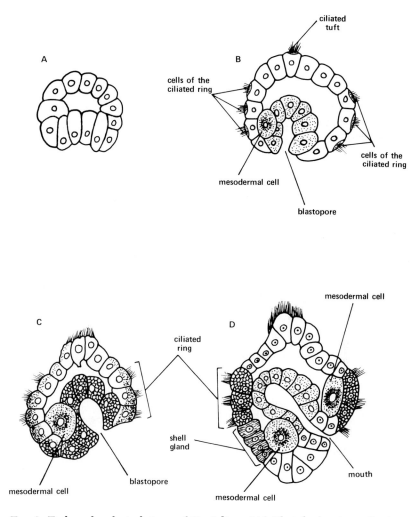

FIG. 6. Early embryological stages of *Dentalium*. (A) Blastula showing cell orientation. (B) Early gastrula showing mesoderm development. (C) Free-swimming gastrula. (D) Fourteen-hour larva. (After Kowalevsky, 1883.)

stage may develop as early as 3 hours after initial cleavage (Verdonk, 1968a) or as late as between 24 and 37 hours after fertilization (Lacaze-Duthiers, 1856–1857): Kowalevsky, 1883). Since the different authors studied different species, the time at which the trochophore develops may be species-dependent. To my knowledge there are no studies indicating what part environmental factors play in affecting the developmental stages of scaphopods.

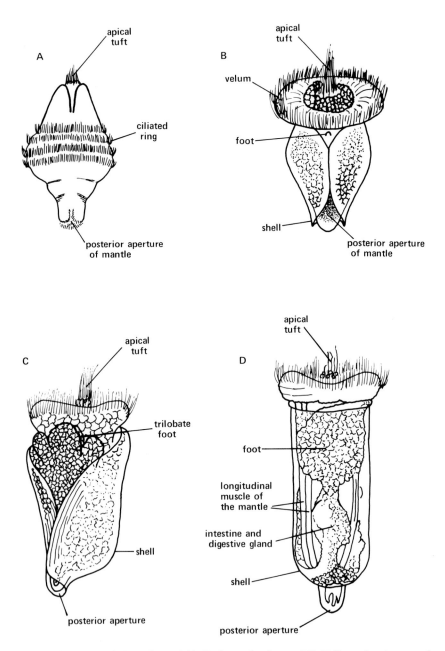

FIG. 7. Larvae of *Dentalium*. (A) Early trochophore. (B) Veliger showing mantle development. (C) Later veliger showing development of the shell and trilobate foot. (D) Prodissoconch stage. (After Lacaze-Duthiers, 1856–1857.)

In a trochophore the pretrochal portion is completely covered with short cilia except for the apical tuft. The tuft is a group of "long, flexible but non-vibratile sense hairs" found at the animal pole (Raven, 1958).

Before shell development the posttrochal region lengthens as the pretrochal portion decreases in size (Fig. 6D). Subsequently the rudiments of the mantle develop and the two folds grow laterally, meeting ventrally first at the posterior of the animal (Fig. 7A). The folds of the developing mantle enclose a swelling that will soon become a trilobate foot (Lacaze-Duthiers, 1856–1857).

Concurrently the nervous system develops. First, two ectodermal depressions, the presumptive cephalic plates, appear near the ciliated tuft and invaginate. As the invaginations become deeper, they meet the walls of the stomodeum, thicken, and detach to become ganglia (Korschelt and Heider, 1900; Raven, 1958).

As the ciliated tuft and ciliated ring are reduced, a definitive velum becomes evident (Fig. 7B). The veliger continues to swim freely for 1–3 days as the prodissoconch develops from the mantle in a manner similar to bivalve shell development. In scaphopods, however, the shell joins to become cylindrical (Fig. 7B–D).

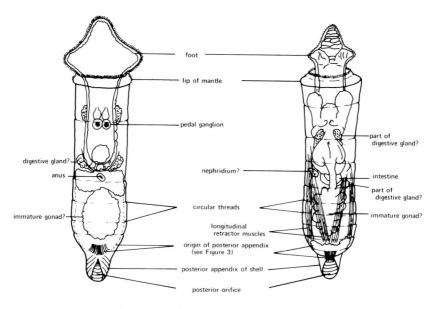

Fig. 8. Left: Juvenile of *Dentalium* shortly after settling. Right: Thirty-five-day-old juvenile of *Dentalium*. No dimensions given in original plate. (After Lacaze-Duthiers, 1856–1857.)

The foot grows, and the tentacular apparatus (the captacula) develops from an ectodermal outgrowth posterior to the mouth. The anus develops late in the Scaphopoda.

As veliger metamorphosis slows, the velum is gradually lost; its cells decreasing in size and losing nuclei. Before the cilia are completely shed, the animal drops to the sea floor to move about sluggishly, partly by use of the foot and partly aided by the velum. Soon the cilia are shed, and the organism takes on external adult characteristics (Fig. 8A), thus ending its free-swimming stage, which lasted from 2 to 6 days (Lacaze-Duthiers, 1856–1857).

It is noteworthy that, although Lacaze-Duthiers (1856–1857) followed scaphopod development to the thirty-fifth day after fertilization, he did not identify any development of the gonads. However, his drawings appear to show immature gonads (Fig. 8). The last 29–30 days of his study demonstrated growth and specialization of the organs of the anterior portion of the body (Fig. 8B) and total loss of the velum with consequent adult crawling and burrowing behavior.

Acknowledgments

This study was made possible in part by a grant from the National Capital Shell Club. I would like to take this opportunity to thank William W. Carter for extensive technical assistance and Stanford University for the research facilities and ship privileges necessary for completion of this study. I am also grateful to Dr. Melbourne Carriker for his critical review of the manuscript. This is University of Delaware College of Marine Studies Contribution No. 114.

References

Arvy, L. (1950). Données histologiques sur l'ovogénèse chez *Dentalium entale* Deshayes. *Arch. Biol.* **61**, 187–195.

Crampton, H. E. (1896). Experimental studies on gastropod development. *Arch. Entwicklungsmech. Org.* **3**, 1.

Dinamani, P. (1964). Burrowing behavior in *Dentalium. Biol. Bull. (Woods Hole, Mass.)* **126**, 28–32.

Gainey, L. F., Jr. (1972). The use of the foot and the captacula in the feeding of *Dentalium* (Mollusca: Scaphopoda). *Veliger* **15**, 29–34.

Geilenkirchen, W. L. M., Verdonk, N. H., and Timmermans, L. P. M. (1970). Experimental studies on morphogenetic factors localized in the first and the second polar lobe of *Dentalium* eggs. *J. Embryol. Exp. Morphol.* **23**, 237–243.

Keen, M. (1971). "Sea Shells of Tropical West America: Marine Mollusks from Baja California to Peru." Stanford Univ. Press, Stanford, California.

Korschelt, E., and Heider, K. (1900). "Textbook of Embryology of Invertebrates" (M. F. Woodward, ed.; transl. by M. Bernard). Macmillan, New York.

Kowalevsky, A. (1883). Etude sur l'embryogénie du Dentale. Ann. Mus. Hist. Nat. Marseille, Zool. 1(7).

Lacaze-Duthiers, F. J. H. (1856–1857). Histoire de l'organisation et du développemente du Dentale. Ann. Sci. Nat. Zool. 46/7), 5–51, 171–255.

Raven, C. P. (1958). "Morphogenesis: The Analysis of Molluscan Development." Pergamon, New York.

Reverberi, G. (1971). Dentalium. In "Experimental Embryology of Marine and Freshwater Invertebrates" (G. Reverberi, ed.), pp. 248–264. North-Holland Publ., Amsterdam.

Rokop, F. J. (1974). Reproductive patterns in the deep-sea benthos. Science 186, 743–745.

Rokop, F. J. (1977). Seasonal reproduction of the brachiopod Frieleia halli and the scaphopod Cadulus californicus in the deep sea. Mar. Biol. 43, 237–246.

Timmermans, L. P. M., Geilenkirchen, W. L. M., and Verdonk, N. H. (1970). The accumulation of Feulgen-positive granules in the egg cortex of Dentalium dentale L. J. Embryol. Exp. Morphol. 23, 245–252.

Tyler, A., and Scheer, B. T. (1937). Inhibition of fertilization in eggs of marine animals by means of acid. J. Exp. Zool. 75, 179.

Verdonk, N. H. (1968a). The effect of removing the polar lobe in centrifuged eggs of Dentalium. J. Embryol. Exp. Morphol. 19(1), 33–42.

Verdonk, N. H. (1968b). The relation of the two blastomeres to the polar lobe in Dentalium. J. Embryol. Exp. Morphol. 20(1), 101–105.

Verdonk, N. H., Geilenkirchen, W. L. M., and Timmermans, L. P. M. (1971). The localization of morphogenetic factors in uncleaved eggs of Dentalium. J. Embryol. Exp. Morphol. 25(1), 57–63.

Wilson, E. B. (1904a). Experimental studies on germinal localization. I. The germ-regions in the egg of Dentalium. J. Exp. Zool. 1(1), 1–74.

Wilson, E. B. (1904b). Experimental studies on germinal localization. II. Experiments on the cleavage-mosaic in Patella and Dentalium. J. Exp. Zool. 1, 197–268.

Yonge, C. M. (1937). Circulation of water in the mantle cavity of Dentalium entalis. Proc. Malacol. Soc. London 22(6), 333–336.

Chapter 5

PELECYPODA (EXCLUDING OSTREIDAE)

A. N. Sastry

5.1 Introduction 113
5.2 Asexual Reproduction 114
5.3 Sexual Reproduction 114
    5.3.1 Sexual Dimorphism 114
    5.3.2 Sex Determination and Hermaphroditism 115
    5.3.3 Anatomy of the Reproductive System 120
    5.3.4 Origin of Gonads and Germ Cells 122
    5.3.5 Gametogenesis 126
    5.3.6 Gametogenic Cycles 131
    5.3.7 Factors Influencing Gametogenesis 137
    5.3.8 Reproductive Behavior 200
    5.3.9 Spawning 202
5.4 Development 217
    5.4.1 Embryonic Development 217
    5.4.2 Larvae 232
    5.4.3 Factors Influencing Larval Growth, Settling, and Metamorphosis .. 239
    5.4.4 Juvenile Development 255
    References 265

5.1 Introduction

Pelecypod reproduction has been studied extensively, but much of the literature is concerned with gonad development and reports on breeding periods. Reproduction is cyclical, and it may be annual, semi-annual, or continuous. Reproduction may be divided into three major phases: gonad development, spawning and fertilization, and development and growth. These phases, functioning continually in coordination with seasonal environmental changes, produce the pattern characteristic of a species. The timing and duration of reproductive activity may be determined through an interaction between endogenous and exogenous factors. The synchronization of breeding periods with the environmental conditions most favorable for the development and growth of progeny is obviously significant for reproductive success. In this chapter, an attempt has been made to outline present knowledge of pelecypod reproduction with an ecological approach and relevant physiological data. Reference to the family Ostreidae is made only for continuity of discussion, and a detailed treatment of this group may be found in this volume, Chapter 6. General accounts of reproduction in pelecypods have been given in reviews by Lankester (1900), Franc (1960), Morton (1960), Galtsoff (1961), Allen (1963), Fretter and Graham (1964), and Purchon (1968).

5.2 Asexual Reproduction

Sexual reproduction is the rule in pelecypods.

5.3 Sexual Reproduction

5.3.1 Sexual Dimorphism

The majority of pelecypods are dioecious, gonochoric, or unisexual. Hermaphroditism occurs in only a few species. Sexual dimorphism is very rare. In some pelecypods the sexes can be distinguished by gonad coloration. Except in a very few species, there are no copulatory organs or accessory sex glands. Accessory genital organs functional in the male phase have been described in the protandric hermaphrodites *Cuspidaria (Naera) cuspidata* (Grobben, 1892) and *Xylophaga dorsalis* (Purchon, 1941). In *X. dorsalis*, a pair of seminal vesicles is found in the male phase, and an accessory genital organ is thought to assist in directing the spermatozoa into the opening of the organ (Figs. 1 and 2).

Sexual dimorphism has been reported to occur in the genera *Lampsilis* and *Astarte* by Lankester, (1900). In *Lampsilis*, the females have broader shells than the males. In *Astarte* males, the border of the shell is smooth, while it is crenulated in females. Saleuddin (1964) reported that the shell of females of *Astarte elliptica, A. sulcata,* and *A. borealis* was larger than that of males. In the commensal bivalve *Montacuta percompressa*, females are larger than the parasitic males (Jenner and McCrary, 1968; Chanley and Chanley, 1970). Except in these few cases, the sexes in pelecypods can be distinguished only by examination of the gonads or from observations made during spawning.

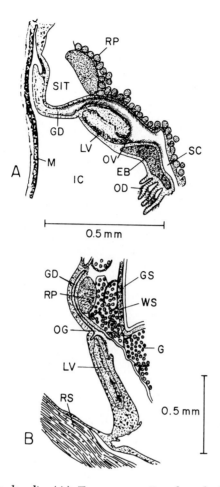

Fig. 1. *Xylophaga dorsalis.* (A) Transverse section through the seminal vesicle. (B) Longitudinal section through the seminal vesicle. EB, Efferent branchial vein; G, gonad tubules; GD, genital duct; GS, gastric shield; IC, infrabranchial cavity; LV, lumen of seminal vesicle; M, mantle; OD, outer demibranch; OG, opening of genital duct; OV, orifice of seminal vesicle; RP, retractor pedis muscle; RS, retractor muscles of siphonal process; SC, supra-branchial cavity; SIT, split in tissues; WS, wall of stomach. (After Purchon, 1941.)

5.3.2 Sex Determination and Hermaphroditism

Pelecypods exhibit wide variation in the expression of sexuality, ranging from strictly gonochoric species to those that are invariably functional hermaphrodites. Sexuality may vary among species belonging

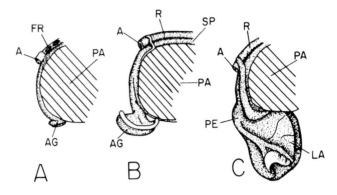

Fig. 2. *Xylophaga dorsalis*. The accessory genital organ in various stages of development. (A) Small, from a female. (B) Medium, from a female or hermaphrodite. (C) Large, from a male. A, Anus; AG, accessory genital organ; FR, feces in rectum; LA, lamella; PA, posterior adductor; PE, peduncle of accessory genital organ; R, rectum; SP, dorsal surface of posterior adductor. (After Purchon, 1941.)

to the same genus and also within populations of the same species. Occasionally, hermaphroditic indivduals are found in species considered strictly gonochoric. According to Coe (1943a), antecedent cells of both sperms and eggs are found in the gonads, except in strictly gonochoric species. If the individuals are to become hermaphrodites, the primary gonad gives rise to both spermatocytes and oocytes. In some dioecious species, the primary gonad may be ambisexual and some oocytes are formed in the early spermary, but they undergo cytolysis during spermatogenesis.

Species reported to be strictly gonochoric are *Modiolus demissus*, *Mya arenaria*, *Barnea truncata*, *Petricola pholadiformis*, *Mytilus califorianus*, *Septifer bifurcatus*, *Donax gouldi* and *Anomia simplex* (Fretter and Graham, 1964), *Placopecten magellanicus* (Drew, 1906; Posgay, 1950; Naidu, 1970), *Chlamys tigerina*, *C. striata*, *C. furtiva* (Reddiah, 1962), *Mulinia lateralis* (Calabrese, 1969a), *Macoma balthica* (Lammens, 1967), and *Brachidontes recurvus* (J. F. Allen, 1962a,b). Occasionally, a few hermaphroditic individuals are found in *Mytilus edulis* (Thorson, 1936; Sugiura, 1962), *Mya arenaria* (Coe and Turner, 1938), *Modiolus modiolus* (Wiborg, 1946), *Placopecten magellanicus* (Merrill and Burch, 1960; Naidu, 1970), *Spisula solidissima* (Ropes, 1968a), and *Modiolus barbatus* (Cahour and Lucas, 1968).

Hermaphroditism in pelecypods has been discussed by Lankester (1900), Pelseneer (1912), Coe (1943a, 1945), Fretter and Graham (1964), and Houtteville and Lubet (1975). Coe (1943a, 1945) proposed

four categories of hermaphroditism in pelecypods: (1) functional hermaphroditism or functional ambisexuality, (2) consecutive sexuality,
(3) rhythmic consecutive sexuality, and (4) alternate sexuality.

The majority of hermaphroditic bivalves are functional hermaphrodites. In these species, spermatozoa and ova are produced simultaneously,
and self-fertilization may occur under the appropriate circumstances.
The best known example in this category is *Argopecten (= Aequipecten)*
irradians in which both types of gametes are produced in distinct parts
of the gonad and reach maturity at about the same time. In *A. irradians,*
the spermatozoa mature slightly earlier than the oocytes, but spawning
begins only after both types of gametes have matured. During spawning,
the discharge of spermatozoa and eggs is normally separated by a brief
period, thereby reducing the chances of self-fertilization (Sastry, 1963).
In *Teredo diegensis* most of the mature individuals are functionally
hermaphroditic following a protandric phase (Coe, 1943a). A small
percent of individuals in this species never change sex and remain true
males. In these individuals, the oocytes are aborted. In *Pecten maximus*
and *Chlamys opercularis* few individuals are distinctly protandric, and
at each spawning the testicular portion of the gonad is spent before
the eggs mature in the ovarian region (Fretter and Graham, 1964). In
Pecten latiauritus var. *monotimeris* on the southern California coast, a
period of sexual inactivity separates the initial male and female phases
(Coe, 1943a). In functional hermaphrodites there appears to be a
tendency toward protandry.

According to Fretter and Graham (1964) functional hermaphroditism
occurs in the genera *Pecten, Chalamys, Tridacna, Tivela, Thracia,*
Cardium, Poromya, Montacuta, Lasaea, Teredo, and *Cetoconcha,* but not
all species belonging to these genera fall into this category. For example,
Cerastoderma (= Cardium) edule has separate sexes, and *Montacuta*
ferruginosa is a consecutive hermaphrodite, whereas *M. substriata* is
a functional hermaphrodite.

In consecutive hermaphrodites, there is a single sex reversal in life.
The sex change normally takes place from a younger male to an older
female phase. Sometimes, a reverse sequence occurs with a regular
proterogynic condition, as in the case of *Kellia suborbicularis* and
Montacuta ferruginosa (Oldfield, 1961). Consecutive sexuality may be
expressed during a single spawning season, with one sexual phase
following the other without an intervening period of recuperation, or
when the spawning period occurs only once a year the sexual phases
may be separated by several months. In *Mercenaria mercenaria,* there is
initially only a functional male phase which is followed by a strictly
male and female phase in the adult. Although the majority of individuals

are protandric and a few are proterogynic, the primary gonads in young adults are distinctly ambisexual, forming a graded series from predominantly male to female. A small number of purely gonochoric individuals also occur in *Mercenaria mercenaria*. In *Bankia setacea*, the protandric female may experience a second sexual phase if the length of life of the individual is prolonged (Coe, 1945). Sometimes the male and female phases in *B. setacea* overlap, resulting in self-fertilization. A small portion of the population may consist of true males that never transform into females and females that have omitted the male phase.

Some species exhibit rhythmic consecutive sexuality, such that sex reversal is repeated, either annually or at closer intervals, and leads to a more-or-less equal number of male and female phases. This occurs in the larviparous oysters *Ostrea lurida, O. edulis,* and *O. equestris* (Gutsell, 1926; Orton, 1927a; Coe, 1932; see also this volume, Chapter 6). In *Teredo navalis,* the sex change from a primary male phase to a female phase is sometimes followed by a second sequence of male and female phases, although mortality during the second male phase is high and only a few individuals reach the second female phase. In a few individuals of this species, the initial sexual phase is female (Grave and Smith, 1936; Coe, 1936).

In alternate sexuality, a sex change occurs in adults that function seasonally as separate sexes. In these species, the sex change and its timing are erratic, hence it is not possible to predict the next phase an individual will assume during the reproductive season. In *Crassostrea virginica* about 70% of the young are functionally male at the first spawning season, while at the second spawning season the numbers of males and females are equal. Females are found in greater numbers among older individuals. A small portion of true males and females also occurs in this species (see this volume, Chapter 6).

The causes underlying the hermaphroditic state in pelecypods have been discussed by Orton (1927b), Coe (1943a, 1945), Tranter (1958a), Franc (1960), and Fretter and Graham (1964). From discussions by various authors, it appears that the hermaphroditic state might be determined by an interaction of genetic and environmental factors. However, there are no experimental studies showing the phenotypic or genetic bases of sex determination in pelecypods. Coe (1943a, 1945) has suggested that there are hereditary modifying factors for activation or suppression of either the male- or female-determining components at certain stages in development, and that these components are responsive to environmental conditions. Coe (1945) has also indicated that a specific hormone activating spermatogonia and another activating oogonia may be released when the appropriate physiological state is reached.

A simultaneous release of both hormones is suggested to result in functional hermaphrodites. The other types of sexuality may depend upon the sequence of hormones produced and the time elapsing between successive periods of production.

Orton (1927b) suggested that metabolic rather than genetic factors determined the sex in bivalves that experience a sex change. The sex change in *Ostrea edulis* was thought to be due to the particular metabolic pathway followed by the animal during its two sex phases. Maleness was associated with carbohydrate metabolism and femaleness with protein metabolism. It was suggested that the accumulation of by-products of one type of metabolism above a certain concentration stimulated the sex change.

Tranter (1958a) has suggested that *Pinctada albina,* a protandric hermaphrodite, inherits the ability to respond differentially to the environment in order to achieve the sexual phase suited for its nutritional condition. Individuals with low food reserves may reproduce less efficiently as females than as males. Germ cell rudiments might respond to the food reserve level, developing toward maleness if at a lower level and toward femaleness if at a higher level. In *Argopecten* (= *Aequipecten*) *irradians* the development of spermatozoa and occytes is influenced by temperature and food level (Sastry, 1966a, 1968). When animals are maintained at 15°C with food, oocytes fail to grow, but spermatozoa develop. At 20°C, however, both spermatozoa and oocytes develop in the presence of food, although oocyte growth is suppressed if food is absent. Awati and Rai (1931) found that the proportion of males in *Crassostrea cucullata* increased when these animals were infested with the pea crab, *Pinnotheres.* The influence of environmental conditions on sex determination in *Anodonta cygnea* is indicated by sex ratios and by hermaphroditic races in populations from different geographic locations (Bloomer, 1939). In *Tivela stultorum* both unisexual and bisexual races are found (Coe, 1945).

The genetic basis of hermaphroditism and gonochorism has been discussed by Montalenti and Bacci (1951) and Montalenti (1960). These authors hypothesize that various combinations of four pairs of genes concerned with sex determination produce male and female phases which are long and short, respectively, or equally long. Gonochoric males and females occur as a result of crossing-over and recombination. Ahmed and Sparks (1970) suggested that crossing-over and recombination might occur in *Mytilus edulis* and *M. californianus,* since both species reproduce as true males and females. A few cytological investigations have been made of the chromosomes in pelecypods, but in none of the species examined so far does there appear to be any evidence of sex chromosomes (Wada, 1978). The bivalents of *Mercenaria mercenaria* and *M. campechiensis* and their hybrids (Menzel, 1968; Menzel and Menzel, 1965) and *Mytilus edulis* and *M. califorianus* (Ahmed and Sparks, 1970) are all homomorphic, isopycnotic, and autosomal. A detailed discussion of problems in sex determination in marine animals is given in the review by Montalenti (1960) and in the books by Bacci (1965) and Reinboth (1975).

According to Fretter and Graham (1964) the hermaphroditic state in molluscs is a more primitive condition than gonochorism. Ghiselin (1969) suggested that hermaphroditism in animals may evolve under conditions where it is difficult to find a mate, when one sex benefits from being larger than the other, and where there are genetically isolated populations. He pointed out that in the deep sea, where the diversity of species is high and the population density is low, simultaneous hermaphroditism might be quite common. Ockelmann (1958) found that between 22 and 58% of the total number of lamellibranch species in the arctic region of eastern Greenland were hermaphrodites of one type or another.

5.3.3 Anatomy of the Reproductive System

A general description of the reproductive system in pelecypods is given by Lankester (1900), Morton (1960), Galtsoff (1961), and Purchon (1968). The gonad of pelecypods is a paired structure and, in some species, is fused in the middle. The reproductive system is very simple, consisting of paired gonoducts, main genital canals, and numerous minor canals which terminate in a network of follicles or alveoli. In many species, the gonad attains maximum bulk before spawning and occupies a large part of the visceral mass. The gonads are diffuse in many species and may penetrate into other body organs. In *Mytilus*, *Anomia*, *Placuna*, and *Brachidontes* the gonads penetrate into the mantle (Lankester, 1900; White, 1937; Sastry, 1955; J. F. Allen, 1962a). In *Kellia*, *Lasaea*, and *Montacuta* (Oldfield, 1961) and *Astarte* (Saleuddin, 1965) the gonads are lobular and are interspersed among the digestive tubules. The gonad in Pecten is a tonguelike mass attached to the anterior side of the adductor muscle (Dakin, 1909).

The reproductive system of *Pecten* has been described in detail by Dakin (1909). In the hermaphroditic species of *Pecten* the gonads are differentiated into an anterior testicular portion and a posterior ovarian region. The gonad surrounds part of the alimentary canal and consists of numerous genital canals which terminate as a fine network of follicles. The genital canals become confluent, forming the gonoducts which open

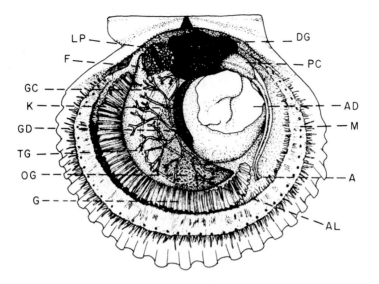

FIG. 3. Reproductive system of *Argopecten* (= *Aequipecten*) *irradians*. A, Anus; AD, adductor muscle; AL, alimentary canal; DG, digestive gland; F, foot; G, gills; GC, genital canals; GD, gonoduct; K, kidney; LP, labial palps; M, mantle; OG, ovarian portion of the gonad; PC, pericardium; TG, testicular portion of the gonad.

into the kidney. The main genital canals are lined by an epithelial layer of ciliated columnar cells. The epithelial layer is supported by connective tissue which surrounds the follicles and the genital ducts. The ciliated columnar epithelium passes into a flattened germinal epithelium which lines the follicles. The main genital canals and the gonoduct in the reproductive system of *Argopecten* (= *Aequipecten*) *irradians* are shown in Fig. 3. The reproductive system in *Mytilus edulis* has been described in detail by White (1937). In *M. edulis*, the gonad consists of paired gonoducts, five major genital canals on each side of the body, and numerous minor canals which terminate as follicles (Fig. 4). The minor genital canals become confluent and become major canals which unite at a point below the pericardium to form the gonoduct. According to White (1937), "the genital duct passes through the mantle and then turning posteriorly runs along the ventral side of the body and parallel with the axis of its inner gill." The gonoduct opens on a prominent papilla located across the ventral side of one of the posterior byssus retractor muscles.

The histology of the alveoli and gonoducts in the reproductive system of *Mya arenaria* has been described by Coe and Turner (1938) and Stickney (1963). The gonoducts and their branches in many respects

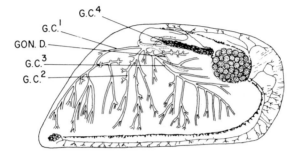

F<small>IG</small>. 4. The reproductive system of *Mytilus edulis* showing the genital canals. G.C.[1], Major genital canal from the anterior region of the mantle and liver; G.C.[2], from the middle region of the mantle; G.C.[3], from the posterior region of the mantle; G.C.[4], on the floor of the pericardium; GON.D., gonoduct. (From White, 1937; modified.)

resemble the alveoli in their histological structure. Stickney (1963) described a genital vesicle in *M. arenaria* and an enlarged oviduct in *Spisula solidissima* that serves as an egg reservoir prior to spawning. Descriptions of the reproductive system and of the histological structure of the gonad have been given for *Argopecten (= Aequipecten) irradians* (Gutsell, 1930), *Mercenaria mercenaria* (Loosanoff, 1937a), *Protothaca (= Paphia) staminea* (Quayle, 1943), *Mytilus edulis*, *Chlamys varia* and *Pecten maximus* (Lubet, 1959), *Pinctada albina*, *P. margaritifera*, and *P. fucata* (Tranter, 1958b,c), *Macoma balthica* (Lammens, 1967), and *Placopecten magellanicus* (Naidu, 1970).

5.3.4 Origin of Gonads and Germ Cells

5.3.4.1 O<small>RIGIN OF</small> G<small>ONADS</small>

The gonads originate from a group of mesodermal cells located in the posterior portion of the body near the visceral ganglion and the ventral side of the pericardium (Wada, 1968). The primordial germ cells multiply and become separated into two groups situated symmetrically on the two sides of the body. Each group of cells continues to multiply and extend anteriorly into the surrounding mesenchyme or vesicular connective tissue to form tubular follicles (Coe, 1943b). In *Mercenaria mercenaria* (Loosanoff, 1937a) and *Venus striatula* (Ansell, 1961a) the gonad first develops as a thin layer of cells anterior to the kidney, between the pericardium and stomach, and proliferates anteriorly to the visceral mass. In *Pinctada albina*, the primordia that gave rise to the gonad are located posterior to the urogenital papilla (Tranter, 1958b). Primary gonad development has been described in *M. mercenaria*

(Loosanoff, 1937a), *Mya arenaria* (Coe and Turner, 1938), *Prothothaca* (= *Paphia) staminea* (Quayle, 1943), *Mytilus edulis* (Lubet, 1959), *Pinctada albina* (Tranter, 1958b), and *Macoma balthica* (Lammens, 1967). For a detailed discussion of the history of primordial germ cells and their continuity, the papers by Woods (1931, 1932) on the freshwater mussel *Sphaerium striatinum* should be consulted.

The formation of follicles and the pattern of gonad development have been described in *Teredo navalis* (Coe, 1943b), *Pinctada albina* (Tranter, 1958b), and *Mytilus edulis* (Lubet, 1959). In *P. albina*, the primordial germ cells spread through the connective tissue and become surrounded with a thin membrane to which the primary germ cells are attached. The follicles enlarge as flattened tubules and branch out and infiltrate the connective tissue between the external epithelium and viscera. As the follicles extend, the primary germ cells are carried to sites of future gonad development. The primary follicles increase in length by spreading as far as possible and by giving off side branches into the connective tissue. The secondary follicles are similar to the primary follicles, except that they have primary germ cells at the tip. When the follicles are established, the primordial germ cells stop multiplying and their derivatives (gonial cells) begin to multiply. In *P. albina*, as the gonad develops and as the gametes are formed, the follicles increase in volume and spread along well-defined paths from the genital pore region on either side of the body (Fig. 5).

According to Coe (1943b), two distinct types of gonads are found in pelecypods. In *Bankia* spp., *Barnea* spp., *Petricola pholadiformis*, and *Teredo navalis*, cells of the gonads differentiate into follicle cells and primary gonial cells. The follicle cells function as nutritive cells and occupy most of the spaces within the follicles, while the primary gonial cells are scattered along the periphery (Fig. 6). The gonads of the genera *Pecten, Mytilus, Modiolus,* and *Paphia* are composed of gonial cells with only minute follicle cells. In this case the gametogenic cells obtain their nutrients from the surrounding connective tissue. A condition intermediary between the above two types is found in *Anomia simplex* in which the follicles of immature gonads are surrounded by vesicular nutritive cells but, when maturity is reached, the nutritive cells are much reduced (Fig. 7).

5.3.4.2 ORIGIN OF GERM CELLS

The formation of germ cells in the hermaphroditic species *Teredo navalis,* and in some other pelecypods, has been discussed by Coe (1943b). After the primordial germ cells branch out to form the follicles, two types of cells, nutritive cells and primary gonial cells, become distinguishable. At the tip of each growing follicle, a group of un-

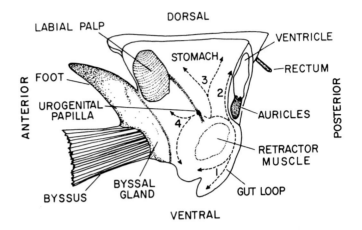

FIG. 5. The spreading of gonad follicles in *Pinctada albina*. The numbered dashed lines indicate the course of spreading of the primary follicles from the genital pore regions on either side of the body. 1, The first path leads around the dorsal edge of the retractor muscle, across the base of the gut loop, and spreads down its length; 2, the second path branches off from the first near the heart and extends up the posterior face of the stomach; 3, the third path, anterior to the heart, bifurcates early and spreads across the flat, lateral face of the stomach; and 4, the fourth path spreads anteriodorsally immediately inside the gills, past the base of the foot and the byssal gland, toward the labial palps. (After Tranter, 1958b.)

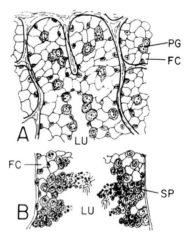

FIG. 6. Portions of primary gonads of *Petricola pholadiformis*. (A) Sexually undifferentiated stage; follicles composed mainly of vacuolated, nutritive cells (FC) with primary gonia (PG) along the central axis or lumen (LU). (B) Later stage in young male; most of the follicle cells have been cytolyzed to supply nourishment for the proliferating spermatogenic cells (SP). (From Coe, 1943b; modified.)

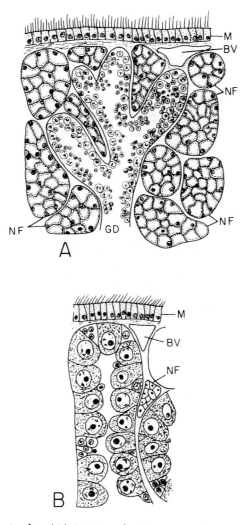

Fɪɢ. 7. *Anomia simplex.* (A) Portion of primary gonad surrounded by nutritive tissue. (B) Follicles of mature ovary in which the nutritive tissue has been almost assimilated. BV, Blood vessels; GD, genital duct; M, mantle; NF, nutritive tissue. (From Coe, 1943b; modified.)

differentiated nuclei is found in a mass of cytoplasm. The nuclei become enclosed to form large, vacuolated cells that function as nutritive cells while the follicles continue to grow. However, a small number of nuclei retain a large portion of cytoplasm and become primary gonial cells. As the follicle continues to grow, the primary gonial cells are left behind on the periphery of the follicle wall. The primary gonial cells multiply as the follicles increase in length. In *T. navalis*, sexual differentiation of the gonial cells becomes apparent when the animals reach a length of about 10 mm. As soon as the primary gonia multiply, they begin to proliferate into groups of nuclei with cytoplasm syncytia. In *T. navalis*, equal numbers of primary gonia differentiate into spermatogonia and oogonia. The formation of germ cells follows the general pattern for other pelecypods, such as that described for *Pinctada albina* (Tranter, 1958d) and *Mytilus edulis* (Lubet, 1959). In *P. albina*, the germ cells of both sexes originate from the primordial cells dispersed around the follicle wall. The primary germ cells are irregular and oval and measure approximately 7–10 μm. One or two nucleoli may be present, but they are not very prominent. The nucleus stains diffusely, showing reticulate filaments, with an indistinct outline which emerges into the surrounding cytoplasm. In *M. edulis*, the primordial germ cells have nuclei with reticulate filaments and cytoplasm rich in RNA (Lubet, 1959).

In addition to primordial germ cells and nutritive cells, a number of other intrafollicular cells also occur in the gonads of pelecypods (Tranter, 1958d; Lubet, 1959; Wilson and Hodgkin, 1967). Tranter (1958d) has described three types of intrafollicular cells in *Pinctada albina* and suggested that they have a resorptive function. Among these, type-A cells are amebocytes and have a phagocytic function (Fig. 8). The phagocytic cells in the follicles have been suggested to contribute nutrients to sex cells during active gametogenesis and to resorb residual cells in the spent follicles (Loosanoff, 1937a,b; Tranter, 1958d; Wilson and Hodgkin, 1967). The function of two other cell types, B and C, is not known. Reviews of bivalve amebocytes and their properties have been published by Takatsuki (1934) and Korringa (1952).

5.3.5 Gametogenesis

There have been many published reports on gametogenesis in pelecypods. Some of these include the work on *Mercenaria mercenaria* and

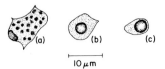

Fig. 8. Intrafollicular nongerm cells. (a) Cell type A. (b) Cell type B. (c) Cell type C. (After Tranter, 1958d.)

Cyprina islandica (Loosanoff, 1937a,b,c, 1953), *Pinctada* spp. (Tranter, 1958c,d, 1959), *Mytilus edulis, Chlamys varia,* and *Pecten maximus* (Lubet, 1959), pectinids (Reddiah, 1962), *Brachidontes recurvus* (J. F. Allen, 1962a,b), *Argopecten (= Aequipecten) irradians* (Sastry, 1963, 1966a, 1968, 1970a), *Chlamys varia, Glycymeris glycymeris, Venus striatula,* and *Scrobicularia plana* (Lucas, 1965), *Macoma balthica* (Lammens, 1967), *Mulinia lateralis* (Calabrese, 1969a), and *Placopecten magellanicus* (Naidu, 1970). In many of these species, gametogenesis follows the same basic pattern.

5.3.5.1 SPERMATOGENESIS

Spermatogenesis has been described in detail from observations of histological sections of gonads of *Mercenaria mercenaria* (Loosanoff, 1937c), *Mytilus edulis* (Lubet, 1959), and *Pinctada albina* (Tranter, 1958d). In *P. albina,* the primary germ cells (stem cells) give rise to a small number of round cells through a series of mitotic divisions. These cells have been termed definitive spermatogonia. The spermatogonial stages between primary germ cell and definitive spermatogonia show a decrease in cytoplasmic volume. The definitive spermatogonia directly give rise to primary spermatocytes which lie free in the follicle as a concentric band just inside the layer of spermatogonia. The primary spermatocytes give rise to secondary spermatocytes, but these are not usually observed because the division rate is very rapid at this stage, the secondary spermatocytes quickly being transformed into spermatids. The spermatids differentiate into spermatozoa which lie as a core in the lumen of the follicle. As the mass of spermatozoa increases in volume, there is a corresponding decrease in the number of cells in the earlier stages, so that the follicles are eventually filled with spermatozoa. The sperm head in *P. albina* is about 1.5 μm in width and spade-shaped, and the tail is about 30 μm long. The acrosome is located in the anterior region of the head as a cone-shaped structure. The sequence of spermatogenic stages is shown in Fig. 9. A similar sequence of spermatogenic stages has been observed in a number of other species (Loosanoff, 1937b,c, 1953; Quayle, 1943; Shaw, 1965; Ropes and Stickney, 1965; Sastry, 1963, 1966a; Calabrese, 1969a; Ropes, 1968b; Naidu, 1970).

Retzius (1904) and Franzén (1955) investigated spermatogenesis in a number of pelecypods and found no morphological differences in spermatozoa of different species except that the nucleus might be elongated in primitive forms. In *Nucula* and *Dosina,* the spermatozoa are usually elongated or conical, but in other species they are either oval or spherical. The acrosome has varied configurations. The acrosome is a long, pointed rod in front of the nucleus in *Mytilus edulis,* whereas in other species a morphologically demarcated acrosome is missing, as

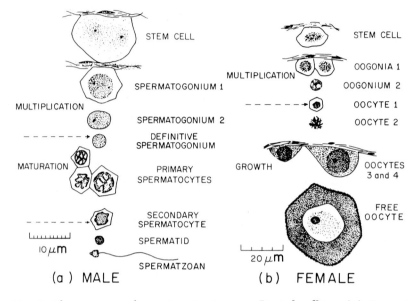

FIG. 9. The sequence of gametogenic stages in *Pinctada albina*. (a) Spermatogenesis. (b) Oogenesis. (After Tranter, 1958d.)

in the case of *Thracia*. Longo and Dornfeld (1967) studied the fine
structure of spermatids in *M. edulis* during differentiation and found
that separate Golgi bodies individually produced numerous proacrosomal
granules which progressively coalesced to form the acrosomal vesicle.
Anderson and Personne (1976) reviewed sperm types in molluscs and
also discussed glycogen metabolism in relation to sperm motility. Bourcart
et al. (1965) described the ultrastructure of spermatozoa of *Mytulis
perna*. The ultrastructure of the midpiece of molluscan spermatozoa has
been described by Anderson and Personne (1976).

Atypical, apyrene, and vermiform spermatozoa have been reported to
occur in *Montacuta tenella* in addition to the normal spermatozoa
(Ockelmann, 1965). Atypical and normal spermatozoa are formed in the
same gonad. Atypical spermatozoa are reported also in the bathyal-
abyssal bivalve *Kellia symmetros* (Ockelmann, 1965). The significance
of atypical spermatozoa is not clear, but they are thought to have a
secretory function concerned either with the formation of typical
spermatozoa or with spawning processes. Oldfield (1961) found that
spermatids in *Montacuta substriata* and *M. ferruginosa* entered larger
cells, called cytophores, which also originate from the germinal layer.
The cytophores are thought to serve a nutritive function. The relationship

between cytophores and spermatozoa in these species appears to continue until after transfer of sperm to the females. Deroux (1961) also observed oblong and round capsules in the gills of *Mysella bidentata* and thought them to be cytophores.

Loosanoff (1937b) in *Mercenaria mercenaria* and Coe and Turner (1938) in *Mya arenaria* have reported atypical spermatogenesis. The follicles contain multinucleated spherical bodies close to the periphery, which develop from the same type of spermatogonia as the normal ones. These multinucleate cells undergo several nuclear divisions while still surrounded by the original mass of cytoplasm contained within the cell membrane. Coe and Turner (1938) found in some cases that these cells broke apart and continued development to normal spermatozoa, but usually they became pycnotic and cytolyzed.

There are no detailed investigations on the time sequence or amount of synchrony during spermatogenesis within an individual. However, observations on spermatogenesis in a number of species indicate synchrony during early stages of differentiation and again when the gonads are approaching maturity. During active spermatogenesis, however, there is considerable overlap of different stages. In *Pinctada albina*, the male follicles in early stages of development contain only primary germ cells and spermatogonia (Tranter, 1958d). A pause appears to occur in the spermatogenic cycle when differentiation has proceeded to the stage of definitive spermatogonia, the gonads resting at this stage. When spermatogenesis proper begins, rapid mitotic division produces spermatogonia in large numbers, and the entire follicle may be filled with spermatocytes and spermatogonia. Later, spermatids and spermatozoa appear in the center of the follicle lumen, while the earlier stages form a peripheral band. For a certain period, the formation of spermatocytes and spermatozoa keep pace with each other. As the follicle begins to fill with spermatozoa, spermatocyte formation slows down. In the mature state, spermatozoa predominate, while there are only a few spermatocytes and primary germ cells. In a number of other species considerable overlap of different spermatogenic stages during the period between spermatogonia formation and development of spermatozoa has been reported (Loosanoff, 1937b; Shaw, 1965; Ropes and Stickney, 1965; Sastry, 1966a).

5.3.5.2 Oogenesis

Oogenesis has been described in detail from histological and histochemical investigations of gonads of *Pinctada albina* (Tranter, 1958d) and *Mytilus edulis* (Lubet, 1959). The same basic pattern appears to be followed during oogenesis in these and many other species. The primary germ cells undergo mitotic division and give rise to primary oogonia. The primary oogonia are attached to the wall of the follicle, and some of them rest in this stage, while others may divide immediately.

The oogonia growing along the wall of the follicle enter meiosis; however, a certain number of primary and secondary oogonia may remain in the resting stage. The oocytes formed following the multiplication stage of the secondary oogonia begin to grow. This continues until the end of oogenesis, although oocyte growth may be separated into previtellogenesis and vitellogenesis stages (Lubet, 1959; Raven, 1961). During the previtellogenesis stage, the leptotene stage of meiotic prophase marks the beginning of the oocyte growth phase (Fig. 9). However, oocyte growth during the period of previtellogenesis is not rapid.

During vitellogenesis, the nucleus enlarges and the cytoplasm accumulates yolk platelets, lipid droplets, and mucopolysaccharide granules (Lubet, 1959). The mitochondria are numerous, and at the end of vitellogenesis they disperse toward the cell periphery. Golgi bodies are also numerous. Oocyte shape changes with the accumulation of reserve products in the cytoplasm. Oocytes are attached by slender stalks to the follicle wall. The lipid inclusions give the characteristic oocyte coloration. The lipid granules in oocytes of *Mytilus edulis* appear to be similar to those in connective tissue (Lubet, 1959). The oocytes completing the growth phase are surrounded by a polysaccharide "chorion" membrane. After ovulation, the follicles are filled with oocytes, and they may be separated by a nongranular substance which appears to be a component of the cell membrane (Tang, 1941; Merrill and Burch, 1960; Lammens, 1967). Further details on oogenesis can be obtained from Lubet (1959) and Raven (1961). Hubner and Anderson (1976) gave a comparative overview of the structural aspects of spiralian oogenesis. Cytological investigations including electron microscopic studies of oocytes in vitellogenesis and unfertilized eggs of *M. edulis* have been published by Humphreys (1962, 1967), Reverberi and Mancuso (1961), and Reverberi (1966, 1967). The cytochemistry of oocyte development in bivalves has been reviewed by Cowden (1976).

There are no detailed studies on the synchrony of oogenesis within individuals in pelecypods. Although the sequence of stages is the same, the duration of successive stages and the amount of synchrony may differ among species. Most published observations on oogenesis in pelecypods have been made on samples of animals, hence they provide only limited information on synchrony within individuals. In a number of species considerable overlap of different stages usually appears within individuals during oogenesis (Loosanoff, 1937b, 1953; Quayle, 1943; Tranter, 1958d; Sastry, 1966a, 1968, 1970a; Lammens, 1967; Allen, 1969;

Naidu, 1970). Generally, different follicles are out of phase with each other. Follicles are lined with primary germ cells and oogonia in the early phases of oogenesis, but later all stages of oogenesis can be found. In *Argopecten (= Aequipecten) irradians* a greater degree of synchrony is observed, especially before the beginning of the oocyte growth phase (Sastry, 1966a, 1968, 1970a). As oocytes begin the growth phase, a few are seen growing into the lumen of the follicle. Tranter (1958d) indicated that a break in oogenesis separated the events before and after the formation of primary oocytes. In *Mercenaria mercenaria*, some oocytes grow, while others remain small (Loosanoff, 1937b). Oogonia and oocytes may be found with primary germ cells interspersed between them. As development proceeds, primary germ cells, oogania, and oocytes in various stages are found in the follicles. Tranter (1958d) has suggested that the stretching of follicles in the early stages of oogenesis keeps pace, but that later in development the growth of young oocytes is delayed until space becomes available after the liberation of oocytes into the lumen of the follicle. It has also been suggested that an arrest of meiotic activity may occur in growing oocytes. In *A. irradians*, all stages in oocyte development occur during the growth phase, but only a few of the earlier stages are found when the gonads are mature and the population is spawning (Sastry, 1966a, 1970a).

5.3.6 Gametogenic Cycles

Generally, the reproductive cycle of a population includes a series of events, namely, activation, growth and gametogenesis, ripening of gametes, spawning, and a quiescent or resting period (Giese, 1959a, 1976; Giese and Pearse, 1974; Sastry, 1966a, 1975). Underlying the reproductive cycle is the gametogenic cycle within the indivduals of a population. The time course of the series of events in the reproductive cycle of a population may be synchronous, so that all individuals breed simultaneously, or asynchronous with gametogenic cycles within the individuals of a population out of phase, so that a proportion of the population may breed at any time. In populations exhibiting an asynchronous breeding period, the individuals may go through successive gametogenic cycles, for example, some tropical and deep-sea animals (Giese and Pearse, 1974; Rokop, 1974).

There have been many studies on the reproductive cycle of pelecypods. In many of these, the reproductive cycle has been divided into a number of stages based on microscopic examination of histological sections of gonads. However, there is little agreement as to the number of stages and the criteria by which these stages have been separated by different authors. Tranter (1958a,c, 1959) determined the seasonal cycle of gonad activity from gonad histology and the macroscopic appearance of gonads of *Pinctada* spp. collected at intervals throughout the year. The extent of gonad development compared to the condition in ripe animals was also determined for the same samples for a quantitative measurement of gonad volume index (Fig. 10). The two expressions of gonadal growth show a basic similarity, indicating that the same phenomenon is being considered. Sastry (1966a, 1970a) determined the gonad index [(gonad weight × 100)/body weight] for samples of

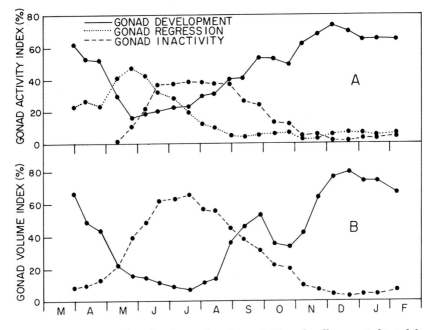

Fig. 10. (A) Seasonal cycle of gonad activity of *Pinctada albina* as indicated by the gonad histology of samples taken at intervals from March 1954 to February 1955. The plotted values for each fortnight represent the extent of development, regression, and inactivity of the gonad relative to a maximal value (100%) for each of the three phases. (B) Seasonal cycle of gonad development as indicated by the macroscopic gonad appearance of the same series of samples. Regressing gonads are grouped with developing ones to yield an overall "gonad volume index" for each sample; each plotted value on the development curve represents the volume occupied by the gonads of the sample relative to that when all gonads are full (100%). The plotted values on the inactivity curve are the percentage frequencies of empty gonads. All curves have been smoothed using a moving average of three. (From Tranter, 1958a.)

Argopecten (= Aequipecten) irradians populations collected at intervals during the year and described the trend of events in the reproductive cycle. Gonadal tissue of the same samples was examined to obtain quantitative and qualitative data on gametogenic activity and to classify the gametogenic cycle into various stages. The data so obtained were used to determine the timing and duration of successive events in the reproductive cycle. The population of *A. irradians* from Beaufort, North Carolina, exhibited an annual reproductive cycle (Sastry, 1966a) which included the following events: a vegetative phase, growth and gameto-genesis, maturation, spawning, and a resting stage. The gonad index for the population of *A. irradians* increased following the vegetative phase and reached a maximum just before spawning (Fig. 11). With spawning, the gonad index decreased, to remain subsequently at a sustained low level during the resting period. Variation in the gonad index of samples examined at intervals during the reproductive period may give an indication of deviation from synchrony in the extent of gonad growth within a population.

Primary germ cells and gonial cells develop in *Argopecten (= Aequi-pecten) irradians* during the vegetative phase. This is followed by the oocyte growth phase and an increase in the gonad index. During gonad

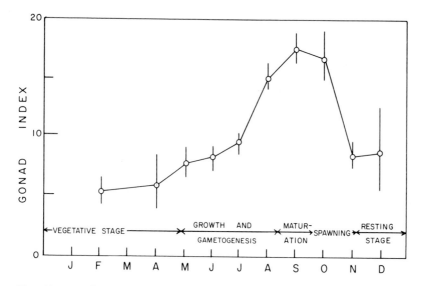

FIG. 11. Annual reproductive cycle of *Argopecten (= Aequipecten) irradians* from Beaufort, North Carolina. The mean gonad index [(gonad weight × 100)/body weight] and the 95% confidence intervals (vertical lines) are shown in relation to histological stages. Samples collected during 1965.

development, oocytes in different stages are found within the gonad
of an individual. When the gonad index reaches a peak value during
the reproductive cycle, full-grown oocytes and spermatozoa predominate
in the gonads. Mean oocyte size in samples of animals collected at
intervals, correlated with the stages in the gametogenic cycle, has been
used to assess the average stage of gonad development (Fig. 12). From
such data, the timing and duration of the gametogenic cycle has been
determined for a population of A. irradians (Sastry, 1966a).

Lammens (1967) determined the growth rate of oocytes in *Macoma
balthica* by measuring the volume of oocyte nuclei during autumn and
winter. The volume of oocyte nuclei shows considerable variation among
different specimens in monthly samples and, as well, among separate
oocytes within each ovary. The variation in these two respects increases
with the progress in oocyte development during the entire period of
oogenesis, suggesting that the members of the population are found in
different stages of gonad development. The oocytes in specimens with
delayed gonad development grow rapidly and mature in spring, since
all the members in the population are mature before spawning. In *M.
balthica* nuclear growth shows a contrasting relationship to cytoplasmic
volume depending on the season. The increase in cytoplasm predominates
in autumn, while nuclear growth is relatively faster than that of cytoplasm in winter. During the first phase of oogenesis, the nucleocyto-
plasmic ratio decreases, to be followed by a considerable increase in
the month before spawning. The changes in ooplasmic RNA and nucle-

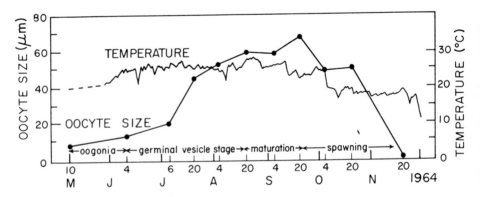

Fig. 12. Changes in average oocyte size of *Argopecten* (= *Aequipecten*) *irradians*
collected at biweekly intervals during the reproductive cycle from Beaufort, North
Carolina. The stages in oocyte development are shown on the bottom of the graph.
The temperature line shows the daily noontime sea temperature in the vicinity
where scallops have been collected. (From Sastry, 1966a.)

olar RNA in growing oocytes of *Spisula solidissima* have been studied by Swift *et al.* (1956).

Spermatogenic cycles have also been determined for a number of species. They follow the same general pattern as in the females of a population (Loosanoff, 1937b, 1953; Ropes and Stickney, 1965; Naidu, 1970). In dioecious species, gonadal development follows a similar pattern in both sexes, and they reach a mature state at about the same time. In a functionally hermaphroditic species such as *Argopecten* (= *Aequipecten*) *irradians*, the primary germ cells and spermatogonia develop during the vegetative phase; then, further differentiation takes place with an increase in the gonad index (Sastry, 1966a, 1970a).

Gametogenic cycles have been described for populations of a number of species. Some of these include the work on *Mercenaria mercenaria* and *Cyprina islandica* (Loosanoff, 1937b, 1953), *Protothaca* (= *Paphia*) *staminea* (Quayle, 1943), *Venus striatula* (Ansell, 1961b), *Brachidontes recurvus* (J. F. Allen, 1962a,b), *Pinctada albina*, *P. margaritifera*, and *P. fucata* (Tranter, 1958a,c, 1959), *Mytilus edulis*, *Chlamys varia*, and *Pecten maximus* (Lubet, 1959), *Mya arenaria* (Ropes and Stickney, 1965), *Spisula solidissima* (Ropes, 1968b), *Argopecten* (= *Aequipecten*) *irradians* (Sastry, 1963), *Chlamys varia* and *C. distorta*, *Glycymeris glycymeris*, *Venus striatula*, *Scrobicularia plana*, and *Donax trunculus* (Lucas, 1965), *D. cuneatus* (Rao, 1967), *D. faba* (Alagarswami, 1966), *Chlamys varia* and *C. distorta* (Reddiah, 1962), *Mytilus edulis planulatus*, *brachidontes* cf. *variabilis*, *Septifer bilocularis*, *Xenostrobus pulex*, and *Amygdalum glaberrimum* (Wilson and Hodgkin, 1967), *Zirphaea crispata* (Allen, 1969), *Cerastoderma* (= *Cardium*) *edule* and *C. glaucum* (Boyden, 1971), *Mytilus edulis* and *M. galloprovincialis* (Brenko, 1971), *Cerastoderma* (= *Cardium*) *edule* (Gimazane, 1971), *Katelysia opima* and *Mytilus viridis* (Nagabhushanam and Mane, 1975a,b), and *Macoma secta* and *M. nasuta* (Rae, 1978).

Gametogenic cycles may occur in populations of a species on an annual, semiannual, or continuous basis. The timing and duration of the gametogenic cycle and the number of cycles within a year may be characteristic of specific populations. In populations of *Mercenaria mercenaria* and *Cyprina islandica* (Loosanoff, 1937b, 1953), *Mytilus edulis* (Chipperfield, 1953; Sugiura, 1959; Moore and Reish, 1969), *Macoma balthica* (Lammens, 1967), *Argopecten* (= *Aequipecten*) *irradians* (Sastry, 1966a, 1970a), *Spisula solidissima* (Ropes, 1968b), *Placopecten magellanicus* (Naidu, 1970), and others, the gametogenic cycle occurs on an annual basis. In the population of *Mercenaria mercenaria* from Long Island Sound, primary germ cells and gonial cells begin development immediately after a brief recuperative period following spawning in early

September. Gametogenesis occurs in autumn and spring and even proceeds slowly during the winter (Loosanoff, 1937b). The population of *C. islandica* from Point Judith, Rhode Island, begins gametogenesis immediately after spawning in late October, but vigorous development only occurs after the winter (Loosanoff, 1953). The populations of *A. irradians* from Nantucket Sound, Massachusetts, and Beaufort, North Carolinia, exhibit gametogenic cycles on an annual basis during the first year of life. After spawning, the second-year animals in the Beaufort population remain in a resting stage through late autumn and winter and then resume gametogenic activity in early spring, perhaps reaching maturity by April (Sastry, 1966a, 1970a; Kirby-Smith, 1970; see also Figs. 11 and 12. In contrast, the population in Nantucket Sound shows asynchrony in gametogenic activity after spawning at the end of the first year. However, the second-year animals of this population mature at the same time as the first-year animals.

In some species, successive cycles in gametogenesis may be interrupted by a brief or a prolonged period of gonad inactivity. In *Mytilus edulis* (Chipperfield, 1953), *Barnea candida* (Duval, 1962), and *Macoma balthica* (Lammens, 1967) gonads of both sexes are similar in appearance during the resting period, as they are before gamete proliferation. After spawning, the gonads are filled with vacuolated cells, and occasionally residual gametes are retained. Subsequently, the gonads regress in both sexes. In others, for example, *Argopecten (= Aequipecten) irradians*, the follicles shink after the discharge of gametes (Sastry, 1966a, 1970a). Even shrinkage of follicles does not occur in *Cyprina islandica*, and they are filled with vacuolated cells (Loosanoff, 1953). In *Mercenaria mercenaria* (Coe and Turner, 1938) and *Protothaca staminea* (Quayle, 1943), after a period of shrinkage and collapse, the gonads are filled with new growing follicles.

Gametogenesis in some tropical species appears to be restricted to certain months in the year, while in others it occurs continuously throughout the year. In *Donax cuneatus* from the Madras coast, India, gametogenesis takes place between September and December, and it is followed by spawning until June. The gonads are inactive for a brief period between July and August (Rao, 1967). *Donax faba* from the Mandapam coast, India, are found in different stages of gametogenesis throughout most of the year, although a greater percentage of the population has inactive gonads in January and August (Alagarswami, 1966). In *Pinctada albina* from Thursday Island, Australia, the highest frequency of ripe gonads occurs during the warmer months and spent gonads predominate during the cooler months (Tranter, 1958a). In a second species, *Pinctada margaritifera*, there are two distinct gametogenic cycles; one in spring and a second one in autumn, each followed by spawning (Tranter, 1958c). In a third species, *Pinctada fucata*, gonad development takes place between September and November, and spawning occurs between December and May. The population has inactive gonads from June to August (Tranter, 1959). In all three species, some individuals with ripe gonads are found outside the major spawning period.

The timing, duration, and number of gametogenic cycles may vary in populations of a species occurring in different parts of the geographic range. The populations of *Mytilus edulis* from the southern California coast (Moore and Reish, 1969) and the Japanese coast (Sugiura, 1959) have mature gametes for a longer period and recover much faster than the populations on the British coast (Chipperfield, 1953). The populations of *M. edulis* in Alamitos Bay, southern California, exhibit the gametogenic cycle on an annual basis but both sexes in populations from different stations vary in terms of gamete maturity and the average length of time required to reach a given stage of gamete development (Moore and Reish, 1969). Renzoni and Giusti (1972) reported variations in the reproductive cycle of *Mytilus galloprovincialis* at 17 stations along the northern shore of the Tyrrhenian Sea in the south of Italy. The populations of *Mya arenaria* in areas north of Cape Cod have an annual gametogenic cycle, while the populations in southern areas exhibit a semiannual cycle of gametogenesis (Ropes and Stickney, 1965). The population of *Argopecten (= Aequipecten) irradians* in Alligator Harbor, Florida, exhibits an annual gametogenic cycle during the first year of life, but after spawning at the end of the first year they rapidly recover, resume gametogenesis, and become mature by late autumn (Sastry, 1963).

5.3.7 Factors Influencing Gametogenesis

The reproductive cycle of a species is a genetically controlled response to the environment (Sastry, 1970a). The pattern of the reproductive cycle in a species is apparently determined through the coordination of successive reproductive events with changes in the external environment. In species occurring in several climatic zones, the reproductive cycle may vary in relation to the local environment as a phenotypic response of a single genotype, or it could be distinctly genetic, or both (Sastry, 1970a). Giese (1959a) and Giese and Pearse (1974) reviewed the influence of exogenous and endogenous factors on the annual reproductive cycle of marine invertebrates including pelecypods. Recent studies indicate that a reproductive response is produced through an interaction of environmental factors—especially temperature, salinity, light, and food—and endogenous factors within an organism. After attaining a certain physiological state, an organism exposed to the required environmental prerequisites begins gonad growth and gametogenesis. The influence of some environmental factors on gonad growth and gametogenesis has been determined experimentally, but the mechanisms coordinating the physiological processes underlying the events within the organism and the changes in the environment are not clearly understood. It is likely that neuroendocrine activity plays a significant role in coordinating the physiological processes, the reproductive events within the organism, to produce a reproductive response relative to changes in the external environment (Sastry, 1970b, 1975). The successive events in the reproductive cycle may be affected differently by various factors, and a detailed analysis requires a separate evaluation of each phase. Only a few studies have been conducted to determine the influence of environmental factors on successive events in the reproductive cycle of any marine invertebrate, hence our knowledge of the environmental interactions and mechanisms controlling the pattern of reproductive activity in a species is still fragmentary (Sastry, 1963, 1966a,b, 1968, 1970a,b; Sastry and Blake, 1971 and unpublished data). With the control of reproductive activity and stimulation of gametogenesis in animals maintained under controlled conditions, it should be possible to determine the effect of various environmental factors singly and in combination, and to study their interaction with endogenous factors to elucidate the mechanisms coordinating the reproductive response at the whole-organism level.

5.3.7.1 Temperature

The influence of temperature on the reproduction of marine invertebrates including pelecypods has been discussed in reviews by Orton (1920), Runnström (1927, 1936), Gunter (1957), Giese (1959a), Vernberg (1962), Kinne (1963, 1964, 1970), McWhinnie (1967), Hedgpeth and Gonor (1969), Loosanoff (1971), and Giese and Pearse (1974).

5.3.7.1.1 *Gametogenesis in Relation to Seasonal Changes in Temperature.* The period of gonad growth and gametogenesis in a number of species has been correlated with seasonal changes in temperature. In many species, gonad growth and gametogenesis seem to occur with declining temperatures in the fall or with increasing temperatures in spring and summer. In *Mercenaria mercenaria* from Long Island Sound, gametogenesis begins immediately after completion of spawning in September, and gametes develop nearly to maturity by December (Loosanoff, 1937b). The population goes through a winter hibernation period, when gametogenic activity is slow, and then resumes vigorous development of spermatozoa and oocytes again in late spring when the temperature reaches about 15°C. In spring, the undersized oocytes grow rapidly and the whole population reaches maturity by May. They remain in this condition until the temperatures reach the summer maximum of 23°–25°C and commence spawning in late July and early August. Gametogenesis in *Cyprina islandica* also begins with declining temperatures in late fall and resumes again in the spring after a period of winter hibernation (Loosanoff, 1953). In *Protothaca (= Paphia) staminea* from Vancouver, British Columbia, the proliferation of follicles and the formation of gonial cells occurs immediately after spawning in October, but the animals remain in this state until February and then begin oogenic growth, reaching maturity in April (Quayle, 1943). Gamete differentiation in this species seems to occur when the temperature is low, and oogenic growth takes place with increasing temperatures in spring. Calabrese (1969a) reported that gametogenesis occurred in *Mulinia lateralis* from Long Island Sound throughout the year, but that it proceeded at a slower rate during the winter.

In comparison, the American oyster, *Crassostrea virginica*, from Long Island Sound enters a prolonged resting period immediately after spawning in summer and remains inactive until the following spring (Loosanoff, 1965). Saleuddin (1964) has reported that *Astarte sulcata* from Millport, Scotland, passes through a prolonged resting period and begins gametogenesis in January when temperatures are at a minimum. Gamete development is accelerated later in spring. Gametes in this species grow slowly in summer, and spawning takes place in fall when the temperature declines from 15° to 10°C.

In the population of *Argopecten (= Aequipecten) irradians* from Massachusetts, primary germ cells and gonial cells develop in winter and early spring, and gamete differentiation begins in April. The population reaches maturity in July with an increase in temperature (Sastry, 1970a). The population of *A. irradians* from Beaufort, North Carolina, however, does not develop primary germ cells and gonial cells until spring, and gamete differentiation takes place with increasing temperature in May (Sastry, 1966a, 1970a). In some tropical species exposed to a slight annual variation in temperature gametogenesis also appears to be restricted to certain periods in the year (Tranter, 1958a,c, 1959; see also Section 5.3.6). These examples illustrate that reproductive timing and duration of gametogenic activity may be characteristic of a species under natural conditions.

In populations of tropical species extending into colder climatic zones of higher latitudes, gametogenesis seems to occur in warmer months, while in high-latitude species extending into warmer climatic zones, gametogenesis seems to occur outside the summer months. Von Oertzen (1972) reported that arctic bivalves in the Baltic Sea had ripe eggs and sperm for prolonged periods and shifted their breeding period to winter or early spring. Wilson and Hodgkin (1967) have determined the annual reproductive cycle of five species of mytilids from Fremantle, Western Australia, and correlated them with seasonal changes in temperature. These species show differences in histological changes accompanying the reproductive cycle, and in the time and duration of spawning. Among these, the boreal and temperate species *Mytilus edulis planulatus* and *Xenostrobus pulex* begin gametogenesis in autumn when temperatures are decreasing and remain in a mature state for a prolonged period before spawning in winter (Fig. 13). These two species remain reproductively inactive during the summer months. In comparison, the tropical species *Brachidontes* cf. *variabilis* and *Septifer bilocularis* initiate gametogenesis in early summer when temperatures are increasing and commence spawning when temperatures are at a maximum in the summer (Fig. 14). In *Amygdalum glaberrimum*, gametogenesis initiated in the autumn continues through winter and spring, and spawning takes place in the summer (Fig. 14). Apparently the pattern of the reproductive cycle and the accompanying gametogenic changes are influenced by the temperature, depending on the thermal history of the species and the geographic distribution of the population.

The pattern of the reproductive cycle of a species within the same geographic area may also vary in relation to differences in environmental temperatures that may occur from year to year. Ropes (1968b) determined the reproductive cycle of *Spisula solidissima* for a number of successive years off New Jersey. Gametogenesis begins in S. *solidissima* immediately after the completion of spawning in December, and the animals mature by May–June when the temperature is about 10°C. Spawning occurs in July and August before a rapid temperature rise from 9.6° to 18.9°C during the early part of September. The clams initiate a second gametogenic cycle in September and spawn again in October–November when the temperature is decreasing from 18° to 10°C. In years when fall temperatures are lower than usual, spawning is delayed and the second cycle does not occur.

5.3.7.1.2 *Stimulation of Gametogenesis.* In some pelecypods, gametogenesis can be induced outside the normal reproductive period and, in animals already undergoing gametogenesis, gamete development can be accelerated by exposing them to suitable temperatures. Loosanoff and Davis (1950, 1963) exposed *Mercenaria mercenaria* collected in winter to elevated temperatures from 5° to 20°C, which induced gametogenesis outside the normal breeding period. Clams brought from the field during the spring could be directly exposed to 20°C and required less time to reach the mature state. Females of *M. mercenaria* apparently require a longer exposure to elevated temperatures than

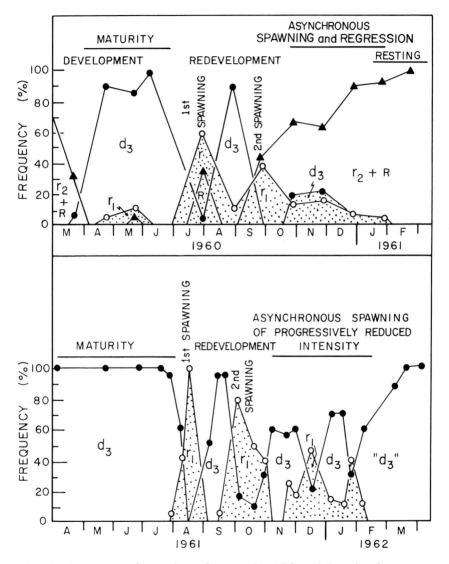

Fɪɢ. 13. Frequency polygon of gonad stages of mytilids with boreal and temperate affinities occurring in the Fremantle area (latitude 32° S), Western Australia. Top: *Mytilus edulis planulatus*; bottom: *Xenostrobus pulex*. R, d_1, d_2, d_3, r_1 and r_2 represent arbitrary stages of development and regression in the gonads. (From Wilson and Hodgkin, 1967; modified.)

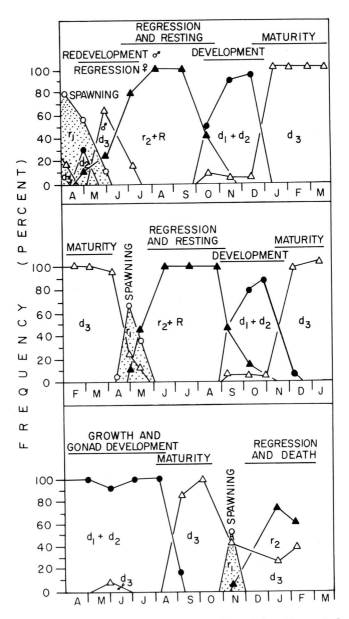

Fig. 14. Frequency polygon of gonad stages of mytilids with tropical affinities occurring in the Fremantle area (latitude 32° S), Western Australia. Top: *Brachidontes* cf. *variabilis*; middle, *Septifer bilocularis*; bottom, *Amygdalum glaberrimum*. R, d_1, d_2, d_3, r_1 and r_2 represent arbitrary stages of development and regression in the gonads. (From Wilson and Hodgkin, 1967; modified.)

males. Sastry (1963) induced the mature state in *Argopecten (= Aequipecten) irradians* from Florida by maintaining them at elevated temperatures in winter, while the population in the field normally does not mature until August. Acceleration of gamete development to spawning at elevated temperatures has also been achieved in *Patinopecten yessoensis* (Yamamoto, 1951b), *A. irradians* from Massachusetts (Turner and Hanks, 1960), *Spisula solidissima* (Ropes, 1968b), *Mytilus edulis* (Lubet, 1957, 1959; Bayne, 1965), *M. galloprovincialis* (Lubet and Bourcart, 1963), and *M. lateralis* (Calabrese, 1969a).

Sastry (1963) has suggested that gamete development to maturity in *Argopecten (= Aequipecten) irradians* can be accelerated after gametogenesis has been intiated and that the rate of development to maturation is dependent on temperature. Chipperfield (1953) has indicated that the rate of gametogenesis in *Mytilus edulis* is approximately proportional to the rate of temperature increase. Acceleration of gamete development up to spawning by elevating temperatures appears to be successful only after the animals have completed postspawning activity. Postspawning recovery involves complex physiological processes leading to the accumulation of nutrient reserves. Loosanoff and Davis (1952) and Bayne (1965) found that the technique of artificially inducing a mature condition was successful only with animals that had accumulated nutrient reserves and developed gonial cells at least to the multiplication stage. It is possible, however, as in the case of *Mercenaria mercenaria* and *A. irradians,* to induce gametogenesis several times by controlling environmental conditions, provided the animals can recuperate from each of the postspawning activities (Loosanoff and Davis, 1963; Sastry, 1966a, 1968).

Colder temperatures delayed gamete development to maturation in *Mercenaria mercenaria* (Loosanoff and Davis, 1963) and in *Mytilus edulis* (Bayne, 1965). *Mercenaria mercenaria* taken from Long Island Sound in May before the beginning of natural spawning and transplanted to colder waters in Maine (about 7°C lower) experienced a delay in gonad maturation and spawning. Bayne (1965) has reported that holding *Mytilus edulis* at 5°C for 3–4 weeks before natural spawning delays maturation and spawning. In other species, however, oocyte growth continues in animals maintained at colder temperatures, but after vitellogenesis the oocytes are cytolyzed, dissolved, and later resorbed (Sastry, 1966a, 1970b).

5.3.7.1.3 *Temperature Effects on Gametogenesis.* Sastry (1966a) determined the effect of different temperatures on successive events in the reproductive cycle of *Argopecten (= Aequipecten) irradians* from Beaufort, North Carolina (Table I). This population initiates gamete differ-

TABLE 1

Changes in Gonad Index, Digestive Gland Index, and Gametogenic Activity in the Reproductive Cycle of *Argopecten* (= *Aequipecten*) *irradians* [a]

Date collected	Size, mean ± S.D. (mm)	Average weight (gm)	Average body weight (gm)	Gonad index, mean ± S.D.	Digestive gland index, mean ± S.D.	Gonad condition		
						Average oocyte diameter (μm)	Color	Gametogenic stage
5/10/64	33.84 ± 4.49	11.33	3.39	5.5 ± 1.1	10.75 ± 1.72	8.0, range 5.2–13.0	Dark	Spermatocytes, oogonia, a few early oocytes
6/11/64 [b]	39.69 ± 3.31	16.86	5.52	9.6 ± 3.2	12.47 ± 1.29	13.0, range 7.8–15.0	Dark	Sperm, oogonia, oocytes
7/6/64 [b]	51.14 ± 4.71	32.64	12.28	15.1 ± 1.84	10.13 ± 1.14	20.3, range 10.4–45.0	Cream, light orange	Sperm, oocytes in germinal vesicle stage
7/20/64	56.33 ± 4.16	37.69	16.53	15.1 ± 1.86	9.80 ± 1.21	45.5, range 20.8–72.8	Cream, light orange	Sperm oocytes in germinal vesicle stage
8/4/64	56.20 ± 4.20	43.19	17.24	14.6 ± 1.30	8.50 ± 1.00	53.4, range 20.8–72.8	Cream, orange	Sperm, oocytes with orange pigmentation
8/20/64 [b]	59.36 ± 2.09	49.99	21.01	16.66 ± 4.41	9.93 ± 1.10	59.5, range 20.8–91.0	Cream, orange	Sperm, oocytes with germinal vesicle shrinkage

Date							
9/4/64	61.39 ± 3.74	52.85	22.02	15.52 ± 2.26	11.41 ± 0.9	58.5, range 26.0–104.0	Cream orange	Sperm, oocytes with germinal vesicle breakdown, a few eggs
9/20/64	64.85 ± 3.36	63.29	26.95	20.95 ± 2.9	10.26 ± 1.40	67.6, range 20.8–104.0	Cream, bright orange	Sperm, eggs, oocytes with germinal vesicle breakdown
10/4/64	65.39 ± 1.39	64.77	23.08	15.9 ± 2.9	11.46 ± 1.47	49.38, range 20.8–98.8	Cream, pale orange	Sperm, a few eggs, oocytes in germinal vesicle stage
10/20/64 [b]	63.00 ± 2.98	54.25	17.31	17.31 ± 1.8	12.05 ± 1.08	50.31, range 18.2–104.0	Cream, pale orange	Sperm, a few eggs, oocytes with germinal vesicle breakdown
11/20/64 [b]	64.75 ± 4.09	63.04	26.22	9.46 ± 2.5	14.2 ± 0.85	—	Pale brown	Residual sperm, inactive ovary
12/20/64	66.20 ± 3.8	69.20	31.26	10.71 ± 0.76	13.75 ± 1.36	—	Pale brown	Residual sperm, inactive ovary
1/20/65	71.01 ± 4.52	81.71	38.87	11.46 ± 1.23	13.35 ± 1.39	—	Pale brown	Inactive

[a] From Beaufort, North Carolina. From Sastry (1966a); modified.
[b] The effect of temperature and starvation on gonad growth and gametogenesis was determined for animals in different stages in the reproductive cycle.

entiation when the gonads have accumulated a minimum amount of body reserves during May–June and when the temperatures rise above 20°C (see Figs. 11 and 12). They reach maturity by the middle of September. Scallops exposed to 10°, 20°, and 30°C before the initiation of gonad growth and gametogenesis and kept without food resorbed the gonial cells. However, animals with a minimum amount of gonadal reserves and with oocytes in the beginning growth stages developed to maturity within the species-specific range of 20°–30°C. In scallops with accumulated gonad reserves and with oocytes in the growth phase, the time required for completing of gametogenesis decreased proportionately with the increase in temperature (Fig. 15). Animals held at 15° and 10°C failed to complete gametogenesis. At these low temperatures, the oocytes cytolyzed after completion of vitellogenesis and dissolution of the germinal vesicle.

Immediately after spawning, starved scallops with resting gonads showed a decrease in gonad growth and failed to initiate gametogenesis when exposed to a series of temperatures between 10° and 30°C. Sastry

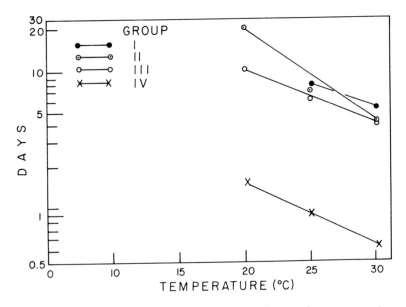

Fig. 15. Time–temperature relationship for completion of gametogenesis and spawning of *Argopecten* (= *Aequipecten*) *irradians* from Beaufort, North Carolina. Groups I–IV represent animals in different stages of gonad development collected at intervals during the reproductive period. Collection dates: group I, June 11, 1964; group II, July 6, 1964; group III, August 20, 1964; group IV, October 20, 1964 (see Table I). (From Sastry, 1966a.)

(1966a, 1968) has suggested that gonad growth and gametogenesis take place under temperature conditions at which nutrient mobilization for the gonads is permitted, which occurs only after the basic metabolic needs of the animal have been met. In scallops with resting gonads, gametogenesis begins upon exposure to a minimum threshold temperature of 20°C, presuming an abundant food supply (Fig. 16). Scallops exposed to a subthreshold temperature of 15°C develop oogonia, but oocyte growth does not occur even though they are supplied with food. When scallops held at 15°C with food are subsequently transferred to higher temperatures (20° and 25°C), oocyte growth begins immediately. Apparently the scallops require exposure to a minimum threshold temperature for activation of the oocyte growth phase, for at subthreshold temperatures the oogonia are developed but further differentiation is suppressed. However, if scallops with oocytes already in the cytoplasmic growth phase are maintained at temperatures below those necessary for triggering growth, oocyte development continues until dissolution of the germinal vesicle (Sastry, 1970b). Sastry (1968, 1970b) has suggested that temperature acts as a triggering stimulus for initiation of the oocyte growth phase. Scallops maintained at temperatures exceeding a certain maximum (25° and 30°C) also fail to initiate gonad growth and gametogenesis. However, gamete growth can proceed at supranormal temperatures if the gonads have accumulated nutrient reserves and the oocytes are in the growth phase. Sastry (1966a, 1970b) indicated that the temperature requirements for activating oocyte growth

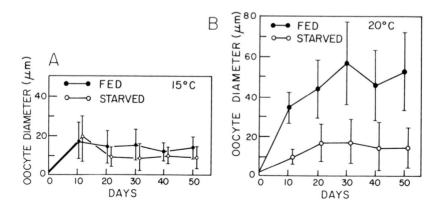

FIG. 16. Oocyte growth response in resting stage animals of Argopecten (= Aequipecten) irradians collected on December 20, 1964, at Beaufort, North Carolina, and maintained at 15°C (A), and 20°C (B). Vertical lines represent the standard deviation. (From Sastry, 1968.)

at the beginning of oogenesis and for attaining maturity could limit the annual period of gonad growth and gametogenesis in the natural environment.

In a further series of experiments, Sastry and Blake (1971), working with a Massachusetts population of scallops, showed that temperature influenced the initiation of oocyte growth by regulating the transfer of nutrient reserves to the gonads (e.g., Section 5.3.7.5.3). The rate of nutrient transfer to the gonads seemed to depend on the stage of gametogenesis and the temperature to which the animals were exposed. Scallops with resting gonads, injected with [^{14}C]leucine into the digestive gland, incorporated greater amounts of labeled amino acid into the gonads at 15°C than those held at 5°C.

The influence of temperature on spermatogenesis has been examined in only a few species. In *Argopecten (= Aequipecten) irradians*, the differentiation of spermatogonia to spermatozoa proceeds more rapidly than oogenesis, and the spermatozoa are already developed several weeks before the oocytes complete vitellogenesis (Sastry, 1966a). Also, spermatogenesis in *A. irradians* takes place at temperatures lower than those required for initiating oocyte growth. Sastry (1966a) determined the effect of different temperatures on spermatogenesis during successive stages of gonad development (Table I; see Section 5.3.7.5.7). Between 10° and 30°C, starved scallops with spermatogonia and spermatocytes predominately resorbed the gonads and no further differentiation occurred. However, scallops with accumulated gonad reserves and in a later stage of spermatogenesis retained spermatozoa at 10° and 15°C but spawned them at 20°, 25°, and 30°C.

5.3.7.1.4. *Variation in Geographically Separated Populations.* The time and duration of gametogenic activity in populations within a species may vary in different parts of the geographic range. Differences in the period of gametogenesis and breeding activity are reported for populations of *Mytilus edulis* from British waters (Chipperfield, 1953), the Japanese coast (Sugiura, 1959), and the southern California coast (Moore and Reish, 1969); *Mya arenaria* from eastern Maine to Chesapeake Bay (Ropes and Stickney, 1965; Pfitzenmeyer, 1965); *Mercenaria mercenaria* from Long Island Sound and Core Sound, North Carolina (Loosanoff, 1937b; Porter, 1964); *Placuna placenta* from the Indo-Pacific region (Sastry, 1955); and *Pinctada margaritifera* from Australia and the Japanese coast (Tranter, 1958c).

Sastry (1963, 1970a) has determined the reproductive cycle of *Argopecten (= Aequipecten) irradians* from Massachusetts, North Carolina, and Florida. The Massachusetts and North Carolina populations exhibit an annual reproductive cycle out of phase with each other (Fig. 17).

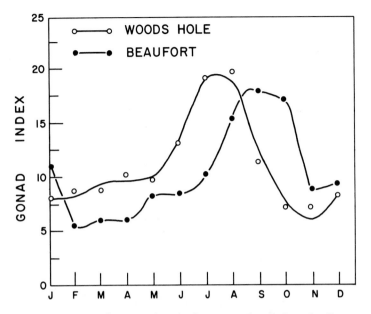

FIG. 17. Variation in the annual reproductive cycle of latitudinally separated populations of *Argopecten* (= *Aequipecten*) *irradians*, which suggests genetic differences. Monthly mean gonad index values fitted with the three-term Fourier curves show a cyclical trend in gonad activity response within the 12-month period. (From Sastry, 1970a.)

Gonad growth and gametogenesis in the Massachusetts population begins in April when water temperatures are about 10°C, and the gametes mature in early summer when temperatures reach approximately 23°C. In the North Carolina population, gonad growth and gametogenesis begin later in June when the temperature is above 20°C, and the gametes do not develop to maturity until the end of summer when temperatures are about 26°–28°C. It is clear that successive events in the reproductive cycle of these two populations occur at different times in the year and also at different environmental temperatures. Sastry (1970a) suggested that such variation between populations may be an adaptive response to the geographic differences in temperature and food production (e.g., Section 5.3.7.5.3).

Sastry (1966b and unpublished data) acclimated scallops in the vegetative stage representing these two populations to a similar series of temperatures and to a constant photoperiod of 12 hr light and 12 hr darkness in the laboratory with an abundant food supply. In the Massachusetts population, gonad growth and gametogenesis occurred at 15°C, whereas the North Carolina population required 20°C. Gonad development did not occur at 10°, 20°, and 25°C in the Massachusetts population and at 15°, 25°, and 30°C in the North Carolina population. Postspawning adults of both populations with resting gonads showed a similar difference in temperature requirement for gonad growth and gametogenesis (Sastry, 1968; Sastry and Blake, 1971). Gonad growth and gametogenesis were suppressed at both subthreshold temperatures and at temperatures exceeding a maximum. At warm temperatures (20° and 25°C, Massachusetts population; 25° and 30°C, Beaufort population), the animals died without initiating gonad growth and gametogenesis. The population of scallops from Massachusetts and North Carolina had different minimum threshold temperatures for the initiation of gonad growth and gametogenesis, which were not altered by thermal acclimation. Massachusetts scallops with resting gonads transplanted to Beaufort, North Carolina, during the winter survived for a period of about 8 weeks but failed to initiate gonad growth and gametogenesis (Sastry, 1966b and unpublished data). The variation in temperature requirements for gonad growth and gametogenesis suggests genetic divergence of populations in the two geographic areas.

Temperature requirements for gonad growth and gametogenesis for the population of *Argopecten* (= *Aequipecten*) *irradians* from Alligator Harbor, western Florida, are similar to those for the Beaufort, North Carolina, population (Sastry, 1961, 1963, 1966a). In the Florida population, gametogenesis begins in the late fall and the gametes develop to maturity by early August. Rapid growth of gonads and intensive gametogenic activity take place in the Florida population between May and July. The Florida population differs from the North Carolina population in that it initiates gametogenesis and commences spawning earlier in the year. The annual temperature cycle in these two areas is similar in many respects, with the exception that late winter and spring temperatures are slightly higher in Alligator Harbor (Sastry, 1961, 1966a). The Florida population also requires higher temperatures for gonad growth and gametogenesis, similar to the North Carolina population. Gonad development is suppressed at 14°C (Sastry, 1963). These differences between the two populations appear to be nongenetic reproductive adaptations to environmental differences at Beaufort, North Carolina, and Alligator Harbor, Florida (Sastry, 1961, 1966a, and unpublished data).

Ropes and Stickney (1965) and Pfitzenmeyer (1965) have reported that the soft-shell clam, *Mya arenaria*, develops gametes both in fall and spring in areas south of Cape Cod, while in the north there is only one period of reproductive activity, in summer. Summer temperatures in the south appear to exceed the optimum required for gametogenesis, while winter temperatures fall below this range, thus producing spring and autumn optima for gametogenesis. Temperature requirements and their interspecific as well as intraspecific influence on reproduction vary when different geographic regions are compared.

5.3.7.2 LIGHT

There are no detailed studies examining the influence of light on gametogenesis in pelecypods. The influence of light on reproduction in some marine invertebrates has been discussed in reviews by Giese (1959a,b), Segal (1970), and Giese and Pearse (1974).

Gonad growth and gametogenesis in *Argopecten* (= *Aequipecten*) *irradians* from Massachusetts are initiated in correlation with increasing day length during spring. Maturity is attained when day length is maximal (Sastry, 1970a). The population of *A. irradians* from Beaufort, North Carolina, however, initiates gonad growth and gametogenesis when the day length is about maximal and mature and spawn with decreasing day length (Sastry, 1970a). Gimazane (1971) found no significant effect of photoperiod on gametogenesis in *Cerastoderma* (=*Cardium*) *edule* with resting-stage gonads.

5.3.7.3 SALINITY

The influence of salinity on reproduction in pelecypods has been little studied, and there are few experimental studies reporting the effect of salinity on gametogenesis. Loosanoff (1948, 1952) reported that *Crassostrea virginica* from Long Island Sound developed spermatozoa at a salinity of 7.5‰, but that the eggs did not develop normally. Butler (1949) found that gametogenesis was inhibited in oysters until the salinity level was increased to above 6‰.

Reproduction in some pelecypods from Madras Harbor, India, has been correlated with salinity changes (Panikkar and Aiyyar, 1939; Paul, 1942). In *Placuna placenta* from Kakinada Bay, India, gonad development takes place during periods of high salinity, and spawning begins with the dilution of seawater by monsoonal rains (Sastry, 1955). Durve (1964) determined the gametogenic cycle in *Meretrix casta* from fish ponds near Mandapam, India, and found that the period of sexual rest between July and October coincided with hypersaline conditions (45‰), while gametogenic activity took place during fairly stable temperatures, specifically between 27° and 33°C, and at a salinity of 30‰. In *Donax faba* from the east coast of India, gametogenesis takes place with increasing salinity following the monsoonal rains (Alagarswami, 1966). In *Donax cuneatus* on the Madras coast, a single annual reproductive cycle occurs and gametogenesis takes place between September and December when the temperature and salinity are low (Rao, 1967). The wood-boring bivalve *Martesia striata*, exposed to a salinity range between 0.5 and 35‰ in the estuarine waters near Cochin, India, breeds for the 8 months in the year when the salinity is high. This activity is interrupted with a decrease in salinity during the period of monsoonal rains (Balasubramanyan, 1970). At Cochin, where the salinity ranges between 0 and 32‰, Nair and Saraswathy (1970) have reported that the shipworm, *Nausitora hedleyi*, breeds when the salinity is low and passes through a resting period when the salinity is high. In another species from the same area, *Teredo furcillatus*, gonads mature when the salinity is high. Nagabhushanam (1970) has reported that developing young of *T. furcillatus* are incubated in the gills throughout the year at Vishakapatnam Harbor, India, where salinity does not decrease below 14.65‰. In areas where salinity decreases to 4.65 and 5.09‰, adults and larvae are killed. According to Purchon (1968), the bivave *Egerina radiata* (Donacidae) inhabits only relatively short lengths of certain South African rivers immediately above the main region of saltwater penetration. Possibly this species depends on increased salinity for breeding purposes. Wilson (1968, 1969) has reported that the reproductive activity of the mussel *Xenostrobus securis* is limited by salinity rather than temperature in the Swan estuary, Western Australia. Prolonged exposure to low salinities causes gonadal resorption, and at very low salinities gametogenesis is inhibited.

5.3.7.4 TIDES

The influence of tides on gonad growth and gametogenesis in bivalves has not been studied in detail. Lammens (1967) reported that first-year *Macoma balthica*, which remain in the surface layers of the substratum, begin to spawn earlier than older animals living in the deeper layers. Apparently the animals in surface layers which are exposed to air are slightly warmer than those at a depth, hence the difference in the populations might be more a temperature effect caused indirectly by tidal action. Campbell (1969), reporting the relationship between carotenoid content and the reproductive cycle in *Mytilus edulis*, states that the rate of feeding affects pigment concentration in the body. The actual amount of carotenoid content is controlled in part by the number of hours the animal is in water during each tidal cycle and is related to feeding. It therefore appears that the extent of gonad growth and intensity of gametogenesis can be influenced by the position of the animal in the littoral zone and the amount of food and time available for feeding, factors which are influenced by the tidal cycle (see also Section 5.3.7.5.2).

5.3.7.5 FOOD

Food availability and the mobilization of nutrients for the gonads is an important aspect of pelecypod reproduction. Considerable work has been done on food, feeding, and digestion, and much of this literature has been reviewed in detail by van Weel (1961), Jørgensen (1966), Owen (1966a,b, 1974), Purchon (1968), Winter (1969, 1970), and Newell (1970). However, many problems remain unanswered on the relationship between food and reproductive activity. Especially, questions related to the fate of products of digestion and their transport to sites of storage and utilization relative to changes in the environment have been examined for only a few species. It is evident that nutrients from ingested food must be mobilized from the gut for other body tissues, especially for the gonads for utilization by the developing gametes. The digestive system of pelecypods is specialized for dealing rapidly and continuously with fine particles of food (Owen, 1966b). Ingested food is mixed with enzymes present in the gastric fluid secreted by the crystalline style and digestive diverticula. The products of digestion are transported into the blind tubules of the digestive diverticula. Digested material is absorbed, and particles are phagocytosed by the digestive cells which give rise to spheres containing spherules of ingested food. These spherules are ultimately conveyed to the intestine by way of intestinal grooves. However, not much is known of the transfer of nutrients and distribution to body tissues, especially gonads. For further discussion on digestion in pelecypods consult Owen (1974).

5.3.7.5.1 *Food Sources.* The majority of pelecypods are suspension feeders, and the source of their food includes phytoplankton, bacteria, fungi, flagellates, dissolved organic matter, and organic aggregates (Jørgensen, 1966). Deposit feeding occurs in *Nucula, Leida, Yoldia, Portlandia, Macoma,* and other genera (Ockelmann, 1958; Newell, 1970). The boring bivalves Teredinidae and Xylophagidae utilize the wood they are boring (Potts, 1923) and supplement their diet with large amounts of organic matter (Jørgensen, 1966). The suctoral scavengers *Lyinsiella* and *Cuspidaria* ingest large particles of decaying animal remains (Yonge, 1928; J. A. Allen, 1962). *Entovalva,* in the holothurian *Synapta,* and *Scioberita,* in an asteriod, are endoparasitic and absorb predigested food from the host through their body walls (Lankester, 1900; Popham, 1940).

Many attempts have been made to determine the relative value of various food organisms and detritus as food for suspension-feeding animals (Blegvad, 1914; Coe, 1948; Coe and Fox, 1944; Verwey, 1952; Sastry, 1955; Davis and Marshall, 1961; Reid, 1969). Phytoplankton is the ultimate source of food for most suspension-feeding animals, and it may be supplemented by organic matter, dissolved organic matter, and bacteria. Qualitative selection of food occurs in suspension-feeding bivalves (Grave, 1916; Fox, 1936). Dinoflagellates have been reported to be selected preferentially for food by *Mytilus californianus* (Buley, 1936) and *M. edulis* (Jensen and Sakshaug, 1970a,b). The adequacy of phytoplankton as a food source depends upon its concentration, rate of production, and availability to the animal. Jørgensen (1966) estimated that typical suspension feeders require about 0.1 mg dry organic matter of phytoplankton per liter of seawater filtered to meet their maintenance requirements, and 0.5 mg/liter for growth and reproduction. Estimates of phytoplankton production in coastal waters of most parts of the world fall within the range needed for meeting typical suspension feeder requirements during the major part of the year (Ockelmann, 1958; Jørgensen, 1966).

Detritus has been suggested to be an important source of food for suspension feeders, but few studies have determined its food value (Fox and Coe, 1943; Coe, 1948; Filatova, 1957; Jørgensen, 1966; Kirby-Smith, 1976). The importance of bacteria in the nutrition of *Mytilus californianus* (Zobell and Landon, 1937; Zobell and Feltham, 1938) and bacteria on detritus particles as a supplemental food source for *Argopecten (= Aequipecten) irradians* (Kirby-Smith, 1970, 1976) has been suggested. Organic detritus and associated microorganisms have been reported to be important in the diet of *Macoma balthica* (Newell, 1965). Baier (1935) indicated that inorganic particulate materials encrusted with bacteria had a higher nutritional value than organic detritus alone.

Stephens and Schinske (1961) and Stephens (1968) reported the uptake of amino acids from seawater by some pelecypods. Fox (1957) has suggested that practically all organic matter in seawater is present in a colloidal state, which is available for suspension-feeding pelecypods. However, it is thought to be only slightly accessible to undisturbed and normally feeding animals (Jørgensen, 1966, 1976). To evaluate the relative importance of different food sources in meeting the requirements of suspension-feeding bivalves, it would be of interest to determine the scope for growth (difference between energy of assimilated food and energy lost in respiration) of animals fed on different diets.

5.3.7.5.2 *Food Consumption and Assimilation.* Feeding and digestion have been reviewed by Owen (1974). The results reported by different investigators on the rhythmicity of water transport and feeding in pelecypods are not in agreement. In undisturbed suspension feeders, the rates of water transport are maintained continuously, with feeding occurring, except when they are exposed by the tide (Loosanoff, 1939, 1942; Jørgensen, 1966). No endogenous tidal rhythmicity in water filtration was found in *Mytilus edulis* (Jørgensen, 1960; Theede, 1963; Davids, 1964) or *Argopecten (= Aequipecten) irradians* (Kirby-Smith, 1970). No endogenous rhythm in pumping, nor any periodicity following alternate periods of light and darkness, was found in *Scrobicularia plana* by Hughes (1969). However, van Dam (1954), Rao (1954), and Verwey (1975) reported rhythmic pumping activity in fully immersed bivalves. In *Mytilus edulis* and *M. californianus* tidal rhythmicity in feeding is reported, with maximal rates of water transport coinciding with high tide and low rates with low tide (Rao, 1954). Winter (1969, 1970) found two exclusive phases of high food consumption alternating with two exclusive phases of digestion within a 24-hr period in *Cyprina islandica* and *Modiolus modiolus*. Differences in filtration rates are also reported in *Mytilus californianus* (Segal *et al.*, 1953) and *Lasea rubra* (Ballantine and Morton, 1956; Morton, 1956; Morton *et al.*, 1957), occurring at different tidal levels. The optimum temperatures for water transport vary in populations of *Mytilus californianus* in different parts of the geographic range (Rao, 1953).

The rates of water filtration in suspension-feeding pelecypods are adapted to the concentration of particles in the water (Jørgensen, 1966). In *Cyprina islandica*, filtration rates increase with decreasing food concentrations (Winter, 1969, 1970). At concentrations above a certain level, food organisms interfere with the feeding process. The percentage of phagocytosed algae in relation to the total amount of algae filtered can be very high when small amounts are ingested.

Winter (1970) determined the influence of temperature on filtration rates and food utilization in *Cyprina islandica*. Filtration rates and also the amount of phagocytosed algae decreased with a decrease in temperature from 12° to 4°C. Although the filtration rates did not change appreciably when the temperature increased from 12° to 20°C, the amount of phagocytosed algae increased from 108 mg (dry weight) at 12°C to 144 mg at 20°C during a 24-hr period. A similar relationship between filtration, food concentration, and utilization of ingested algae was found in *Modiolus modiolus* (Winter, 1969, 1970). The extent of food utilization has been determined in *Modiolus demissus* (Kuenzler, 1961), *Mya arenaria* and *Venus striatula* (J. A. Allen, 1962), *Dreissena polymorpha* (Mikheyev and Sorokin, 1966), and *Mytilus edulis* (Bayne, 1975) with the use of ^{14}C- and ^{32}P-labeled algae.

In *Argopecten (= Aequipecten) irradians*, filtration rates are compensated between 10° and 26°C (Kirby-Smith, 1970), and all reproductive activity occurs within this range (Sastry, 1966a, 1968). In *Mytilus edulis*, there is a complete temperature acclimation of filtration rate and metabolic rate between 5° and 20°C and, as a result, the scope for growth (difference between energy of assimilated food and energy lost in respiration) is relatively independent of temperature over this range (Bayne *et al.*, 1975; Widdows, 1973, 1978a,b). Winter (1970) found a correlation between filtration rate, phagocytosed algae, and body size in *Cyprina islandica* and *Modiolus modiolus*. The efficiency of assimilation of ingested food is also dependent upon size and food concentration in *Mytilus edulis* (Widdows, 1973). Recent studies investigating the relationship among food concentration, temperature, and food consumption and assimilation in *Mytilus edulis* have been reviewed (Bayne, 1976a,b).

A number of factors such as temperature, tidal level, latitude, particle size, quantity of food, and filtration rates may influence food consumption, digestion, and assimilation in an organism. Therefore the energetics and adaptation to the environment may influence the timing of gametogenesis and the extent of gonad growth in a population at a given geographic location.

5.3.7.5.3 *Relationship between Food and Gonad Development.* Gonad development is an energy-demanding process. The mobilization of nutrients from the ingested food for the gonads is essential for gamete development. The relationship between food and gonad development has been studied in detail for only a few species (Sastry, 1966a, 1968, 1970a, 1975; Sastry and Blake, 1971; Gimazane, 1972; Bayne, 1975, 1976a). Nutrients from ingested food are distributed to various body organs for assimilation and storage. For many species it is still unclear whether gonad development depends on food ingested directly from the surrounding water, on stored reserves, or on both. In high latitudes where food production is seasonal, some species may store nutrient reserves in body organs when food availability is maximal and subsequently utilize them for gonad growth, or supplement food intake to meet the reproductive requirements (Ockelmann, 1958; Ansell *et al.*, 1964; Ansell and Trevallion, 1967; Ansell and Lander, 1967; Bayne, 1976a). The relationship among food availability in the environment, storage, and reproductive activity varies among species.

In *Argopecten (=Aequipecten) irradians* (Sastry, 1966a, 1968, 1970a), *Tellina tenuis* (Trevallion, 1971), *Abra alba* and *Chlamys septemradiata* (Ansell, 1974a,b), and *Placopecten megellanicus* (Thompson, 1977; Ehinger, 1978; Sastry and Ehinger, unpublished data), the periods of food abundance and of gonad development the are nearly coincident. In *Mercenaria mercenaria* (Loosanoff, 1937b) and *Venus striatula* (Ansell, 1961b), the recovery period is very brief and gametogenesis begins immediately after spawning. In these species, the initiation of gonad development coincides with the increase in phytoplankton after a winter delay. While food abundance and reproductive activity can be correlated in some species, only a few studies have been undertaken to determine the influence of food on reproductive timing and the extent of gonad development (Lubet, 1959; Sastry, 1966a, 1968, 1970a; Ansell and Trevallion, 1967; Campbell, 1969). The relationship between food abundance and seasonal gonad development varies for populations of *Argopecten* (= *Aequipecten*) *irradians* from Alligator Harbor, Florida, Beaufort, North Carolina, and Nantucket Sound, Massachusetts (Sastry, 1961, 1963, 1966a, 1970a). Gonad growth and gametogenesis in the population from Alligator Harbor correlate with seasonal changes in phytoplankton production. Chlorophyll concentrations show spring and summer oscillations, low values in the middle of summer, a series of peaks throughout autumn, and a decrease to low values during winter (Grice, 1953; Marshall, 1956). Rapid gonad growth in this population occurs in May–July, and postspawning adults recover to maturity in autumn, which coincides with phytoplankton blooms. In the Beaufort population, gonad growth and gametogenesis begins in May–June when the temperature increases above 20°C. This coincides with peak phytoplankton production. The gonad index increases rapidly between May and July and remains at a steady level through midsummer (Fig. 18; see also Fig. 11 and Table I). The digestive gland size remains constant. The time and extent of gonad growth coincide with increasing and decreasing temperatures and pulses in high phytoplankton production. Although food concentrations are high in summer (Fig. 19), the gonad index shows no increase until early fall. The amount of nutrients mobilized for the gonads appears then to depend on the food concentration, temperature, and basic metabolic requirements of the animal. Gonad growth and gametogenesis are dependent upon direct intake of food during the period of gonad development (Sastry, 1966a, 1968). Gonad growth and gametogenesis in the northern population of *A. irradians* from Nantucket Sound occur between April and early July and differ from those of the Beaufort and Alligator Harbor populations (Sastry, 1961, 1963, 1966a, 1970a; see also Section 5.3.7.1.4). Phytoplankton blooms occur in the northern areas in midwinter and early spring (Fish, 1925; Smayda, 1957; Ryther and Yentsch, 1958; Pratt, 1959; Yentsch, 1963; Marshall and Wheeler, 1965). In the northern populations of *A. irradians* the period of gonad development does not

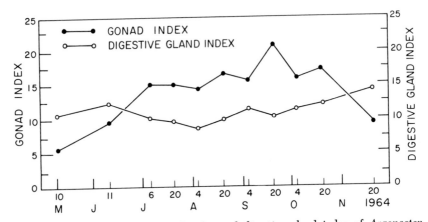

FIG. 18. Changes in mean gonad index and digestive gland index of *Argopecten* (= *Aequipecten*) *irradians* collected at biweekly intervals during the reproductive period at Beaufort, North Carolina. Compare with Fig. 19 for correlation between gonad growth activity of scallops and phytoplankton production in Beaufort Channel. (From Sastry, 1966a.)

coincide with peak phytoplankton production, suggesting differences in adaptation (Bullock, 1955; Prosser, 1955; Vernberg, 1962) for timing reproduction between populations in the northern and southern climatic zones. Population differences in details of the relationship between food abundance and gametogenesis also occur in *Tellina tenuis* (Ansell and Trevallion, 1967; Trevallion, 1971) and *Macoma balthica* (Caddy, 1967; Lammens, 1967; Gilbert, 1973).

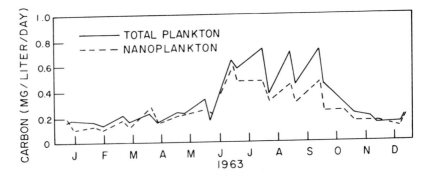

FIG. 19. Seasonal differences in phytoplankton production in Beaufort Channel, North Carolina, where *Argopecten* (= *Aequipecten*) *irradians* have been collected for determination of the changes in gonad index shown in Fig. 18. (From Williams and Murdoch, 1966; modified.)

In other pelecypods, seasonal gonad development is linked with the storage and utilization of reserves accumulated in the body during the period of phytoplankton maximum (Ockelmann, 1958). The quantity of gametes developed in *Crassostrea virginica* during spring is influenced by the amount of food ingested and the reserves accumulated during the preceding recovery period (Loosanoff, 1965). In *Mytilus edulis* (Chipperfield, 1953), *Pecten maximum* (Comely, 1974), and *Macoma balthica* (De Wilde, 1975) nutrients accumulated during the summer are stored and utilized for gametogenesis during the autumn and winter. The interrelationships among food and storage reserves, energy metabolism, and gametogenic activity in *Mytilus edulis* have been recently studied in considerable detail, and this information has been extensively reviewed (Bayne, 1975, 1976b; Gabbott, 1975, 1976). In the deposit-feeding bivalve *Macoma balthica*, there is a distinct recovery period, but this species is not entirely dependent upon accumulated nutrient reserves for subsequent gamete development (Lammens, 1967). During midsummer, the follicles are filled with connective tissue elements which gradually disappear with the development of gametes in autumn. Gamete development continues in winter through utilization of the food ingested.

5.3.7.5.4 *Changes in Body Weight.* Seasonal changes in body weight in relation to reproductive activity have been reported for *Tellina tenuis* by Ansell and Trevallion (1967). Body weight is minimal in February and March and increases greatly in May, corresponding to gonad proliferation. A gain in body weight occurs through immediately increased utilization of the food supply in the water for gonad development and building up of reserves. Body weight begins to decline during June and fluctuates during July and August. It has been suggested that these fluctuations occur because of a changing balance of shell growth, growth of body and gonads, and loss of gametes in spawning. Body weight continues to decline from September to March, and the sex of the animals cannot be distinguished when the food supply is limited in the environment and reserves are utilized. The overall decrease in body weight represents the net deficit of food available over the metabolic requirements of the animal.

Changes in shell length, total weight (including shell), and body weight in relation to the gonadal cycle of the Beaufort, North Carolina, population of *Argopecten (= Aequipecten) irradians* have been determined by Sastry (1966a). Growth is rapid between May and July and slow in summer. The relationship among growth, food concentrations, and temperature have been determined by Kirby-Smith (1970). Changes in total weight and body weight parallel each other throughout the year. Body weight and the gonad index increase simultaneously during

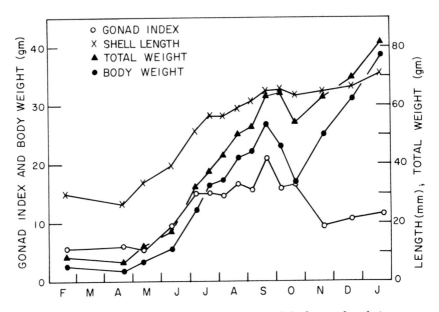

Fɪɢ. 20. Changes in shell length, total weight, and body weight of *Argopecten* (= *Aequipecten*) *irradians* from Beaufort, North Carolina, in relation to the reproductive cycle. The average values were determined for 10 individuals each month. (From A. N. Sastry, unpublished data.)

May and early July. During the remainder of summer, the gonad index remains steady, while the body weight continues to increase until late September (Fig. 20). The gonad index increases in late August and early September and declines with a corresponding decrease in body weight during the spawning period. After spawning, the body weight increases, while the gonad index remains steady throughout late autumn and early winter. Nutrient reserves may accumulate in the gonads independently of other body organs, depending on food concentrations and environmental temperatures. The increase in body weight after spawning represents an accumulation of nutrient reserves in the body tissues and/or a gain in water content.

5.3.7.5.5 *Changes in Body Component Index.* Seasonal changes in body components have been determined for some species throughout the year to learn whether they vary in relation to the reproductive cycle in terms of possibly supplying nutrients to the gonads for growth and gametogenesis. In *Tivela stultorum*, the body component indexes (the ratio of the component wet weight to the total weight) remained at a steady level throughout the year, except for a high gonad index during the breeding period when the body fluid index decreased (Giese et al., 1967). Since all body components including the digestive gland maintained a steady proportion throughout the year, it has been suggested that none of the tissues serve as a source of nutrient supply for gametogenesis (Fig. 21). The accessory tissue of gonads in T. stultorum is thought to serve as its own storage organ.

Sastry (1966a, 1970a, and unpublished data) determined seasonal changes in the body component indexes of populations of Argopecten (= Aequipecten) irradians from Beaufort, North Carolina, and Nantucket Sound, Massachusetts. The gonad index of the North Carolina population showed an annual cycle with a maximum in September, a decline with spawning, and a steady level during late autumn and early winter (see Fig. 11). The digestive gland index decreased with the beginning of gonad growth in May–June and remained at a minimal plateau until early winter (see Fig. 18). The adductor index remained at a steady high level until mid-June and then declined steadily until early winter. The pooled index for the remaining body tissues (mantle, gills, kidney, and heart) remained at a steady high plateau until July, declined until September, and increased again during autumn and early winter (Fig. 22). In the Massachusetts population, the gonad index increased until July–August and decreased between August and October with spawning; it remained at a steady level during autumn and early winter. The digestive gland index decreased with the beginning of gonad growth in April and remained at a low level until September; it then showed a slight increase in late autumn. The adductor index increased until June and later declined during autumn and early winter. The pooled tissue index for the remaining body components was steady until May and then decreased until August. The pooled tissue index increased dramatically between August and December (Fig. 23).

The relative changes in the gonad index and digestive gland index in the two populations show variation perhaps reflecting differences in their adaptation to food availability in the environment, food storage, and energy allocation for gonad development. The decline in the digestive gland index in the Massachusetts population during the period of gonad growth is more pronounced than that in the North Carolina population (Fig. 24). Sastry (1966a) has indicated that a decline in the digestive gland index occurs during the period of gonad growth as a result of the rapid transfer of nutrients from ingested food. The digestive gland index is higher than the gonad index during vegetative and resting stages in the reproductive cycle of Argopecten (=Aequipecten) irradians. This coincides with the colder environmental tempera-

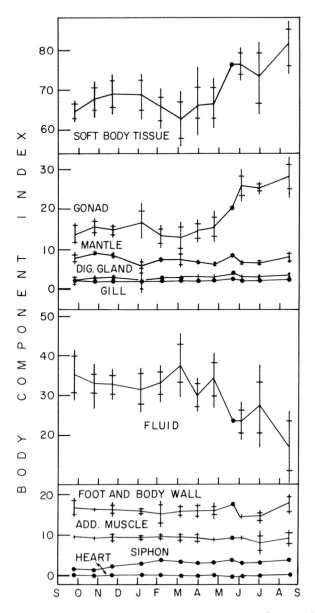

FIG. 21. Body component indexes [(body component wet weight × 100)/body wet weight] of *Tivela stultorum* for the year 1965–1966. The vertical lines through the points are the standard deviations. The horizontal markings on these lines are twice the standard error and mark the 95% confidence limits. (From Giese *et al.*, 1967; modified.)

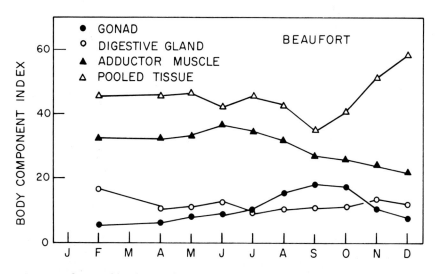

Fig. 22. The monthly mean body component indexes [(body component wet weight × 100)/body wet weight] of *Argopecten (= Aequipecten) irradians* from Beaufort, North Carolina. The pooled tissue index included the mantle, gills, kidney, and heart. The average values were determined for 10 individuals each month. (From Sastry, unpublished data.)

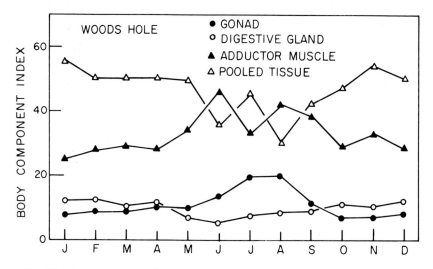

Fig. 23. The monthly mean body component indexes [(body component wet weight × 100)/body wet weight] of *Argopecten (= Aequipecten) irradians* from Woods Hole, Massachusetts. The pooled tissue index included the mantle, gills, kidney, and heart. The average values were determined for 10 individuals each month. (From A. N. Sastry, unpublished data.)

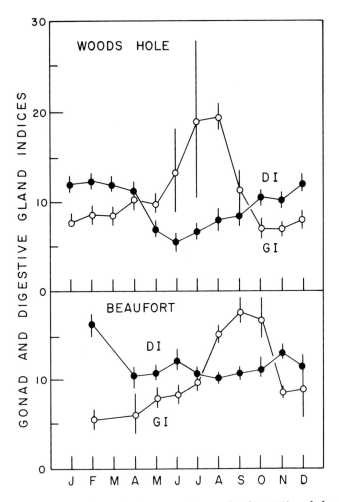

Fig. 24. The inverse relationship between the gonad index (GI) and the digestive gland index (DI) of latitudinally separated populations of *Argopecten* (= *Aequipecten*) *irradians* collected during 1965. Vertical lines represent the 95% confidence intervals. (From Sastry, 1970a.)

tures. Nutrients from ingested food may be stored in the digestive gland and transferred to body organs when the gonads are in a quiescent state. During the period of gonad development, however, nutrients may accumulate in the gonads at the expense of other body organs.

Fuji and Hashizume (1974) determined the seasonal changes in body components of 1-, 2-, and 3-year-old scallops, *Patinopecten yessoensis*, in Matusi Bay, Japan, in terms of dry weight and calorific value.

Rapid gonadal growth during winter was followed by the growth of nongonadal tissues in late winter and spring. Gonad weight in each age group decreased with spawning. Shell growth occurred in the months following spawning.

5.3.7.5.6 *Distribution of Nutrients in the Body.* Nutrient distribution and sites of storage have been examined in only a few species with the use of radiotracer methods, and many questions still remain as to how nutrient transfer from one body organ to another affects reproductive activity. The amount of nutrients mobilized for the gonads could be affected by energy requirements for somatic growth and basic metabolism. These relationships may change according to age and reproductive stage and also according to changes in food concentrations and temperature conditions in the environment.

J. A. Allen (1962) fed *Phaeodactylum* labeled with ^{32}P to *Mya arenaria, Venus striatula,* and *Cuspidaria cuspidata* and determined the distribution of ^{32}P within the body and its rate of excretion from the body. The digestive gland had the highest count per milligram of dry tissue in *M. arenaria* and *V. striatula,* with no drop in count with time (Table II and III). The gonads accumulated only moderate amounts of ^{32}P and showed no evidence of a rapid buildup. The second highest concentration of ^{32}P was found in kidney tissue. The concentration of ^{32}P in muscle and shell was much lower than in the other tissues. The labial palps and gills contained moderate amounts of radiophosphorus. The ^{32}P content of gills showed an increase with time, suggesting storage. In *C. cuspidata,* the distribution of radiophosphorus was more uniform, suggesting only absorption (Table IV).

Allen (1970) reported results of further experiments on the uptake and distribution of radiophosphorus in *Venus striatula, Mya arenaria,* and *Mytilus edulis.* These studies were carried out for varying periods up to a maximum of 6 weeks for comparison with the short-term experiments reported above. In *Mya arenaria,* the digestive gland accounted for nearly 60% of the ^{32}P assimilated a day after feeding ^{32}P-labeled *Phaeodactylum.* Over the next 6 days the percentage of radiophosphorus decreased. After 1 day, the amount of ^{32}P in the kidney was about 22% of the total amount in the body, but this percentage decreased to 14% by the end of the first week. The percentage of ^{32}P at the same time increased in the gills, palps, and mantle (Table V).

The digestive gland retained a considerably higher concentration of ^{32}P than in other body organs of both *Venus striatula* and *Mytulis edulis* even 6 weeks after feeding. The mantle and adductor muscle showed no marked increase in concentration of ^{32}P over the values initially re-

TABLE II

The Distribution of ^{32}P in Mya arenaria [a]

	Controls	1	2	3	4	5	6
				Counts/min/mg of tissue (dry weight)			
Digestive gland	51	898	2592	2132	2208	4812	4938
Gonad	2	44	74	104	160	222	135
Kidney	8	260	892	460	483	1220	1304
Siphon	—	—	—	—	—	132	—
Mantle epithelium	2	17	90	149	—	291	897
Adductor muscle	0	5	22	21	—	88	—
Palp	19	43	156	64	—	614	926
Gill	4	51	253	131	—	772	1258
Shell edge and periostracum	0	13	62	—	—	42	—
Time from start of experiment to tissue estimation (hr)	30	19	24	30	74	115	158
Time in algal suspension (hr)	30 [b]	19	24	30	5½	6	86
Length of shell (mm)	79	70	75	82	75	73	57
Estimated counts/min of ^{32}P extracted (millions)	0.004	2.30	2.55	2.10	4.43	4.42	2.55
Sex	Ripe ♀	Ripe ♀	Ripe ♀	Ripe ♂	Ripe ♀	Ripe ♀	Ripe ♀

[a] From J. A. Allen (1962).
[b] ^{32}P in solution only.

TABLE III

DISTRIBUTION OF ^{32}P IN *Venus striatula* [a]

		Counts/min/mg of tissue (dry weight)			
	Controls	1	2	3	4
Digestive gland	52	33,475	5,940	39,990	48,823
Gonad	7	1,717	169	2 180	12,437
Kidney	9	4,447	846	5,760	18,403
Siphon	—	—	—	—	15,292
Mantle epithelium	—	—	—	—	5,341
Adductor muscle	—	—	—	—	1,398
Foot	—	—	—	—	3,022
Time from start of experiment to tissue estimation (hr)	210	72	114	186	234
Time in algal suspension (hr)	2 [b]	2	2	2	2
Length of shell (mm)	22	21	21	19	19
Estimated counts/min of ^{32}P extracted (millions)	—	0.58	0.21	0.63	1.65
Sex, all spent	♀	♂	♀	♂	♀

[a] From J. A. Allen (1962).
[b] [32]P in solution only.

corded 3 days after feeding. The kidney and gill showed an increase to about half the ^{32}P concentration in the digestive gland. The gill concentrated a fair amount of radiophosphorus, which increased with time after assimilation. From these experiments it was concluded that

TABLE IV

THE DISTRIBUTION OF ^{32}P IN *Cuspidaria cuspidata* [a]

	Counts/min/mg of tissue		
	1	2	3
Digestive gland	35	19	26
Kidney	20	16	17
Mantle epithelium	26	30	9
Gonad	0	4	6
Time from start of experiment to tissue estimation (hr)	52	52	52
Time in algal suspension (hr)	41	41	41
Length of shell (mm)	14	12	14
Sex, all ripe	♂	♀	♀

[a] From J. A. Allen (1962).

TABLE V

Proportional Concentrations of ^{32}P in Selected Tissues of Bivalves Fed on Labeled *Phaeodactylum* [a,b]

Tissue	*Mya arenaria*						*Venus striatula* Immature female	Half-mature female	Mature female	*Mytilus edulis* [c]
Digestive gland	68.1	63.6	69.7	60.0	52.0	65.0	42.6	42.0	64.6	54.8
Gonad	3.3	1.8	3.4	2.8	1.4	3.3	2.6	17.5	2.7	5.3
Kidney	19.7	21.9	15.0	15.2	13.8	8.6	15.0	10.6	19.4	18.4
Mantle	1.3	2.2	4.9	3.6	9.5	2.7	5.6	4.7	No count	3.4
Adductor muscle	0.4	0.5	0.7	0.7	No count	1.7	3.3	0.3	1.9	No count
Palp	3.3	3.8	2.1	7.7	9.7	10.2	2.2	2.8	1.6	4.8
Gill	3.9	6.2	4.3	9.6	13.3	9.8	28.7	22.2	9.7	12.4
Days	¾	1	1½	5	7	3	36	36	36	36

[a] Calculated from the average results of counts of the activity in 1 μg dry tissue. From Allen (1970).

[b] Animals were allowed to feed on labeled algae for 2 hr and were then transferred to fresh seawater; the radioactivity in tissues was determined at the time intervals; the reproductive stage of the animals, except for *V. striatula* on day 36, was not reported.

[c] 17°C and 34‰.

the digestive gland was the main site of assimilation of labeled phosphorus and that its concentration decreased with time as it was dispersed to other tissues. The gonads accumulated little ^{32}P. However, it has been suggested that the state of maturity influences the distribution of radiophosphorus within the body. In *V. striatula*, animals with maturing gametes showed a slightly higher amount of ^{32}P in the gonad when compared to those with either immature or fully developed gametes (Table V). There was no difference in the proportion of ^{32}P distributed in the tissues of *M. edulis* exposed to 17 and 34‰ salinity (Table VI).

5.3.7.5.7 *Regulation of Gonad Growth.* The accumulation of nutrient reserves for utilization by developing gametes and the decline in gonad size or weight with the release of gametes during spawning are responsible for cyclical changes in the gonad index. The mechanisms controlling the transfer of nutrient reserves to gonads and the effects of environmental factors on this process have been examined for only a few species, and many problems still remain to be resolved (see Section 5.3.7.6). The relationship between food supply and gonad growth in the population of *Argopecten (= Aequipecten) irradians* from Beaufort, North Carolina, has been determined by exposing field-collected animals representing successive stages in the reproductive cycle to various temperatures in the laboratory without supplying food (Sastry, 1966a). At 10°, 20°, and 30°C, starvation of animals, at the beginning of gonad growth and gametogenesis resulted in a decrease

TABLE VI

PROPORTIONAL CONCENTRATION OF ^{32}P IN SELECTED TISSUES OF *Mytilus edulis* AT FULL AND HALF-SALINITY [a,b]

	Full salinity, 34‰			Half-salinity, 17‰		
Digestive gland	55.4	56.2	39.2	72.0	49.6	36.2
Gonad	4.8	16.2	16.7	12.4	12.8	19.8
Mantle	16.0	9.3	18.5	2.1	10.1	14.9
Adductor muscle	4.3	2.9	6.9	0.3	6.3	3.1
Gill	19.5	15.2	27.3	12.5	21.1	26.0
Tissue from completion of uptake (days)	7	21	51	7	21	51

[a] Calculated from the average results of counts in 1 μg dry tissue. Uptake of ^{32}P was from a solution in seawater. From Allen (1970); modified.

[b] Animals were maintained at the respective salinities throughout the experiment; animals from labeled seawater were transferred to fresh, filtered seawater after 7 days, and the radioactivity in tissues was determined at time intervals.

TABLE VII

EFFECT OF TEMPERATURE AND STARVATION ON *Argopecten* (= *Aequipecten*) *irradians* AT THE BEGINNING OF GONAD GROWTH AND GAMETOGENESIS [a,b]

Temperature (°C)	Number of animals	Average length (mm)	Average weight (gm)	Average gonad index	Average digestive gland index	At the time of 50% survival [c]		Gonad condition	
						Spawned	Average oocyte diameter (μm)	Color	Spermatogenic stage
10	8	34.5	12.81	3.3	10.64	No	Resorbed	Whitish	Spermatocytes
20	8	35.6	14.12	3.2	4.96	No	Resorbed	Pale brown	Spermatocytes
30	8	35.2	13.50	3.2	3.70	No	Resorbed	Pale brown	Spermatocytes

[a] From Sastry (1966a).

[b] The gonad index and digestive gland index of the sample analyzed at the beginning of the experiment (June 11, 1964) were 9.36 and 12.47. Few spermatozoa, oogonia, and oocytes prior to the growth phase (mean diameter, 13 μm) were present in the gonads.

[c] Time for 50% survival: 152 days at 10°C, 75 days at 20°C, and 20 days at 30°C.

in tissue indexes. Also, the oogonia and primary oocytes were resorbed (Table VII). Animals held at 25° and 30°C and with minimal gonad reserves and oocytes in the cytoplasmic growth phase developed and released gametes (Table VIII). At 15° and 20°C, the digestive gland and gonad indexes decreased, and the oocytes were resorbed. At 10°C, the gonad and digestive gland indexes remained the same, but the oocytes failed to grow. Similar results were obtained at 10°C with starved animals having accumulated reserves and oocytes in the vitellogenesis growth phase, as well as with animals having full-grown oocytes. At 20°C, the latter released gametes. Animals with resting gonads showed a decrease in both organ indexes at all experimental temperatures and failed to initiate gametogenesis (Table IX). Organ indexes decreased more markedly in animals at higher temperatures than those at lower temperatures. Sastry (1966a) has indicated that an aboundant food supply is necessary for gonad growth and gametogenesis in A. irradians, since the nutrient reserves accumulated in the body are not adequate to support gonad growth and gametogenesis. Under the stress of starvation, the nutrient reserves from both the digestive gland and the gonad are utilized to meet the basic metabolic requirements. When the gonads have accumulated a minimal supply of nutrient reserves, however, the gametes develop to maturity even though the animals receive no more food. It appears that gonad growth and gametogenesis in A. irradians can take place in the presence of abundant food and under temperature conditions that permit nutrient mobilization for the gonads after the animal's basic metabolic requirements have been met (Sastry, 1966a).

The relationship among food, temperature, and gonad development was determined for the population of A. irradians from Beaufort, North Carolina (Sastry, 1968). Resting-stage animals were exposed to 15° and 20°C, and one group at each temperature was provided with food, while a second group was starved. The gonad index in animals receiving food at 20°C increased, but the digestive gland index decreased (Fig. 25). A slight increase in the gonad index and a decrease in the digestive gland index was shown in animals receiving food at 15°C. Both gonad and digestive gland indexes decreased in starved animals at 15° and 20°C (Fig. 25). Apparently, postspawning adults of this population require an abundant food supply and exposure to a minimum threshold temperature (20°C) for the initiation of gonad growth and gametogenesis (see Fig. 16). Nutrient reserves accumulated in the body seem inadequate to support gonad development under these conditions (e.g., Section 5.3.7.1.3). Animals held at a subthreshold temperature (15°C) did not accumulate an appreciable amount of gonad reserves; they may

TABLE VIII

EFFECT OF TEMPERATURE AND STARVATION ON *Argopecten* (= *Aequipecten*) *irradians* WITH ACCUMULATED GONAD RESERVES AND OOCYTES AT THE BEGINNING OF THE GROWTH PHASE [a,b]

Temperature (°C)	Number of animals	Average length (mm)	Average weight (gm)	Average gonad index	Average digestive gland index	At the time of 50% survival [c]		Gonad condition	
						Spawned	Average oocyte (μm)	Color	Gametogenic stage
10	10	53.4	51.6	14.2	9.3	No	23.0	Cream, pale orange	Sperm, oocytes
15	10	53.0	41.6	11.1	5.1	No	Absorbed	Pale brown	Spermatocytes
20	10	53.1	38.0	6.0	9.1	No	Absorbed	Pale brown	Spermatocytes
25	10	48.6	34.0	6.9	7.8	8th day sperm, eggs	Absent	Pale brown	Residual sperm
30	10	53.3	39.6	10.3	8.1	5th day sperm, eggs	Absent	Pale brown	Residual sperm

[a] From Sastry (1966a).

[b] The gonad index and digestive gland index of the sample analyzed at the beginning of the experiment (July, 6, 1964) were 15.1 and 10.13. Spermatozoa and oocytes (mean diameter, 20.3 μm) were present in the gonads.

[c] Time for 50% survival: 5 days at 10°C, 120 days at 15°C, 90 days at 20°C, 22 days at 25°C, and 7 days at 30°C.

TABLE IX

Effect of Temperature and Starvation on *Argopecten* (= *Aequipecten*) *irradians* in the Resting Stage of the Reproductive Cycle [a,b]

Temperature (°C)	Number of animals	Average length (mm)	Average weight (gm)	Average gonad index	Average digestive gland index	At the time of 50% survival [c]				
						Spawned	Average oocyte diameter (μm)	Gonad condition		
								Color	Spermatogenic stage	
10	6	65.5	69.39	8.57	12.57	No	Absent	Pale brown	Spermatocytes
15	6	64.0	64.52	9.17	11.00	No	Absent	Pale brown	Spermatocytes
20	6	64.5	67.26	6.52	11.58	No	Absent	Pale brown	Spermatocytes
25	6	65.7	65.96	5.58	6.30	No	Absent	Pale brown	Spermatocytes
30	6	66.1	69.29	4.92	8.45	No	Absent	Pale brown	Spermatocytes

[a] From Sastry (1966a).

[b] The gonad index and digestive gland index of the sample analyzed at the beginning of the experiment (November 20, 1964) were 9.46 and 14.2. The gonads were inactive except for the presence of few residual spermatozoa.

[c] Time for 50% survival: 25 days at 10°C, 26 days at 15°C, 20 days at 20°C, 10 days at 25°C, and 7 days at 30°C. No oocytes were observed, nor was spawning evident.

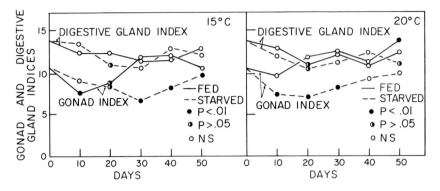

FIG. 25. Changes in the gonad index and the digestive gland index in resting-stage animals of *Argopecten* (= *Aequipecten*) *irradians* from Beaufort, North Carolina, maintained at 15° and 20°C. (From Sastry, 1968.)

have been stored in the digestive gland or transferred to other body tissues. However, neither gonad growth nor gametogenesis occurred in fed animals held at 25°C.

Gonad growth occurs in *Argopecten* (= *Aequipecten*) *irradians* within a narrow range of temperatures ($> 15°$ and $< 25°C$) when oocytes begin cytoplasmic growth phase. It is suggested that nutrients are transferred to gonads upon initiation of the cytoplasmic growth phase of the oocytes (Sastry, 1968). After the initiation of oocyte growth, the gametes continue to develop in animals exposed to warmer temperatures (25°C), suggesting nutrient transfer to the gonads. Animals with oocytes in the cytoplasmic growth phase transferred to colder temperatures (10° and 15°C) also continue development to completion of the vitellogenesis growth phase (Sastry, 1970b).

Evidence of nutrient transfer to gonads with the initiation of gametogenesis in *Argopecten* (=*Aequipecten*) *irradians* was obtained by injecting [¹⁴C]leucine into the digestive gland of scallops with inactive gonads collected during winter from Buzzards Bay, Massachusetts, and following its distribution in the body organs (Sastry and Blake, 1971 and unpublished data). The scallops were maintained at sub-threshold (5°C) and threshold (15°C) temperatures for gametoegenesis and provided with an ample supply of *Monochrysis lutheri* as food during the course of the experiment. The gonad index for animals at 5°C, remained steady, while at 15°C it showed a slight increase after 30 days of feeding (Fig. 26). The digestive gland index remained at a near-constant level in animals held at 5°C, while in 15°C animals it initially fluctuated and later increased with gonad growth. The digestive

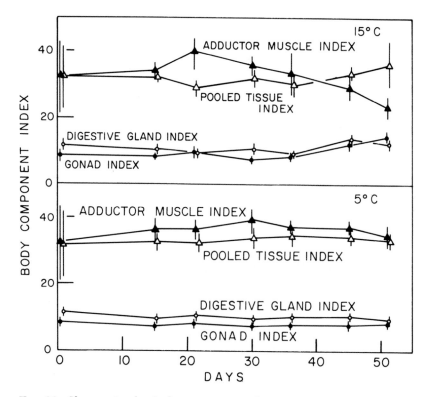

FIG. 26. Changes in the body component indexes in resting-stage animals of *Argopecten* (= *Aequipecten*) *irradians* from Buzzards Bay, Massachusetts, maintained at 5° and 15°C. The scallops were provided with *Monochrysis lutheri* as food. The pooled tissue index included the mantle, gills, kidney, and heart. Vertical lines represent the standard deviation. (From Sastry and Blake, 1971, modified; also unpublished data.)

gland index of laboratory-fed animals did not decrease with gonad growth as in field animals. At 15°C, the adductor muscle index increased during the first 20 days and then declined, while at 5°C it remained at a steady level after an initial increase. At 15°C, the pooled index for the mantle, kidney, gills, and heart fluctuated in the beginning and later increased when the adductor muscle index was decreasing. In animals held at 5°C, the pooled tissue index for the remaining body components remained at a steady level.

The uptake of [^{14}C]leucine into the mantle, kidney, gill, and heart (pooled) of 5° and 15°C animals was higher than into any other body component when measured after 1 day and again after 1 week. In animals

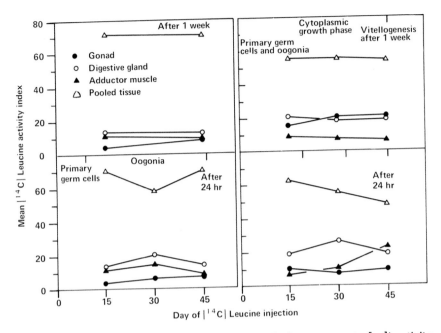

FIG. 27. Incorporation of [¹⁴C]leucine into the body components [radioactivity index = (total dpm/organ × 100)/(total dpm/animal)] of *Argopecten* (= *Aequipecten*) *irradians* collected in the resting stage from Buzzards Bay, Massachusetts, during the last week of November and maintained at 5° (left) and 15°C (right) with *Monochrysis lutheri* as food. [¹⁴C]Leucine was injected into the digestive gland of scallops on days 15, 30, and 45 after the beginning of the experiment; and radioactivity of the body components was measured 24 hr and 1 week after injection. The pooled tissue included gills, mantle, kidney, and heart. The oogenic stage of scallops sampled at each time interval from the two experimental temperatures is shown at the top of the figure. See Fig. 28 for oocyte growth responses of the same animals. (From Sastry and Blake, 1971, modified; also unpublished data.)

held at 5°C, the uptake of labeled amino acid into the gonad was less than into any other body component (Fig. 27). The incorporation of [¹⁴C]leucine into the adductor muscle was higher than into the gonad and less than into the digestive gland when measured after 1 day and again after 1 week. At 15°C, incorporation of [¹⁴C]leucine into the gonad was less than into the digestive gland and adductor muscle when measured 1 day after injection. One week after injection, however, uptake into the gonad was higher than into the digestive gland and adductor muscle (Fig. 27).

The cytoplasmic growth phase of the oocytes was initiated in animals exposed to 15°C, but at 5°C only oogonia developed (Fig. 28). Animals

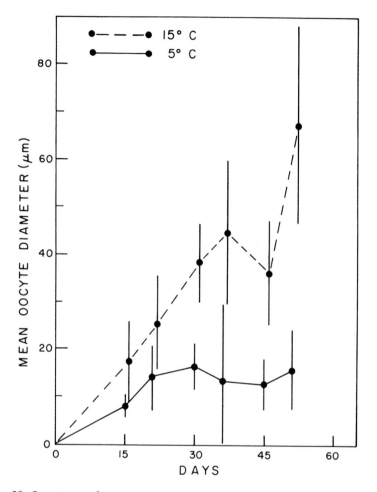

Fig. 28. Oocyte growth response in resting-stage animals of *Argopecten* (= *Aequipecten) irradians* from Buzzards Bay, Massachusetts, maintained at 5° and 15°C with *Monochrysis lutheri* as food. (From Sastry and Blake, 1971.)

injected with [14C]leucine and measured 24 hr later showed greater incorporation of labeled amino acid into both the digestive gland and gonad in 15°C animals than into 5°C animals (Fig. 27) but the difference in uptake between the two groups was not statistically significant. The incorporation of [14C]leucine into the gonads of animals measured 1 week after injection was significantly higher at 15°C than at 5°C. Nutrient mobilization for the gonads occurs slowly, and the rate of transfer is influenced by the stage of gametogenesis and the environmental temperature.

The relative uptake of [^{14}C]leucine into the gonad digestive gland, and adductor muscle seems to vary in relation to reproductive activity. The incorporation of labeled amino acid into the other body components remained approximately the same. In animals with quiescent gonads, at 5°C, the amount of [^{14}C]leucine incorporated into the adductor muscle showed no change at different time intervals. In animals with developing oocytes, at 15°C, the uptake of [^{14}C]leucine into the gonad increased with time, while it decreased in the adductor muscle and digestive gland. Since the uptake into the adductor muscle is less than into any other body component in animals with developing oocytes, it appears that the gonadal increase may occur at the expense of the adductor muscle. The relative amount of nutrients transferred into the adductor muscle and the gonad seems to be affected by the gametogenic activity and metabolism of the animals; for example, with an increase in demands during gonad development, nutrients may be channeled more into the gonads than for somatic growth. It is not clear, however, whether the nutrient reserves are utilized driectly from the adductor muscle to meet the increasing demands of gonads during development.

Vassallo (1973) reported that [1–^{14}C]acetate activity in the digestive gland of *Chlamys hericia* decreased, while it increased in the gonad, suggesting the transfer of nutrients from the digestive gland to the gonads. Recent radiotracer experiments with *Mytilus edulis* have also demonstrated that the digestive gland controls the distribution of assimilated food to the body organs (Bayne, 1975, 1976a; Gabbott, 1975, 1976), and the rate of transfer varies with the season.

5.3.7.5.8 *Functions of the Connective Tissue.* The connective tissue around the follicles of the gonads of *Ostrea lurida* (Coe, 1932; Loosanoff, 1962), *Mercenaria mercenaria* (Loosanoff, 1937b), *Crassostrea angulata* (Bargeton-Couteaux, 1942), *Mytilus edulis* and *M. galloprovincialis* (Lubet, 1959), *Chlamys varia, C. distorta, C. tigerina,* and *C. furtiva* (Reddiah, 1962), *Barnea candida* Duval, 1962), and *Crassostrea virginica* (Kennedy and Battle, 1964; Loosanoff, 1965) disappears with the growth of follicles and then appears again after spawning. It has been suggested that lipids and glycogen accumulated in these cells during the recovery period serve as nutrients for the developing gametes (Bargeton-Couteaux, 1942; Chipperfield, 1953). This gonadal connective tissue strongly resembles follicle cells in *Macoma balthica,* and it also has been suggested to have a nutritive function in the development of sex cells (Lammens, 1967). Coe and Turner (1938) found inclusions of lipid and globular albuminous nature in the follicle cells of *Mya arenaria* and suggested that they were nutritive in function.

5.3.7.5.9 *Biochemical Changes.* Biochemical changes associated with the reproductive cycle have been reported for a number of species, but the relative influence of gonad development on the distribution and storage of biochemical constitutents in various body components has been examined in only a few cases. Much of the work on biochemical changes in pelecypods has been discussed in reviews by Giese (1969), Gabbott (1975, 1976), and Bayne (1976a). A review of lipids in marine invertebrates including pelecypods is given by Giese (1966) and Lawrence (1976). A detailed discussion of the methods for studying the biochemical composition of marine invertebrates is also given by Giese (1967). Seasonal changes in the soft body weight, water content, ions, and major biochemical constituents, namely, proteins, lipids, and carbohydrates, have been determined for a number of species. The soft body weight increases with gonad development to a maximum just before spawning and then decreases with the release of gametes. Water content is minimum when the gonads are fully developed and increases during the spawning period. Ansell and Trevallion (1967) have reported that the water content in *Tellina tenuis* is minimal just before spawning when the body weight is maximal. The body weight decreases during the spawning period, and the water level increases. During autumn, the water content in this species remains steady, while the body weight continues to decrease. In *Mercenaria mercenaria*, similar changes in the body weight and water content also occur during gonad development and spawning, but the water content in this species increases steadily during winter (Ansell *et al.*, 1964; Ansell and Lander, 1967). In *Mytilus edulis*, the water content increases in winter and spring and is lowest in autumn and early winter, coinciding with spawning and gonad development (Williams, 1969). Nair and Saraswathy (1970) found a high water content in *Nausitora hedleyi* during the breeding period. Water content in the soft body also showed a reciprocal relationship with the ambient salinity.

Changes in ash content have been determined for some species throughout the year. In *Mytilus edulis*, ash content increases with gonad development and decreases with the release of gametes (Williams, 1969.) Tanaka and Hatano (1952) found that ash levels varied between 12.1 and 5.7% of the body weight, with the lowest level in winter. Nair and Saraswathy (1970) determined changes in ash content in *Nausitora hedleyi* in monthly samples collected throughout the year and found a distinct fall in the ash level with the onset of the breeding period. In *Tivela stultorum*, *M. edulis*, and *Modiolus demissus*, the ash level is highest in the digestive gland (Giese, 1969). Ash levels remain low in the testis and ovary of *T. stultorum* for much of the year and increase with the increase in gonad index. Ash levels are low in both immature testis, ovary, and indeterminate gonads.

Changes in inorganic constituents in the ash have been reported for some species. In *Pinctada martensii*, changes in the concentration of chloride, phosphate, sulfate, sodium, magnesium, and iron have been reported, which may correlate with the reproductive condition of the animals (Tanaka and Hatano, 1952). Nair and Saraswathy (1970) have reported values for chloride, phosphorus, and calcium as a percent of the body weight for animals collected throughout the year. Concentrations of phosphorus and calcium are also reported in percent of ash weight. Calcium and phosphorus contents are high during July, August, and September and low in October and November. This species breeds between June and December and passes through a resting period between January and May.

The biochemical composition of the entire soft body has been analyzed for a number of species, as discussed in the review by Giese (1969). A few examples are discussed here. Ansell and Trevallion (1967) determined the biochemical composition of *Tellina tenuis* in relationship to seasonal changes in reproductive activity and food. Total carbohydrates and nitrogen increase with an increase in body weight when gonadal development is progressing rapidly. During this period, carbohydrates, especially glycogen, increase in various body parts. During spawning, the biochemical constituents decrease with the decline in body weight. When the gonads are mature and when spawning takes place, there is a high proportion of carbohydrates. This could indicate that following maturity, the carbohydrates present are stored reserves. The body weight declines between September and February, with a corresponding decrease in carbohydrates, but proteins and lipids remain more or less constant. Carbohydrate reserves are utilized by *T. tenuis* during September–February to supplement the metabolic deficit caused by the limited food during this period. Ansell *et al.* (1964) and Ansell and Lander (1967) determined the seasonal biochemical cycle in a *Mercenaria mercenaria* population introduced in waters off Southampton, England. The condition index (the mean flesh weight as a percentage of the total wet weight) showed a distinct annual cycle following the reproductive cycle, with a maximum occurring when the gonads were developing to maturity in spring and summer. The index later decreased with spawning. The body weight remained at a low level in February and March. The lipid fraction showed no seasonal variation. Proteins decreased from February to September and later increased. Carbohydrates, however, increased from February until July and later decreased. Carbohydrates and proteins showed a reciprocal relationship in their ratios, which closely followed the cycle of the condition index. The net gain in stored reserves at the end of the spawning period is accounted for by the increase in carbohydrates. In winter, the body weight decreases, apparently entirely because of the utilization of carbohydrate reserves.

Seasonal changes in the biochemical composition of *Mytilus edulis* in British waters have been reported by Daniell (1920, 1921, 1922), Williams (1969), and Bayne and Thompson (1970). The increase in proteins and lipids, and decrease in carbohydrates, during winter and spring correlates with gonad development. Proteins and lipids decrease during the spawning period, coinciding with a rapid increase in the proportion of carbohydrates. With the redevelopment of the gonads, carbohydrates decrease and other fractions increase. The biochemical changes in *M. edulis* from the coast of Spain differ from those reported for the population from the British waters which seem related to the difference in reproductive cycle between these two areas (Fraga, 1956a,b).

Changes in biochemical composition have also been reported for *Pinctada martensii* (Ashikaga, 1948; Tanaka and Hatano, 1952), *Teredo pedicellata* (Lane et al., 1952; Greenfield, 1953), *Pecten jacobeus* (Lopez-Benito, 1955), oysters and clams (Venkataraman and Chari, 1951), *Martesia striata* (Nagabhushanam, 1961; Srinivasan, 1963; Srinivasan and Krishnaswamy, 1964), *Donax cuneatus* (Rahaman, 1965a,b), *Nausitora hedleyi* (Nair and Saraswathy, 1970), *Patinopecten yessoensis* (Takahashi and Mori, 1971), *D. vittatus, Abra alba, Chlamys septemradiata*, and *Nucula sulcata* (Ansell, 1972, 1974a,b,c), *Mytilus edulis* (De Zwann and Zandee, 1972; Gabbott and Bayne, 1973; Dare and Edwards, 1975), *Argopecten* (= *Aequipecten*) *irradians* (Estabrooks, 1973), *Pecten maximus* (Comely, 1974), *Katelysia opima* (Nagabhushanam and Mane, 1975c), and *Macoma balthica* (Brukema and De Brunn, 1977).

Seasonal variations in the biochemical composition of the body components have been reported for only a few species. Giese et al. (1967) analyzed the biochemical composition of separated body components of *Tivela stultorum* (Fig. 29; see Fig. 21 for the relative sizes of these components). The protein level is high in all body components except the gonads, where a decrease to a little over 20% occurs during July. The carbohydrate level is also high (10–25%) for all body components, except the siphons and gills which contain about 5%. The carbohydrate level is high during the early stages of gametogenesis and decreases

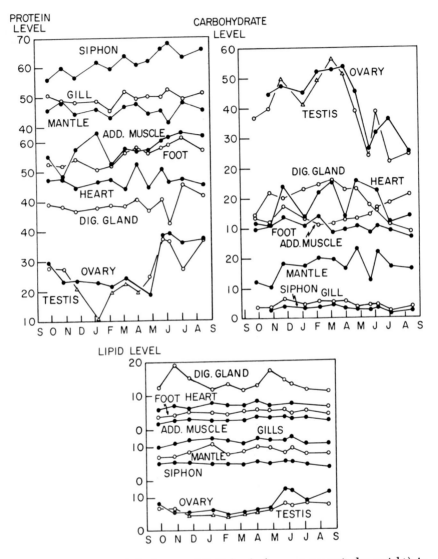

FIG. 29. Protein, carbohydrate, and lipid levels (percent per unit dry weight) in body components of *Tivela stultorum* for the year 1965–1966. The triangles indicate indeterminate gonads. Compare with Fig. 21. (From Giese *et al.*, 1967; modified.)

after the gametes have developed. In comparison, the protein level in the gonad is high when the gametes are present and decreases after their release where there is an accompanying increase in carbohydrates. The decrease in carbohydrates with the increase in protein has been suggested to be due to the conversion of carbohydrates into protein during gametogenesis. The lipids remain at the same level, except in the ovary where they seem to increase slightly, probably because of high lipid levels in the eggs. Large stores of carbohydrates in the gonads of *T. stultorum* as well as in other species, for example, *Teredo pedicellata* and *Mytilus edulis,* indicate that stored reserves may be utilized for gametogenesis (Giese et al., 1967; Gabbott, 1975).

Ansell et al. (1964) determined seasonal changes in the biochemical composition of the adductor muscle, mantle, siphons, visceral mass (gonads), digestive gland, and foot in *Mercenaria mercenaria.* The biochemical composition of these body components showed little seasonal variation, with the exception of the foot, mantle, and siphons. In these organs, carbohydrates decreased with the corresponding increase in proteins during the period of gonad proliferation. Carbohydrates from these organs are apparently utilized at the time of gametogenesis to meet the needs of rapidly proliferating gonads. The lipid content was high in the digestive gland in comparison to other body organs, perhaps because of its function as a site for lipid storage. Gonads contained higher levels of carbohydrates than any other body organ. Since there is no marked seasonal cycle in the biochemical composition of other organs and because the carbohydrate content in gonads is greater than other constituents, the gonadal increase has been suggested to be responsible for seasonal changes in the carbohydrate/protein ratio for the whole animal. Seasonal changes in stored glycogen and lipids have also been reported in the horse clam, *Tresus capax,* by Reid (1969).

Bayne and Thompson (1970) determined the biochemical composition of mantle (germinal) and nonmantle (somatic) tissues of *Mytilus edulis* maintained in the laboratory under different conditions of nutritional and temperature stress. In *M. edulis,* the mantle tissue serves both as a site of storage and of gamete production. Gamete production occurs at the expense of other body reserves when the gametes are not ripe. However, after gametogenesis is completed, animals at sublethal temperatures and under nutritional stress cannot maintain the gametes in a ripe condition and the gonads regress. Animals maintained under stressful conditions initially mobilize carbohydrate reserves from the mantle to meet energy requirements. As stress continues, proteins from mantle tissue and carbohydrates and proteins from nonmantle tissue are also utilized for maintenance. Apparently there is a complex relationship between the development of gametes and the readiness with which carbohydrate and protein reserves are utilized in animals under stress.

The interrelationship among food, sites of energy reserves, and the utilization of reserves relative to gamete development in *Mytilus edulis* has been discussed in detail by Gabbott and Bayne (1973) and Bayne (1975, 1976a). A detailed discussion of energy metabolism in *M. edulis*, the interrelationship among different body tissues, their energy requirements, and metabolic transformation of reserves has been given by Gabbott (1975, 1976).

In *Mytilus edulis*, the carotenoid content correlates with the reproductive cycle (Campbell, 1969). Pigment content increases in April–May, when the animals are mature, and decreases rapidly during summer, reaching a minimum during winter. Mussels stored at 10°C with a limited food supply show no seasonal variation in carotenoid content when compared to fresh animals obtained from the field. Rates of feeding affect pigment concentration, and in starved animals carotenoids are lost more rapidly from the hepatopancreas than from the gonads.

Idler *et al.* (1964) have determined the total lipids, unsaponifiable material, provitamin D, and sterols in monthly samples of male and female *Placopecten magellanicus*. During the prespawning period in August, the total lipids, unsaponifiable material, and sterol levels are low in the adductor muscle. This has been suggested to be due to active transfer of these materials to the gonadal tissue. During the retarded growth period in February and March, the lipid content of the muscle is decreased, but there is a compensating increase in unsaponifiable material and the sterol content of total lipids. Lubet and de Longcamp (1969) reported seasonal variations in sterol esters, cholesterol, and triglycerides in *Mytilus edulis*. Using ^{3}H- and ^{14}C-labeled precursors, de Longcamp *et al.* (1974) found very little conversion of [^{3}H]acetate into any of the sterols and none into cholesterol in the gonads of male and female *M. edulis*. Gabbott (1976) indicated that this may be due to short incubation times of 2 and 6 hr. In *Chlamys hericia*, the lipid in the digestive gland is converted into fatty acid and glycerol and then transferred to the gonad where synthesis of triglycerides and hydrocarbons occurs (Vassallo, 1973).

Thompson (1977) determined the seasonal changes in biochemistry and blood chemistry of *Placopecten magellanicus*. Carbohydrates, lipids, and protein content of somatic and gonadal tissue were measured on a monthly basis. In this species, there was no evidence of mobilization of energy reserves stored in somatic tissues for the gonad during its growth and maturation. There was also no accumulation of reserves in the gonads after spawning. Plasma lipid and protein content increased at the time of gonad growth and maturation, much of it probably going to gamete production. In somatic tissues, lipid varied little during the year, while protein was somewhat lower in fall and winter than in summer and carbohydrate increased rapidly in early summer and then decreased with spawning through winter.

5.3.7.5.10 *Energy Allocation for Gamete Production.* Detailed studies conducted in recent years on the energy flow in bivalve populations subjected to different temperatures and food conditions either in the laboratory (Bayne, 1975; Ehinger, 1978) or seasonal variations in nature (Hughes, 1970; Fuji and Hashizume, 1974) are beginning to provide information as to how energy is allocated for somatic growth, maintenance metabolism, and gamete production. The energy allocated for gamete production is the assimilated energy remaining after requirements for metabolism and somatic growth have been met. In *Macoma balthica,* where there is no somatic growth after the animals have reached adulthood, all the energy available for production after maintenance requirements have been met is utilized for gamete production (Gilbert, 1973). In some species where somatic growth occurs throughout life, the energy requirements for gamete production are also high. Bayne (1975) estimated that gamete production as a percentage of total production increased from 8% for 1-year-old to 94% for 4-year-old *Mytilus edulis.* The energy balance of *M. edulis* subjected to combined nutritional and temperature stress relative to gametogenic activity has been recently reviewed (Bayne, 1975). Fuji and Hashizume (1974) determined the seasonal energy budget for three age groups of *Patinopecten yessoensis* in Matusi Bay, Japan. Most of the energy consumed by this species was utilized for somatic growth during the nonreproductive period and, after the onset of gametogenesis, the major part of the stored energy was utilized for gamete production (Fig. 30). The amount of energy utilized for gamete production relative to the amount of energy assimilated increased markedly for the older age groups. Ehinger (1978) determined the seasonal energy balance of a population of *Placopecten magellanicus* occurring below the euphotic zone in Narragansett Bay, Rhode Island. This population was found to adjust its metabolic energy expenditures relative to the seasonal variation in food availability and temperature. A portion of the assimilated energy was made available for growth (including gonad growth) in all stages of the reproductive cycle except maturity.

5.3.7.6. Neurosecretion

Nervous and hormonal systems in pelecypods are interrelated structurally and functionally, and the nervous system plays a role in neurotransmission as well as in the synthesis and discharge of secretions. Neurons secrete both neurohumors and neurohormones. Studies on neurohumors have involved their chemical nature (Blaschko and Milton, 1960; Milton and Gosselin, 1960; Puppi, 1963; Sweeney, 1963; Zs Nagy et al., 1965; Paparo, 1972; Stefano and Aiello, 1975; Haley et al., 1976), their control of ciliary activity (Aiello, 1957, 1960, 1962, 1965, 1970;

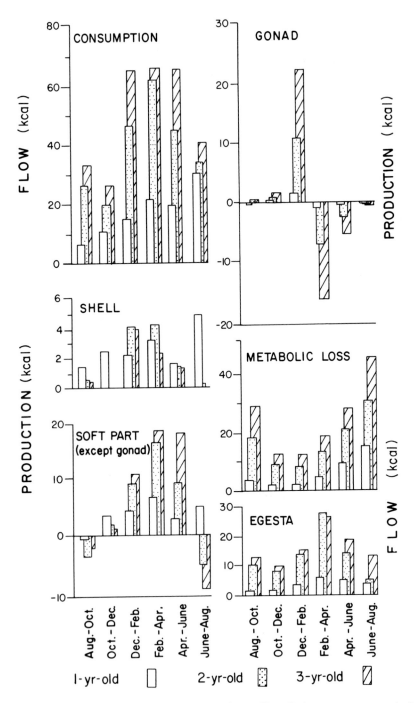

F𝐼G. 30. Components of energy flow for the scallop *Patinopecten yessoensis* in relation to time. (From Fuji and Hashizume, 1974; modified.)

Paparo and Aiello (1970), and their influence on oxygen consumption (Moore et al., 1961; Moore and Gosselin, 1962). The control of cilio-excitation and cilioinhibition by neurotransmitters and their possible role in the regulation of functional processes has been discussed by Sleigh (1974) and Bayne (1976b).

Neurosecretion in pelecypods has been reviewed by Gabe (1965, 1966), Lubet (1966, 1973), Martoja (1972), and Golding (1974). The presence of neurosecretory cells has been demonstrated by histological studies on a number of species. Their number and location vary among species. Neurosecretory cells are located in cerebral (cerebropleural) ganglia and visceral ganglia. In primitive species such as *Nucula nucleus* cells rich in acidophilic secretions from a dense cap on the dorsal aspect of the cerebral ganglia (Gabe, 1966). Neurosecretory cells have been observed in the dorsorostral and dorsocaudal areas of cerebropleural ganglia in *Teredo* (Gabe and Rancurel, 1958). In higher lamellibranchs, neurosecretory cells are less numerous and more localized. They are generally found in the dorsal cell caps of cerebral ganglia and the dorsal cell layer of visceral ganglia (Lubet, 1955a, 1959; Nagabhushanam, 1963, 1969; Nagabhushanam and Mane, 1973; Antheunisse, 1963). Blake (1972) found that the neurosecretory cycle of the large neurons in the cerebral ganglia of *Argopecten (= Aequipecten) irradians* appeared identical to that of the visceral ganglia. The neurosecretory cells were found to be less numerous in the cerebral ganglia than in the visceral ganglia. In bivalves, the neurosecretory cells are not usually found in the pedal ganglia, although they have been reported to occur in all ganglia of the freshwater mussels *Unio tumidus* (Fahrmann, 1961) and *Dreissena polymorpha* (Antheunisse, 1963).

In most species neurosecretory cells are small or medium-sized, with an approximate diameter of 20 μm. Neurosecretory perikarya are ovoid or pyriform. The general histological features are similar to those of plasmochrome cells, but neurosecretory perikarya can be distinguished by the marginal position of the Nissl bodies and the presence of acidophilic secretions in the cytoplasm (Gabe, 1955, 1966). Different categories of neurosecretory cells have been distinguished based on their size and morphology. In *Mytilus edulis* and *Chlamys varia* some neurosecretory cells are pear-shaped, unipolar, and up to 25 μm, while others are small and multipolar (Lubet, 1959). Pear-shaped (type I) and oval-shaped (type II) neurosecretory cells were distinguished in *Crassostrea virginica* and *Meretrix casta* (Nagabhushanam, 1963, 1969) and *Katelysia opima* (Nagabhushanam and Mane, 1973). Different categories of neurosecretory cells have also been reported in the freshwater mussel *Unio tumidus* (Fahrmann, 1961).

The appearance and position of neurosecretory products within the perikarya vary with the stage of the neurosecretory cycle (Lubet, 1955a,b, 1959; Gabe, 1966; Gabe and Rancurel, 1958; Blake, 1972). In some cells neurosecretory granules are few, while in others they are abundant and remain discrete. In still other cells, neurosecretory products are present in lumps or pools. The discharge of neurosecretory products is characterized by cytoplasm and the presence of small quantities of secretory products between the vacuole and axon hillock (Gabe, 1966). Signs of axonal transport are not very distinct in marine bivalves. Neurosecretory products have been observed in the axon hillock and proximal parts of the interganglionic paths of the axon, but they disappear in the neuropile and are not seen in the communicative branches, commissures or nerves leaving the ganglia. Welsh (1961) indicated that electron microscopic studies of *Mercenaria mercenaria* showed an occasional axon filled with electron-dense granules and having ultrastructural characteristics similar to neurosecretory neurons of arthropods and vertebrates. Endocrine glands or neurohemal organs have not been identified in bivalve molluscs. Umiji (1969) has reported that a neurohemal area exists on the cerebral commissure of *Mytilus perna*. The transport of neurosecretory substances by axons, intermediate cells, and possibly glial cells has been suggested by Lubet (1955b) and Umiji (1969). Several authors have suggested that glial cells play a role in storage and transport and that glial cells and epineurons can function as neurohemal organs (Fahrmann, 1961; Antheunisse, 1963). However, the chemical nature, transport, and fate of neurosecretory products is not clearly established.

Cyclical activity in neurosecretory cells was observed by histological studies on *Chlamys varia* and *Mytilus edulis* (Lubet, 1955a, 1959), *Spisula solidissima*, *Meretrix casta*, *Yoldia* sp., *Brachidontes recurvus*, *Modiolus demissus*, and *Mulinia lateralis* (Nagabhushanam, 1963, 1969), *Katelysia opima* (Nagabhushanam and Mane, 1973), *Mytilus perna* (Umiji, 1969), and *Argopecten (= Aequipecten) irradians* (Blake, 1972). Seasonal changes in cyclical activity have been related to the reproductive cycle. However, Welsh (1961) and Antheunisse (1963) have pointed out that the reproductive cycle is usually seasonal and in turn related to such environmental factors as temperature, food abundance, and light. Therefore it is not safe to conclude that neurosecretory substances directly regulate gonad activity.

Temperature and salinity changes have been reported to affect neurosecretory activity within the cerebral ganglia of *Mytilus edulis* (Lubet and Pujol, 1965). A rapid increase in temperature of 10°C and holding animals at the elevated temperature for 1 hr, or a sudden decrease in salinity to 20‰, results in emptying of the neurosecretory cells of the cerebral ganglia. A decrease in temperature or an increase in salinity to 45‰ was reported to result in the accumulation of neurosecretory products. With similar treatment, however, there was no variation in neurosecretory activity in the visceral ganglia. Nagabhushanam (1964) also reported that high temperature, low salinity, and electrical shock stimulated emptying of neurosecretory products in the visceral ganglia of *Crassostrea virginica*.

In *Mytilus edulis*, the pear-shaped cells have a distinct annual cycle, and the small, star-shaped cells show continuous activity (Lubet, 1959). The accumulation of neurosecretion in the pear-shaped cells of the cerebral ganglia of *M. edulis* and *Mytilus galloprovincialis* occurs during the phase of gametogenesis. Emptying of neurosecretion occurs when the gametes are mature. The activity of multipolar neurons and of cells of visceral ganglia is continuous.

Therefore gametogenesis appears to be linked with secretion in the pear-shaped cells (Lubet, 1959). In *Crassostrea virginica*, the activity of type I cells varies during the year; the neurosecretory material accumulates in January and reaches a maximum in March (Nagabhushanam, 1963). The number of cells containing secretion decreases between April and September; the cells are empty between October and December. In the freshwater mussel *Dreissena polymorpha*, the neurosecretory material begins to accumulate in the cerebral ganglia in autumn; maximum activity takes place during winter. A period of inactivity occurs in summer. In the visceral ganglia, the cycle is similar to that in the cerebral ganglia, but emptying takes place during the summer (Antheunisse, 1963).

In the adult population of the tropical bivalve *Katelysia opima* from the west coast of India, gametogenesis, followed by spawning and resting phases, occurs twice each year. Neurosecretory products begin to accumulate in the type I cells with the initiation of gametogenesis and reach a maximum when the animals are mature (Nagabhushanam and Mane, 1973). Secretory granules in neurosecretory cells decrease with spawning and are not seen in resting animals. The neurosecretory cycle and reproductive cycle closely parallel each other (Fig. 31).

Demonstration of the role of neurosecretion in the reproduction of bivalves has been difficult with standard surgical procedures. Bilateral ablation of cerebral ganglia during the resting phase and at the beginning of gametogenesis delays gametogenesis in *Mytilus edulis;* the few gametes formed may undergo lysis before spawning (Lubet, 1959, 1965). Removal of ganglia at latter stages of gametogenesis accelerates gamete maturation. After ablation of ganglia during the resting phase the amount of RNA and lipid is decreased and glycogen is increased in

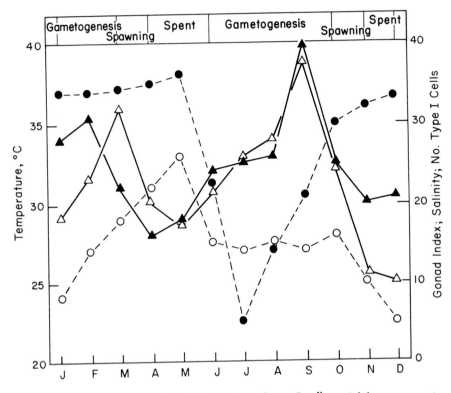

FIG. 31. Reproductive cycle and the number of type I cells containing neurosecretory granules in *Katelysia opima* relative to changes in temperature and salinity. O, temperature (°C); △, gonad index; ●, salinity (‰); ▲, number of type I cells. (From Nagabhushanam and Mane, 1973, 1975a; modified.)

gonadal connective tissues. Umiji (1969) has found that ablation has no effect on follicle development nor on the intrafollicular connective tissue. Although cerebral ganglia appear to secrete neurosecretory substances during gametogenesis, it has been difficult to determine whether their mechanism is nervous or hormonal (Lubet, 1965). The integrity of the cerebral ganglia is, however, essential for progress of the normal course of the reproductive cycle (Lubet, 1955b, 1959, 1966; Nagabhushanam, 1962; Antheunisse, 1963). The cells of visceral ganglia do not appear to be involved in the regulation of reproductive activity in *M. edulis*.

Blake (1972) examined the effects of temperature and starvation on neurosecretory activity and oogenesis in an *Argopecten (= Aequipecten) irradians* population from Massachusetts. The annual cycle of neurosecretory activity of the population was divided into five stages based on changes in size, granulation, and vacuolization of the neurons. The neurosecretory cycle stages were found to coincide with the following stages in oogenesis: stage I, vegetative or resting stage with primary germ cells; stage II, oogonia and early oocytes; stage III, cytoplasmic growth phase; stage IV, vitellogenesis and maturation; and stage V, spawning. For the first 12 months, the neurosecretory cycle and reproductive cycle of the population were highly synchronous. Both cycles showed significant correlation with seasonal changes in temperature but not with each other.

Argopecten (= Aequipecten) irradians collected from the field in different stages of the annual reproductive cycle were maintained at 5° and 15°C with food, and their subsequent responses in neurosecretory activity and oogenesis were followed for about 8 weeks (Blake, 1972). Three-month-old scallops with primary germ cells and in neurosecretory cycle stage I held at the two temperatures showed no change in either neurosecretory cycle stage or stage of oogenesis (see Section 5.3.7.7). Six-month-old scallops with oogonia and early oocytes (23 μm in diameter) held at 5°C showed no increase in oocyte diameter or change in neurosecretory cycle stage. Those held at 15°C and animals transferred from 5° to 15°C after 28 days showed an increase in oocyte diameter with corresponding progress of the neurosecretory cycle to stages III and IV (Fig. 32). When the animals at 15°C were transferred after 28 days to 5°C, the neurosecretory cycle reverted to stage II and oocyte diameter remained the same during the next 4-week period. Scallops in the later stages of oogenesis (e.g., the cytoplasmic growth phase and vitellogenesis) held at 15°C reached maturity with corresponding changes in neurosecretory cycle stage (Fig. 32). Animals held at 5°C showed no change in the stage of oogenesis or neurosecretory cycle. In postspawning animals, the neurosecretory cycle returned to stage I for a brief period and then advanced to stage II with the development of oogonia. The neurosecretory cycle stage and oogenic stage remained the same in animals held at 5°C, but the two cycles progressed to later stages in animals held at 15°C.

Blake (1972) also determined the effects of starvation of Argopecten (= Aequipecten) irradians collected from the field at different stages during the annual reproductive cycle. Starvation of scallops in the vegetative phase resulted in the resorption of germinal epithelium at both 5° and 15°C, but the neurosecretory cycle remained in stage I. Similar treatment of animals with oocytes in the cytoplasmic growth phase and the neurosecretory cycle in stage III showed no change at 5°C, but at 15°C the oocytes grew to vitellogenesis with corresponding

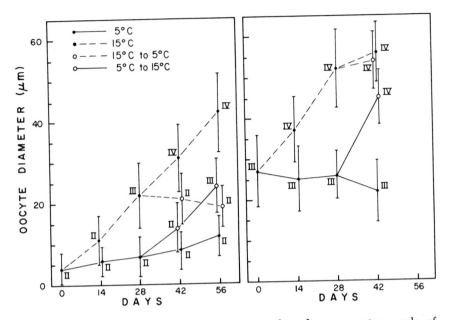

FIG. 32. Effect of temperature on oocyte growth and neurosecretory cycle of *Argopecten* (= *Aequipecten*) *irradians* from Massachusetts maintained at subthreshold (5°C) and threshold (15°C) temperatures for oogenesis with *Monochrysis lutheri* as food. Left, Mean changes in the neurosecretory cells and oocyte diameter of 6-month-old scallops with oogonia and early oocytes at the beginning of the experiment. Right, Mean changes in the neurosecretory cells and oocyte diameter of 10-month-old scallops with oocytes in the cytoplasmic growth phase at the beginning of the experiment. Roman numerals show the predominant neourosecretory stage. After 28 days some individuals were transferred from 15° to 5°C and from 5° to 15°C. (From Blake, 1972; modified.)

changes in the neurosecretory cycle. Starvation of animals undergoing vitellogenesis and neurosecretory cycle stage IV had no effect on oocyte development or neurosecretory stage at either 5° or 15°C. When partially spawned animals were starved, the residual gametes were resorbed and the neurosecretory cycle returned to stage II. Prolonged starvation of these animals resulted in resorption of the ovarian portion of the gonad, while the neurosecretory cycle remained in stage II. A depletion of reserves from the digestive gland was indicated by vacuolization of the epithelium of the tubules and ducts of the digestive diverticulum. Prolonged starvation of *A. irradians* at the time oocytes are initiating the cytoplasmic growth phase, or exposure of animals to subthreshold temperatures in the later stages of oogenesis, seems to result in neuron degeneration and atrophy (Blake, 1972).

In *Argopecten (= Aequipecten) irradians,* the change in the neurosecretory cycle stage from II to III seems to act as a switching mechanism, and oocyte growth may be initiated or delayed depending upon the ambient temperature and food conditions (Sastry, 1968, 1970b, 1975; Sastry and Blake, 1971; Blake, 1972). Continued progress of both cycles occurred only when the animals were exposed to threshold temperatures with food. The change in the neurosecretory cycle from stage II to III may allow mobilization of the nutrient reserves from the digestive gland for the gonads, which is essential for the growth of oocytes (see Section 5.3.7.5; see also Sastry, 1966a, 1968, 1970b, 1975; Sastry and Blake, 1971; Blake, 1972). For a certain time after the neurosecretory cycle has progressed to stage III, however, subsequent exposure of the animals to subthreshold temperatures reverses the neurosecretory cycle to stage II and delays the growth of oocytes. However, if the animals have been at the threshold temperature for a prolonged period of time with the neurosecretory cycle in stage III and the oocytes well into the cytoplasmic growth phase, or in stage IV with oocytes undergoing vitellogenesis, subsequent exposure to subthreshold temperatures does not reverse the neurosecretory cycle stages. Prolonged exposure of animals with oocytes undergoing vitellogenesis to subthreshold temperatures results in the vacuolization of cytoplasm and lysis of oocytes (Sastry, 1966a, 1968). Blake (1972) found that neuronal degeneration in *A. irradians* was correlated with the extent of oocyte disintegration and resorption.

In *Argopecten (= Aequipecten) irradians,* the gametogenic and neurosecretory cycles appear to be dynamically linked, with both cycles regulated by changes in the environmental temperature (Fig. 33). Feedback controls through the mediating neuroendocrine substances may regulate metabolism, accumulation of reserves and mobilization of nutrients for the gonads, and gametogenesis relative to changes in the environment (Sastry, 1975).

The relationship between the nervous system and metabolism and growth has been examined in *Mytilus edulis* (Bourcart and Lubet, 1965; Lubet and Pujol, 1965; Lubet, 1965, 1966) and *Crassostrea virginica* (Nagabhushanam, 1964). Bilateral removal of cerebral ganglia results in a delay of growth, which is reversed quickly in smaller animals, while it causes disorders in lipid metabolism in adults and has little or no effect on glycogen metabolism and storage. Ablation of visceral ganglia causes weakening of muscle tonus, a decrease in the respiration rate, and a loss of reserves from starvation in *M. edulis* (Altman, 1959; Lubet, 1965, 1966). In *C. virginica* (Nagabhushanam, 1964), *Katelysia opima* (Nagabhushanam and Mane, 1973), and *Argopecten (= Aequipecten)*

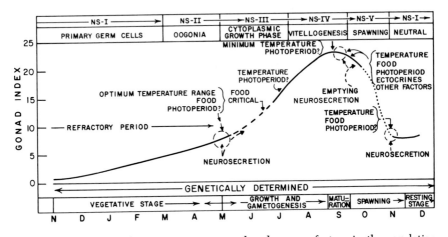

Fɪɢ. 33. Interaction between exogenous and endogenous factors in the regulation of the annual reproductive cycle of *Argopecten* (= *Aequipecten) irradians*. (From Sastry, 1975.)

irradians (Blake, 1972), the water filtration rate is affected on removal of the visceral ganglia. Visceral ganglia of pelecypods contain large amounts of acetylcholine and 5-hydroxytryptamine, and their removal may cause a loss of these chemical mediators affecting the integration of functional processes of the animals (Bayne, 1976b).

In *Mytilus edulis*, synthesis of glycogen takes place in the mantle tissue until gonad development is complete. Glycogen content increases in spring (April–May), but there is no storage in the connective tissue. Glycogen synthesis and breakdown take place on a seasonal basis, hence the metabolic responses are essentially long-term, according to Gabbott (1976). Glycogen metabolism and gametogenesis have been suggested to be regulated by the same neurohormones in *M. edulis* (Gabbott, 1975). Bourcart and Lubet (1965) have examined the effect of neurosecretion on the accumulation and degradation of glycogen reserves in *Mytilus galloprovincialis*. Comparison of the glycogen content of decerebrated and control mussels has shown that cerebral hormones have no direct affect on glycogen metabolism. Houtteville and Lubet (1974, 1975) have distinguished between the effect of cerebral ganglia on gonad development and on the utilization of glycogen and protein reserves in connective tissues and the action of visceral ganglia in the control of accumulation of reserves during the resting period of the reproductive cycle. It has been reported from observations on mantle tissue of *M. edulis* maintained in culture with cerebral or visceral ganglia that visceral ganglia have a neuroendocrine control over the accumulation of glycogen reserves in storage tissue, and cerebropleural ganglia over the release of reserves and gametogenesis (Houtteville and Lubet, 1974, 1975; see also Sastry and Blake, 1971; Sastry, 1975; Bayne, 1975, 1976a; Gabbott, 1975, 1976). Bayne (1976a) reported that glucose was found in the blood of *M. edulis* at high levels during summer after spawning and may help trigger glycogen synthesis. In autumn and early winter, neurosecretory activity has been suggested to provide a hormonal control switching on the cycle of degradation of glycogen and the initiation of gametogenesis. Once initiated, gametogenesis apparently proceeds through winter until spawning, when control over the digestive gland is reestablished.

5.3.7.7 EFFECT OF AGE

Animal size at the beginning of gametogenesis and size at maturity have been reported for a number of species. In many cases, it is not possible to examine the same individual at successive stages in the life cycle. Hence the effect of age on gametogenesis has to be inferred from a series of observations made on a population. In some pelecypods, sexuality can appear precociously in the postlarval stage or juvenile stage. This has been observed in *Chlamys varia*, *Glycymeris glycymeris*, and *Venus striatula* (Lucas, 1966).

The effect of age on the types of gametes developed is most evident in species showing a sex change during their life (e.g., Section 5.3.2). The wood-boring bivalves *Xylophaga dorsalis* (Purchon, 1941), *Teredo navalis* (Coe, 1943b), *T. furcillatus* (Nagabhushanam, 1970), *Nausitora hedleyi* (Nair and Saraswathy, 1970), and others, show a sequence of sex changes that correlate with the size and age of the animals. In *T. navalis*, sexual differentiation begins 3–4 weeks after entering wood. Young animals normally pass through an initial male phase and later change over to a female phase. In *X. dorsalis*, young males change over to a female phase when they are about 3 mm in length. The sex change in some younger individuals may be delayed, resulting in an equal proportion of males and females in this segment of the population. Individuals in the population exceeding 8 mm are all females. In *Mercenaria mercenaria*, the primary bisexual gonad is formed at about 4–6 mm, and its protandric nature is expressed by a proliferation of spermatozoa when the animals are about 5–7 mm. In the second year, the individuals become definitive males and females (Loosanoff, 1937a). A similiar sequence in gonadal development occurs also in *Venus striatula* (Ansell, 1961a).

The pearl oysters *Pinctada albina*, *P. margaritifera*, and *P. fucata* are protandric hermaphrodites. They first mature as males, and then pass

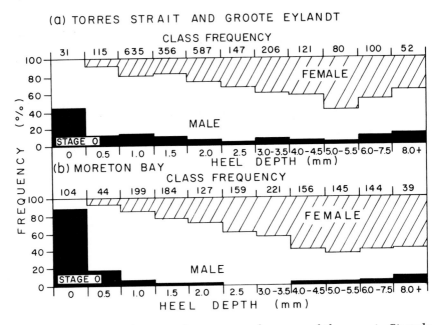

FIG. 34. Variation with size in the percentage frequency of the sexes in *Pinctada albina*, illusrating the protandric sexuality of the species. (From Tranter, 1958a.)

through several sexual phases during the life of an individual (Tranter, 1958a,c, 1959). In these species, the sex ratio varies with size (Fig. 34). The middle age group is composed of an equal proportion of both sexes, but as the population becomes older there is an increase toward femaleness. In *Chlamys varia*, the majority of first-year animals mature as males, and some later change to females any time after spawning (Reddiah, 1962). Sexual differentiation begins in *C. varia* at a length of 13 mm, and spawning at a length of about 20 mm. Males predominate in the smaller size group (13–23 mm), but at about 24 mm the sex ratio becomes equal. The larger animals (25–38 mm) are mostly females. The sex change occurs at about 25 mm after the first spawn. In *Chlamys tigerina*, gametogenesis begins at a shell length of 13–14 mm, and sexual maturity is reached at about 16–17 mm, when the animals are about 8 months of age (Reddiah, 1962). In *Mulinia lateralis*, the gonads develop within 6 weeks after the larvae settle (Calabrese, 1969a).

The age of the individual appears to have some relationship to the extent of gonad development in some species. In *Macoma balthica*, the gonads are less extensive in first-year animals, in comparison to older animals (Lammens, 1967). Coe and Turner (1938) found that the number of eggs in the female correlated with animal size. Brousseau (1978) observed a positive correlation between female body size of *Mya arenaria* and oocyte production. In *Patinopecten yessoensis*, the proportion of energy used for gamete production to energy assimilated increases markedly with increasing age. This ratio is about 2% in 1-year-old, 7% in 2-year-old, and 12% in 3-year-old animals (Fuji and Hashizume, 1974; see also Fig. 30). In *Placopecten magellanicus*, gonad weight and age of the animals show a sigmoid relationship (Naidu, 1970). The testes of any age group in this species are usually heavier than the ovaries.

Populations of *Argopecten (= Aequipecten) irradians* from Nantucket Sound, Massachusetts, Beaufort, North Carolina, and Alligator Harbor, Florida, reach sexual maturity at the age of 12 months and begin spawning (Belding, 1910; Gutsell, 1930; Sastry, 1961, 1963, 1966a, 1970a). This species has a life span of 18–24 months, and only a few individuals survive to spawn for a second time. The cytoplasmic growth phase of oocytes begins in the Alligator Harbor population at the age of 4–5 months, in the Beaufort population at the age of 7–8 months, and in the Nantucket Sound population at the age of 8–9 months (Sastry, 1961, 1963, 1966a, 1970a). In all three populations, spermatozoa develop earlier than eggs. Animals maintained at elevated temperatures develop gametes to maturity and spawn at an earlier age, once certain physiological prerequisites and nutritional requirements are satisfied (e.g., Section 5.3.7.1). The animals collected at the age of 4–5 months from Alligator Harbor and maintained at $23° \pm 1.0°C$ released gametes when they were only 5–6 months of age (Sastry, 1961, 1963). Individuals of the Beaufort population developed to maturity at the age of 9–10 months when exposed to 25° and 30°C (Sastry, 1966a). These observations indicate that, although the population may take 12 months to reach the mature state in nature, the animals can be accelerated to reach maturity earlier by exposing them to higher temperatures. Animals with gonial cells collected from Nantucket Sound at the age of 6–7 months, provided with food, and exposed to 10°, 15°, 20°, and 25°C in the laboratory showed oocyte growth only in those held at 15°C (Fig. 35). At 15°C, the cytoplasmic growth phase of oocytes was initiated after 45 days of feeding at the age of 7.5–8.5 months, and vitellogenesis occurred by the age of 9–10 months. The beginning of the oocyte cytoplasmic growth phase in animals held at 15°C occurred at about the same time as in those in the field. It is likely that animals may have to reach a minimum age before they can initiate oocyte cytoplasmic growth and progress further with gonad development (see Section 5.3.7.6). After initiation of cytoplasmic growth, however, the rate of gamete development to maturity is dependent on temperature within

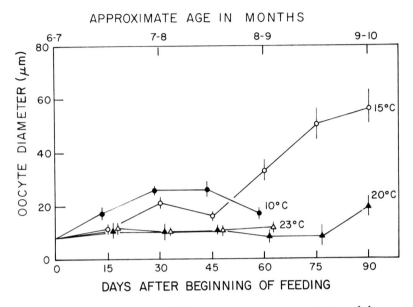

FIG. 35. The influence of age and different temperatures on initiation of the oocyte growth phase in *Argopecten* (= *Aequipecten*) *irradians* from Woods Hole, Massachusetts. (From Sastry, unpublished data.)

the species-specific range, provided nutritional needs are satisfied. Post-spawning adults, however, initiate the cytoplasmic growth phase of oocytes within a week to 10 days after exposure to the threshold temperature with food (e.g., Section 5.3.7.5). After peak spawning, the Alligator Harbor population of *A. irradians* recovers rapidly and reaches a mature state by late autumn and early winter. The Beaufort population remains in an inactive state during late autumn and early winter, but gametogenesis seems to occur in second-year animals which reach the mature state by April (Kirby-Smith, 1970). The population of *A. irradians* from Nantucket Sound initiates gametogenesis during autumn, but they are highly asynchronous. Second-year animals become mature and spawn at the same time as first-year animals. Postspawning adults of *A. irradians* go through successive gametogenic cycles as long as temperatures are within the range for gonad development and food is available (Sastry, 1963, 1968; Sastry and Blake, 1971).

5.3.7.8 EFFECT OF PARASITES

The parasites and diseases of pelecypods have been reviewed in detail by Cheng (1967), Sindermann (1970), and Sniesziko (1970). A variety of protozoan and trematode parasites are reported in pelecypods, some of which are found in gonadal tissue, but their effect on gameto-genesis in the host is not always clear.

Naidu (1971) and Stevenson and South (1975) reported that colonies of zoochlorellae, unicellular algae inhabiting the mantle, gonad, and muscle of *Placopecten magellanicus* in shallow waters, are parasitic and cause general loss of body condition. Field (1922) and Sprague (1965) have reported that the microsporidian *Chytridiopsis mytilovum* is found in eggs of *Mytilus edulis*. Cysts ranging in diameter from 14 to 20 μm are found in egg cytoplasm abutting on the nucleus, sometimes in the nucleolus or in undifferentiated gonadal cells. However, the effect of parasitic infestation on the egg is not known. Cysts of the gregarine *Nematopsis* have been reported in tissues of a number of genera of European bivalves, *Mytilus, Cardium, Donax, Tellina, Spisula, Solen,* and others, but there is no evidence of pathogenicity (Dollfus, 1922).

Infection by trematode parasites has been reported to cause parasitic castration in some pelecypods. The trematode *Bucephalus haimeanus* occurring in tissues of *Cerastoderma (= Cardium) tuberculatum* and *Syndosmya alba* is reported to cause parasitic castration of the host (Cheng, 1967). Sporocysts of *Bucephalus mytili* occurring in the digestive gland and the gonad of *Mytilus edulis* are thought to cause parasitic castration of the host. The bucephalid cercariae, *Cercaria noblei* in *Mytilus californianus* and *Cercaria (Gymnophallus) fulbrighti* in *Cerastoderma (= Cardium) edule,* occupy principally the digestive gland, foot, and gonads. *Cercaria myae* occurring in the digestive gland of *Mya arenaria* has been suggested to cause complete castration of the host (Uzmann, 1952). Palombi (1934) has reported that sporocysts of *Bacciger bacciger* infect both the visceral mass and reproductive organs of *Tapes decussatus, T. pullastra, T. semidecussata, T. aureus, Donax vittatus, Pholas candida,* and *Tellina exigua.* Infestation by sporocysts is reported to cause parasitic castration of the host. Sporocysts of the trematode parasite *Protoeces maculatus* are reported from the sinuses and lymph spaces in the gonad of *Mytilus edulis.* In parasitized animals, sporocysts develop in the vascular system and cause a serious reduction in the circulatory system, including that in the mantle, and prevent normal gametogenesis. Follicular development is either seriously impaired or totally precluded (Uzmann, 1953). Dupouy and Martinez (1973) studied the influence of *Protoeces maculatus* on the reproductive cycle of *Mytilus galloprovincialis.* Parasitic castration was found when miracidia invaded during the resting phase, and maximal proliferation of sporocysts occurred before the development of gonial cells. The sporocysts utilized the storage substances (glycogen) of the intertubular tissue and this later resulted in arrest of the gametogenic cycle.

Sellmer (1967) has reported that sporocysts containing furcocerous cercariae of *Parvatrema borealis* occurring in the gonads of *Gemma gemma* cause parasitic castration or interfere with reproduction. Sporocysts of *Postmonorchis donacis* in *Donax gouldii* (Coe, 1955) and *Cercaria milfordensis* in *Mytilus edulis* (Uzmann, 1953) cause sterilization of the host. In contrast, sporocysts containing microcercous cercariae of *Cercaria adranocerca* are found in the gonads of *G. gemma,* and the infected host does not develop embryos, indicating parasitic castration (Sellmer, 1967).

The copepod parasite *Mytilicola intestinalis* is found in *Mytilus edulis.* The gonads of parasitized animals are reduced by 10–20% compared to those of nonparasitized individuals (Mann, 1956). Williams (1969) has reported that the parasite interferes by delaying breeding in infected animals. It is suggested that parasitic infestation causes short-lived changes in biochemical constituents during the breeding period. Neurosecretion in parasitized individuals may occur more slowly than in nonparasitized animals, which delays the breeding period. *Mytilicola orientalis* is reported to infest *Mytilus crassitesta* in the Inland Sea of Japan (Galtsoff, 1964). Odlaug (1946) found *Mytilicola orientalis* in *Mytilus edulis* and *Protothaca* (= *Paphia*) *staminea* in lower Puget Sound, Washington.

Symbiotic crabs, *Pinnotheres pisum,* infesting *Mytilus edulis* and *Cardium norwegicum,* are reported to cause partial or complete cessation of gamete production in the host (Berner, 1952).

5.3.8 Reproductive Behavior

5.3.8.1 COURTSHIP

Courtship has not been reported to occur in this group.

5.3.8.2 AGGREGATION

Aggregation is well known in some sedentary species such as oysters and mussels, and it may be significant for synchronization and coordination of spawning. It might be expected that aggregation of adults is also valuable in the synchronous release of gametes in both sedentary and free-living species and also increases chances for fertilization of gametes liberated into the surrounding water. Aggregation usually occurs in such forms as boring bivalves and commensal species, but again its significance is not clear in terms of reproduction, although it is probably safe to conclude that it increases success in reproduction.

In *Mytilus edulis,* the larvae settle in shallow waters. After the juveniles grow for a certain period of time, they detach themselves and are transported by currents to resettle in other places. This behavior is apparently repeated several times until the juveniles find a substratum such as a bank of adult mussels (Verwey, 1952, 1954, 1957; de Block and Geelen, 1958). Lent (1969) has reported that juveniles of *Modiolus demissus* crawl freely and, upon contacting a group of mussels, attach themselves with the byssus. A clumped distributional pattern in this species appears to result from thigmotactic movements. The movement of mussels is greatly reduced at sizes greater than 45 mm, the size at which they become sexually mature. However, it is not clear what significance this aspect of aggregation may have in relation to reproduction. Golikov and Scarlato (1970), in studying population densities of various size groups of *Patinopecten yessoensis* and *Spisula sachalinensis,* found aggregation of individuals after they reached sexual maturity. Population density is greatest after settlement, but the juveniles disperse. However, after *Patinopecten yessoensis* and *S. sachalinensis* become sexually mature at the age of 3 years, they aggregate with significantly increasing densities in favorable areas. Obviously this aggregation of mature adults would be advantageous for synchronization of spawning and fertilization of the gametes released. In *Montacuta ferruginosa,* adults as well as young become entangled by byssus threads in burrows and also on the host sea urchin (Gage, 1966a). This species incubates its embryos, releasing them as larvae. Therefore the aggregation of adults would enhance successful fertilization of gametes.

5.3.8.3 MIGRATION

Migratory behavior in pelecypods and its relation to reproduction are not clear. Bean clams, *Donax* spp., are characteristically found on fully exposed beaches throughout the world. They execute surface leaps and can also burrow efficiently in order to migrate up and down the beach with the rising and falling tides (Stoll, 1938; Mori, 1950; Jacobson, 1955; Ansell, 1969; Tiffany, 1971). Whether migration with changing tides has any relationship to reproduction is not known. A number of pectinid species are known to swim, which results in local movement, but there is no convincing evidence of migration (Baird, 1966). Migration is discussed for *Pecten maximus* by Baird and Gibson (1956) and Baird (1958, 1966); for *Argopecten (= Aequipecten) irradians* by Belding (1910), Sastry (1961), and Moore and Marshall (1967). Gibson (1956) and Mason (1957) tagged *P. maximus* and found no evidence of migration. Likewise, Dickie (1955) and Posgay (1953) found no evidence of migration in *Placopecten magellanicus.* The Tasmanian scallop,

Notovola meridionalis, is an active swimmer but seems to swim only locally (Olsen, 1955). However, the importance of such mobility for reproduction remains conjectural.

5.3.9 Spawning

5.3.9.1 FACTORS INDUCING SPAWNING

The factors inducing spawning in pelecypods have been discussed in reviews by Giese (1959a), Loosanoff and Davis (1963), Galtsoff (1961, 1964), Fretter and Graham (1964), Loosanoff (1954, 1971), and Giese and Pearse (1974). A variety of factors—temperature changes, salinity, light, mechanical shock, and chemicals—have been reported to induce spawning. Spawning of both sexes of the oyster *Crassostrea virginica* have been described in detail (Galstoff, 1964; see also this volume, Chapter 6), but similar observations for other pelecypods are not available. In *C. virginica,* the gametes are dispersed through a coordinated action of the adductor muscle, the epithelium of the gills, and the edge of the mantle. In *Spisula solidissima,* spawning is accomplished by rhythmic contractions of the adductor muscle, and the gametes flow out into the water or are emitted by muscular activity (Ropes, 1968b). The spawning behavior of *Argopecten (= Aequipecten) irradians* (Sastry, 1961) and *Scrobicularia plana* (Hughes, 1971) has been described.

5.3.9.1.1 *Temperature.* Temperature has been considered one of the important factors in stimulating spawning in a number of pelecypods. In many species, spawning seems to occur at a definite temperature or following a definite temperature change (Orton, 1920; Nelson, 1928a,b; Berner, 1935; Thorson, 1946; Loosanoff, 1953; Sastry, 1963, 1966a). Spawning occurs with rising temperatures in *Cyprina islandica* (Loosanoff, 1953), at maximum temperatures in *Mercenaria mercenaria,* and with decreasing temperatures in the North Carolina population of *Argopecten (= Aequipecten) irradians* (Sastry, 1963, 1966a) and the Georges Bank and Port au Port Bay, Newfoundland, populations of *Placopecten magellanicus* (Posgay, 1953; Naidu, 1970). Oocyte growth to maturity occurs with increasing temperatures, and spawning occurs at the maximum temperature (Loosanoff, 1937b; Yamamoto, 1951a,b; Ropes, 1968b) or with declining temperatures after the maximum (Posgay and Norman, 1958; Sastry, 1963, 1966a). In temperate latitudes, spawning usually occurs during the warmer months of the year. In tropical species, spawning may occur more intensively in warmer months for some species (Stephenson, 1934), but in others there is no seasonal correlation (Rao, 1951; Tranter, 1958a; Sastry, 1955; Narasimham, 1969).

Population differences in the temperature stimulus necessary to induce spawning also occur in some species. The Massachusetts population of *Argopecten (= Aequipecten) irradians* spawns when water temperatures are rising (14°–16° to 23°C), while those from North Carolina and Florida release gametes with declining temperatures (25° to 17°–18°C) after an initial increase (Belding, 1910; Gutsell, 1930; Sastry, 1963, 1966a, 1970a).

Spawning has been induced by temperature changes in some species. In *Mercenaria mercenaria*, a rapid increase in temperature from 5° to 30°C induces spawning (Loosanoff and Davis, 1963). Calabrese (1969a) reported that spawning occurred in *Mulinia lateralis* exposed to 20°C in January. In the population of *Argopecten (= Aequipecten) irradians* from Alligator Harbor, Florida, spawning was induced by exposing animals to an initial increase in temperature to 30°C and a subsequent decrease in temperature from 27.9° to 22°C (Sastry, 1963). The temperature levels at which the gametes may be released also vary with the season. The Alligator Harbor population of *A. irradians* normally spawns in late summer and autumn, but animals could be induced to mature and stimulated to spawn throughout most of the year (Sastry, 1963). During colder months, animals in general liberated spawn at temperatures lower than those during warmer months. In *A. irradians*, the minimum temperature requirements for spawning decrease with the progress in development of oocytes to maturity. In the laboratory, animals with oocytes at the beginning of the cytoplasmic growth phase fail to release gametes at 20°, 15°, and 10°C. In later stages of oogenesis, animals maintained at 20°C spawn, but those held at 15° and 10°C fail to release gametes (Sastry, 1966a).

Differences in response to temperature stimulus occur between sexes. In *Scrobicularia plana*, threshold temperatures for the spawning of males are lower than those required for females (Hughes, 1971). In *Argopecten (= Aequipecten) irradians*, spermatozoa are released more readily and also at lower temperatures than are ova (Sastry, 1966a).

Spawning may be advanced or delayed, depending upon the temperature conditions to which the animals are exposed. In *Mercenaria mercenaria* (Loosanoff and Davis, 1951) and *Mytilus edulis* (Bayne, 1965), spawning was delayed by maintaining nearly mature animals at colder temperatures. In *Argopecten (= Aequipecten) irradians*, oocyte growth and spawning have been advanced by exposing animals with developing oocytes to 25° and 30°C (Sastry, 1966a). Observations by a number of authors on the spawning of pelecypods in the field indicate that gamete release can be delayed or advanced by variations in temperature conditions from year to year (Loosanoff, 1937b; Lammens, 1967; Ropes, 1968b; Naidu, 1970).

A temperature change failed to induce spawning in *Pitar morrhuana*, *Mytilus edulis*, and *Modiolus demissus* (Loosanoff and Davis, 1963) and *Brachidontes* cf. *variabilis* and *Septifer bilocularis* (Wilson and Hodgkin, 1967). However, storage of animals in a dry state for several hours and subsequent exposure to rising temperatures stimulate spawning in *Mytilus edulis* (Berg and Kutsky, 1951) and *Pecten maximus* (Gruffydd and Beaumont, 1970). A detailed discussion of temperature influence on spawning in marine invertebrates, including pelecypods, has been given by Gunter (1957), Giese (1959a), and Kinne (1970).

5.3.9.1.2 *Salinity.* Decreasing salinity with the onset of the northeast monsoon has been reported to be the stimulus for spawning in a number of bivalves from the Indian coast (Hornell, 1909; Panikkar and Aiyyar, 1939; Paul, 1942, Rao, 1951; Sastry, 1955; Rao *et al.*, 1962; Nair and Saraswathy, 1970). Salinity changes and not temperature have been reported to influence spawning of the mussel *Xenostrobus securis* in Swan estuary, Australia (Wilson, 1968, 1969).

5.3.9.1.3 *Light.* A detailed discussion of the effect of light on the release of gametes in some marine invertebrates is given by Segal (1970). The influence of light as a factor coordinating spawning in pelecypods has not been investigated.

In some species, peaks in spawning during the breeding period have been correlated with the lunar phases. Lunar periodicity in spawning has been suggested for *Cumingia tellinoides* (Grave, 1927a), *Chlamys opercularis* (Amirthalingam, 1928), *Mytilus edulis* (Battle, 1932), and *Pecten maximus* (Tang, 1941). In *Argopecten* (= *Aequipecten) irradians* from Beaufort, North Carolina, two spawning peaks occur within the 8-week breeding period coinciding with the period between the full moon and the last quarter (Sastry, unpublished data). Mileikovsky (1970) reported lunar periodicity in the spawning of some pelecypods in the White Sea. Korringa (1947, 1957) suggested that the semilunar rhythm was linked to tidal rhythm rather than light intensity.

5.3.9.1.4 *Mechanical Stimuli.* A number of other factors, such as wave shock, pressure changes, and tidal cycle have been suggested to induce spawning in some species (Young, 1942, 1946). Rough handling, stretching, and pricking the adductor muscle induces spawning in *Mytilus edulis* (Field, 1922; Loosanoff and Davis, 1963). Removing one valve and placing the mantle in seawater is reported to induce the release of gametes in *M. edulis* (Costello *et al.*, 1957), while Iwata (1950) and

Sugiura (1962) have found that stimulation with electric current induces spawning in this species.

5.3.9.1.5 *Chemicals*. A number of chemicals have also been reported to be effective in stimulating the release of gametes (Iwata, 1951a,b; Sagara, 1958; Breese *et al.*, 1963; Bayne, 1965). Spawning was observed in *Mytilus edulis* within 1–5 hr after the injection of 0.5 ml of 0.5 M potassium chloride into the mantle cavity (Iwata, 1951a). When the mantle was bathed for 5–10 min in 0.5 M potassium chloride, ammonium chloride, or barium chloride, the gametes were released in 30–90 min (Iwata, 1951b). It was suggested that the higher mobility of cations might be the basis for chemical stimulation, with action not directly on the eggs but on the ovarian cells. The excitation of these cells then results in meiosis and subsequent release of eggs. Sagara (1958) showed that spawning could be induced in some pelecypods by placing them in ammoniated seawater or by injecting sodium hydroxide into the gonad. Kraft mill effluents were found to be highly effective in triggering spawning in *M. edulis* and *Mytilus californianus* (Breese *et al.*, 1963).

Gametes of the opposite sex also stimulate spawning. In *Chlamys varia*, spawning is initiated by the gametes of the opposite sex and also by mechanical and thermal stimulation (Lubet, 1951, 1955b). It was found that thermal stimulation was more effective than mechanical stimulation in this species when combined with extracts of sex cells. Enhanced effectiveness of thermal stimulation combined with the addition of gametes of the opposite sex has been reported for a number of species (Loosanoff and Davis, 1963; Bayne, 1965). Wada (1954) has reported that the addition of an egg water suspension or a sperm suspension stimulates spawning in *Tridacna*. In *Mytilus edulis* and *M. galloprovincialis*, males can be stimulated to release gametes by eggs or egg extracts, but fresh sperm or sperm extracts are not effective in the release of eggs by the female (Lubet, 1951, 1955b).

Some species produce substances which stimulate spawning in other members of the same population (Galtsoff, 1961, 1964). Young (1942, 1946) has found that testicular tissue of *Mytilus californianus* contains a substance causing spawning in females. A similar gonad stimulant affects the discharge of sperm in *Tridacna* (Wada, 1954). In *Turtonia minuta*, where fertilization takes place in the mantle cavity prior to the formation of egg capsules, spawning of eggs may be induced by the sperm or by some secretion of a prostatic nature released by the male (Oldfield, 1964). Substances released by one or both sexes and by other species may also induce spawning in members of a population (Nelson, 1936; Wada, 1954; Fretter and Graham, 1964). Galtsoff (1961) has pointed out that males are more sensitive to a heterogeneous collection of stimuli and that this might be valuable in triggering the reaction for spawning of the whole population.

5.3.9.1.6 *Neurosecretion.* Neurosecretory control of spawning in *Mytilus edulis* and *Chlamys varia* has been suggested by Lubet (1955b, 1957). In these species, neurosecretory products of the cerebral ganglia are reduced when response to environmental stimuli is maximal (see Section 5.3.7.6). Just before spawning some of the neurosecretory cells release their secretion. Ablation of the cerebral ganglion accelerates spawning, whereas removal of the visceral ganglion retards spawning. It has been suggested that the neurosecretion of cerebral ganglia has an inhibitory effect on spawning and that only after the disappearance of neurosecretory products does the animal become receptive to environmental factors stimulating spawning. Galtsoff (1961) indicated that a delay in spawning on removal of the visceral ganglion might be due to the isolation of receptors on the mantle and prolongation of the inhibitory action of the cerebral ganglion. Injections of blood from a spawning animal into a mature nonspawning individual indicate that no stimulating substance is present in the blood of the female (Lubet, 1955a; Galtsoff, 1961).

5.3.9.2 SYNCHRONIZATION AND COORDINATION OF SPAWNING

In species releasing gametes to the external environment, fertilization results from random contact between male and female sex cells. However, coordination of spawning results in a synchronous release of gametes of both sexes, thereby increasing the chances for successful fertilization and the production of a large number of larvae. The advantage of timing of spawning within a brief period, when conditions are best for survival and growth and when food is abundant, have been emphasized by Thorson (1950). However, the degree of synchronization of spawning seems to vary among species and within populations of the same species in different parts of their geographic range. In high latitudes, where seasonal changes in the environment are pronounced, a high degree of synchronization in spawning is observed, especially in species with planktotrophic larvae (Ockelmann, 1958). At lower latitudes, where diminished seasonal variations allow spawning and development throughout the year, spawning takes place for a more prolonged period. However, even in the tropics, breeding seasons appear to be limited to certain periods, with spawning peaks, suggesting some synchronization (Yonge, 1930; Stephenson, 1934; Gunter, 1957; Kinne, 1970).

Year-round observations on reproduction in deep-sea pelecypods are few, and the information available on the amount of synchrony or on the factors coordinating spawning is very limited. Sanders and Hessler (1969) reported that breeding occurred in *Nucula cancellata* throughout the year. Rokop (1974) reported that *Nucula pontonia, N. darella,* and *Bathyarca* sp. bred continuously in the bathyal San Diego trough. In these species, gametogenesis is continuous, with only partial spawning occurring at any one time.

The environmental factors triggering spawning have been investigated, but the factors responsible for bringing the population to a mature stage, so that spawning can be coordinated for the synchronous release of gametes, have not received much attention. Environmental factors may synchronize gonad development in members of a population at an early stage in their progress toward maturity, so that spawning can be coordinated to achieve synchronous release of gametes. Results of experimental studies on environmental regulation of the neurosecretory cycle and gonad development in *Argopecten (= Aequipecten) irradians* indicate that the controlling mechanisms within the organism might become responsive to environmental influences at the beginning of the cytoplasmic growth phase of oocytes (Sastry, 1966a, 1968, 1970b, 1975; Sastry and Blake, 1971; Blake, 1972; see also Section 5.3.7). A population responding to environmental influences would initiate gonad development at the same time and progress toward maturity synchronously.

In *Argopecten (= Aequipecten) irradians,* oocytes of different sizes occur within the gonad of an individual throughout the period of gonad development, even to the mature state. This pertains even to animals maintained under uniform conditions in the laboratory. The variation in oocyte size may be due to temporal differences in initiation of the cytoplasmic growth phase. This could be influenced by the amount of nutrients channeled into the gonads. Although different sized oocytes occur within an individual, the members of the population as a whole are synchronous during their development to maturity and during the breeding period. Even after the beginning of spawning, a process that lasts for about 8 weeks, the gonads still contain developing oocytes of different sizes. This is reflected by the stepwise decline in the gonad index as well as by variation in the size of oocytes in the spawning animals (see Figs. 12 and 18). The breeding period of the Beaufort, North Carolina, population terminates in November when environmental temperatures reach about 17°–18°C (Sastry, 1966a). It was found that animals with developing oocytes maintained at 15° and 10°C in the laboratory never released gametes. At 10° and 15°C, the oocytes were cytolyzed and eventually reabsorbed after dissolution of the germinal vesicle (Sastry, 1966a, 1970b).

Once the population has reached maturity, spawning may be triggered through interaction between the organism and external factors that induce gamete release (see Fig. 33). If all members of the population are mature and react simultaneously to the factors inducing spawning, gametes may be released synchronously and a short breeding period results. Therefore synchrony in spawning appears to depend on a critical state of maturity within an individual and among members of the population, as well as their responsiveness to the exogenous factors inducing gamete release. Once the animals are mature, a variety of factors seem to induce spawning (see Section 5.3.9.1).

Spawning is a protracted phenomenon in a number of species. Several peaks within breeding periods lead to an extended breeding season. In *Cyprina islandica*, for example, spawning takes place between July and mid-October (Loosanoff, 1953). At the beginning of the breeding season in July, only 75% of the population is mature. The end of breeding period is marked by a predominance of spent animals (Fig. 36). The

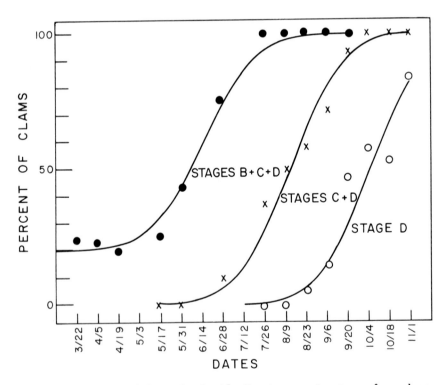

FIG. 36. Percentage of clams, *Cyprina islandica*, in successive stages of sexual maturity at different times of the year. Stage B, Ripe but had not begun to spawn; stage C, partly spawned; stage D, completely spawned. Solid lines represent cumulative percentages of a normal curve. (From Loosanoff, 1953; modified.)

population of *Placopecten magellanicus* in Port au Port Bay, Newfoundland, also has an extended breeding season because of partial or interrupted spawning (Naidu, 1970). In southern latitude populations of *Mya
arenaria* spawning takes place both in spring and autumn (Porter, 1964).
Campbell (1969) found that *Mytilus edulis* high in the intertidal zone
spawned before those of the same size below the tidal level. When
different age groups in a population spawn at different times, an
extended breeding season is observed (Fretter and Graham, 1964;
Alagarswami, 1966). More data on the influence of various exogenous
and endogenous factors controlling the beginning of oocyte growth and
mechanisms coordinating maturation and spawning are needed to
analyze the reproductive cycle effectively.

5.3.9.3 BREEDING PERIOD

The breeding periods of some pelecypods living in different climatic
zones are summarized in Table X. In eastern Greenland a number of
species with planktotrophic larvae spawn during the summer, while
species with lecithotrophic larvae breed in summer and autumn (Ockelmann, 1958). At midlatitudes, the breeding period occurs for most
species in spring, summer, or autumn. In *Macoma balthica* from Den-
Helder, Netherlands, spawning takes place in the spring (Lammens,
1967). *Mercenaria mercenaria* from Long Island Sound (Loosanoff,
1937b) and *Argopecten (= Aequipecten) irradians* from Massachusetts
and Connecticut (Belding, 1910; Marshall, 1960; Sastry, 1970a) spawn
during the summer. *Cyprina islandica* from Point Judith, Rhode Island
(Loosanoff, 1953), *A. irradians* from Beaufort, North Carolina, and
Alligator Harbor, Florida (Sastry, 1963, 1966a), *Spisula solidissima*
from the New Jersey coast (Ropes, 1968b), and *Placopecten magellanicus*
from Port au Port Bay, Newfoundland (Naidu, 1970), spawn during
autumn.

The pearl oyster, *Pinctada albina*, from Thursday Island, Australia,
spawns actively during autumn (Tranter, 1958a). In another species,
Pinctada margaritifera, spawning is limited to early summer and winter
(Tranter, 1958c). A third species, *Pinctada fucata*, spawns during
summer–autumn (Tranter, 1959). But in all three species minor spawning also occurs outside the major spawning period. On the Indian coast,
a number of species have a prolonged breeding season, yet it is restricted
to specific months of the year (Rao, 1952; Abraham, 1953; Nayar, 1955;
Sastry, 1955; Alagarswami, 1966). In other Indian pelecypods, spawning
takes place throughout the year, even though peak spawning is limited
to certain months (Antony Raja, 1963; Narasimham, 1969).

Duration of the breeding period in some species increases with a
decrease in latitude. The breeding period in the population of *Argo-*

TABLE X

SUMMARY OF THE BREEDING PERIODS OF SOME PELECYPODS OCCURRING IN DIFFERENT CLIMATIC ZONES

Species	Location	Latitude	Notes on breeding period	Reference
Arca glacialis [a]	N.E. Greenland	70°–85° N	End of July or beginning of August	Ockelmann (1958)
Astarte borealis [a]	N.E. Greenland	70°–85° N	Very late in autumn	Ockelmann (1958)
Astarte crenata [a]	N.E. Greenland	70°–85° N	Very late in autumn	Ockelmann (1958)
Astarte elliptica [a]	N.E. Greenland	70°–85° N	Very late in autumn	Ockelmann (1958)
Astarte montagui [a]	N.E. Greenland	70°–85° N	Very late in autumn	Ockelmann (1958)
Axinopsis orbiculata [a]	E. Greenland	70°–85° N	Late in summer	Ockelmann (1958)
Cerastoderma (= Cardium) ciliatum	N.E. Greenland	70°–85° N	End of July and beginning of August	Ockelmann (1958)
Hiatella arctica	N.E. Greenland	70°–85° N	May to beginning of July	Ockelmann (1958)
Lyonsia arenosa [a]	N.E. Greenland	70°–85° N	After end of August	Ockelmann (1958)
Macoma calcaria	N.E. Greenland	70°–85° N	End of July or beginning of August	Ockelmann (1958)
Macoma moesta [a]	N.E. Greenland	70°–85° N	End of August or beginning of September	Ockelmann (1958)
Modiolaria discors laevigata [b]	N.E. Greenland	70°–85° N	In summer, greatest intensity in August	Ockelmann (1958)
Modiolaria nigra [b]	N.E. Greenland	70°–85° N	Summer, with maximum in August	Ockelmann (1958)
Mya truncata	N.E. Greenland	70°–85° N	May–July	Ockelmann (1958)
Nucula tenuis [a]	N.E. Greenland	70°–85° N	End of July or beginning of August	Ockelmann (1958)
Pandora glacialis [a]	N.E. Greenland	70°–85° N	After middle of August	Ockelmann (1958)
Pecten islandicus	E. Greenland	70°–85° N	End of July and beginning of August	Ockelmann (1958)
Portlandia artica [a]	N.E. Greenland	70°–85° N	End of July to beginning of August	Ockelmann (1958)
Propeamusium groenlandicum [a]	N.E. Greenland	70°–85° N	Late autumn	Ockelmann (1958)
Propeamusium imbriferum [a]	N.E. Greenland	70°–85° N	Late in autumn	Ockelmann (1958)
Thyasira gouldi [a]	N.E. Greenland	70°–85° N	August–September	Ockelmann (1958)

Yoldia hyperborea [a]	S. E. Greenland	70°–85° N	End of July and later	Ockelmann (1958)
Anomia squamula	Velikaya Salma Sound, White Sea	64°–68° N	July–August	Mileikovsky (1970)
Macoma balthica	Oresund, Denmark	56° N	September–October	Thorson (1946)
	Plymouth, U.K.	50°25′ N	September–October	Lebour (1947)
	Velikaya Salma Sound, White Sea	64°–68° N	June–August	Mileikovsky (1970)
Mya arenaria	Dutch Wadden Sea,	53° N	March–April	Lammens (1967)
	Velikaya Salma Sound, White Sea	64°–68° N	June–August	Mileikovsky (1970)
	Booth Bay Harbor, Maine	43°–47° N	Late June–August	Ropes and Stickney (1965)
	Woods Hole, Mass.	42°–43° N	May–June	Costello *et al.* (1957)
	Chesapeake Bay	37°–39° N	May, October–November	Pfitzenmeyer (1962)
Mytilus edulis	Velikaya Salma Sound, White Sea	64°–68° N	June–August	Mileikovsky (1970)
	Brighton, U.K.	50°80′ N	April–May	Campbell (1969)
	Prince Edward Island, Canada	46°47′ N	May–early August	Sullivan (1948)
	Central California	36° N	Spring and early summer	Graham and Gay (1945)
	Alamitos Bay, California	33°40′ N	Late fall to winter	Moore and Reish (1969)
	Cape Hatteras, North Carolina	35° 30′ N	April–June, October–November	Wells and Gray (1960)
Turtonia minuta [b]	White Sea	64°–68° N	June–August	Matveeva (1953)
	Northumberland, U.K.	55°–56° N	March–July	Oldfield (1955)
Protothaca staminea	Lady Smith Harbor, Canada	60° N	April–October	Quayle (1943)

(continued)

211

TABLE X (Continued)

Species	Location	Latitude	Notes on breeding period	Reference
Patinopecten caurinus	Kodiak, Alaska	56°–58° N	Early June–mid-July	Hennick (1970)
Astarte sulcata [a]	Largs Channel, Scotland	55°75′ N	October–November	Saleuddin (1964)
Venerupis pullastra	Millport, Scotland	55°75′ N	March–July	Quayle (1952)
	North Sea	53°60′ N	September–December	Rees (1950, 1951, 1954)
Venus striatula	Plymouth, U.K.	50°25′ N	June–October	Lebour (1938a)
	Millport, Scotland	55°75′ N	May–late August	Ansell (1961b)
	North Sea	53°60′ N	September–November	Rees (1951, 1954)
Zirphaea crispata	Northumberland Coast, U.K.	55°–56° N	March–October	Allen (1969)
Chlamys distorta	Port Erin, U.K.	54°10′ N	May–August, November–December, lesser spawning February, March, April	Reddiah (1962)
Chlamys furtiva	Port Erin, U.K.	54°10′ N	June, October–November	Reddiah (1962)
Chlamys striata	Port Erin, U.K.	54°10′ N	August–early September	Reddiah (1962)
Chlamys tigerina	Port Erin, U.K.	54°10′ N	June	Reddiah (1962)
Chlamys varia	Port Erin, U.K.	54°10′ N	June, September–November	Reddiah (1962)
	French coast	42°–51° N	April–May, September–October	Dalmon (1935); Lubet (1959)
Pecten maximus	Port Erin, U.K.	54°10′ N	Spring, autumn, and summer	Mason (1958)
Siliqua patula	Cook Inlet, Alaska	53°71′ N	July–early September	McMullen (1967)
	Queen Charlotte Island, Canada	52°–54° N	July–August	Bourne and Quayle (1970)
	Clatsop beaches, Oregon	42°–46° N	Late May–early June	Hirschhorn (1962)

Species	Location	Latitude	Spawning period	Reference
Montacuta ferruginosa	North Sea	53°60′ N	July–October	Rees (1951, 1954)
	Plymouth, U.K.	50°25′ N	July	Oldfield (1964), Gage (1966b)
Montacuta substriata	North Sea	53°60′ N	July–November	Rees (1951, 1954)
	Plymouth, U.K.	50°25′ N	Midsummer	Oldfield (1964), Gage (1966b)
Scrobicularia plana	Conway Bay, U.K.	53°33′ N	June–August	Hughes (1971)
Kellia suborbicularis	Plymouth, U.K.	50°25′ N	May–November	Lebour (1938b)
Lasaea rubra [b]	Plymouth, U.K.	50°25′ N	May–August	Oldfield (1964)
Placopecten magellanicus	Port au Port Bay, Newfoundland	48°30′ N	September–October	Naidu (1970)
Adula californiensis	Newport, Oregon	44°30′ N	June–October	Lough and Conor (1971)
Nucula proxima	Maine	43°–47° N	Summer	Drew (1901)
Cerastoderma edule	French coast	42°–51° N	July–August	Gimazane (1971)
Argopecten irradians	Massachusetts	42°–43° N	Mid July–end of August	Belding (1910)
	Niantic River, Connecticut	42° N	Mid-June to July	Marshall (1960)
	Beaufort, North Carolina	35° N	Mid-September–mid-November	Gutsell (1930); Sastry (1966a)
	Alligator Harbor, Florida	30° N	August–early December	Sastry (1963)
Cyprina islandica	Point Judith, Rhode Island	42°–43° N	Early July–October	Loosanoff (1953)
Cumingia tellinoides	Woods Hole, Mass.	42° N	Mid-June–mid-September	Costello *et al.* (1957)
Yoldia limatula	Woods Hole, Mass.	42° N	Early summer	Costello *et al.* (1957)
Mercenaria mercenaria	Long Island Sound	41° N	July–September	Loosanoff (1937b)
	Core Sound, North Carolina	34°–37° N	June, September–October	Porter (1964)

(continued)

213

TABLE X (Continued)

Species	Location	Latitude	Notes on breeding period	Reference
Mulinia lateralis	Long Island Sound	41° N	July–September	Calabrese (1969a)
Spisula solidissima	New Jersey	39°–41° N	July–August, mid-October–early November	Ropes (1968b)
Pododesmus cepio	Tomales Bay, California	38°33' N	June–August	Leonard (1969)
Pseudopythina rugifera	Tomales Bay and Bodega Bay, California	38°33' N	Contained embryos in April	Narchi (1969)
Brachidontes recurvus	Chesapeake Bay	37°–39° N	June–November	J. F. Allen (1962a, b)
Patinopecten yessoensis	Sea of Japan	35°–45° N	May–August	Golikov and Scarlato (1970)
Spisula sachalinensis	Sea of Japan	35°–45° N	End of June through October	Golikov and Scarlato (1970)
Argopecten gibbus	North Carolina, western Florida	34°–37° N	March–June	Kirby-Smith (1970)
		25°–31° N	August–September	Bullis and Ingle (1958); Sastry (1962)
Cardita ventricosa [b]	Southern California	33°–37° N	Throughout the year	Jones (1963)
Pecten circularis aequisulcatus	Mission Bay, California	33° N	Spring and early summer	Coe (1945)
Pecten latiauritus	La Jolla, California	33° N	Early spring and early autumn	Coe (1945)
Cyclinella tenuis	Biscayne Bay, Florida	26° N	January–May, June–September, October–December	Wright and Moore (1970)
Meretrix casta	Bombay coast, India	18°–20° N	March–June	Rai (1932)
	Adayar estuary, India	12°–14° N	July–August, October–November, March–May	Abraham (1953)

214

Species	Location	Latitude	Spawning period	Reference
Teredo furcillatus	Mandapam, India	10° N	November–June	Durve (1964)
	Vishakaptnam Harbor, India	18° N	Throughout the year	Nagabhushanam (1970)
Anadara granosa	Kakinada Bay, India	17°55′ N	Throughout the year, peak spawning between January and April	Narasimham (1969)
Placuna placenta	Pinang Island, Malaysia	5°30′ N	August–October, some breed throughout the year	Pathansali (1964)
Solen kempi	Kakinada Bay, India	17°55′ N	October–January	Sastry (1955)
	N.W. coast of India	16°–24° N	October–March	Rao et al. (1962)
Paphia laterisulca	Ratnagiri, India	17° N	September–April, with October–November and February–March spawning peaks	Nagabhushanam and Dhamme (1977)
Donax cuneatus	Madras coast, India	12°–14° N	December–June	Rao (1967)
	Palk Bay, India	10° N	January–April	Nayar (1955)
Katelysia opima	Adayar estuary, India	12°–14° N	December–January	Rao (1952)
Modiolus striatulus	Madras Harbor, India	12°–14° N	Throughout the year, peak breeding in March–April and July–August	Antony Raja (1963)
Donax faba	Mandapam, India	10° N	November–June	Alagarswami (1966)
Nausitora hedleyi	Cochin, India	10° N	June–December	Nair and Saraswathy (1970)
Pinctada albina	Thursday Island, Australia	10°30′ S	Throughout the year, peak spawning in April–May	Tranter (1958a)
Pinctada fucata	Thursday Island, Australia	10°30′ S	December–May, lesser spawning in other months	Tranter (1959)
Pinctada margaritifera	Thursday Island, Australia	10°30′ S	Early January and late July, minor spawning in other months	Tranter (1958c)
Pinctada maxima	Thursday Island, Australia	10°30′ S	September–March	Minaur (1969)

215

(continued)

TABLE X (Continued)

Species	Location	Latitude	Notes on breeding period	Reference
Mytilus perna	Santos, Brazil	24° S	April–early June, September	Lunetta (1969)
Amygdalum glaberrimum	Fremantle, Australia	32° S	November	Wilson and Hodgkin (1967)
Brachidontes cf. *variabilis*	Fremantle, Australia	32° S	April	Wilson and Hodgkin (1967)
Mytilus edulis planulatus	Fremantle, Australia	32° S	Mid-July–September, followed by asynchronous spawning	Wilson and Hodgkin (1967)
Septifer bilocularis	Fremantle, Australia	32° S	Mid-April	Wilson and Hodgkin (1967)
Xenostrobus pulex	Fremantle, Australia	32° S	August–September, asynchronous spawning follows	Wilson and Hodgkin (1967)

a Lecithotrophic development.
b Direct development (cf. Table XIV).

pecten (= *Aequipecten*) *irradians* from Alligator Harbor, Florida, extends from late summer to late autumn, while it occurs only in autumn (late September–mid-November) in the population from Beaufort, North Carolina (Sastry, 1963, 1966a). In pearl oysters, *Pinctada* spp., the breeding season is more restricted in the higher latitudes and occurs in warmer months, while in the lower latitudes there is a longer breeding period with major spawning occurring outside the summer months (Tranter, 1958a).

Wilson and Hodgkin (1967) have indicated that the breeding pattern in mytilids from Fremantle, Western Australia, correlates with their latitudinal distribution and zoogeographic affinities. Tropical mytilids, *Brachidontes* cf. *variabilis* and *Septifer bilocularis,* breed for a short period in summer and autumn, while the temperate species, *Mytilus edulis planulatus* and *Xenostrobus pulex,* spawn intensively in winter and spring (Fig. 37). The spawning peaks in *M. edulis planulatus* and *X. pulex* are followed by several months of asynchronous spawning of reduced intensity. However, *Amygdalum glaberrimum* has tropical affinities, but its breeding pattern at Fremantle does not conform to that of either tropical or temperate species (Fig. 37).

5.4 DEVELOPMENT

5.4.1 Embryonic Development

5.4.1.1 Egg Maturation

Raven (1958) gave a detailed description of the maturation process in fully grown oocytes of molluscs, including some pelecypods. The maturation process involves shrinkage of the germinal vesicle and extrusion of nucleoplasm, complete disappearance of nuclear membranes and formation of the first maturation spindle, disappearance of the nucleolus and extrusion of the first polar body, and formation of the second maturation spindle and second polar body. The time relationship between maturation and fertilization varies among species.

In some species, the germinal vesicle of the fully grown oocyte breaks down either before or immediately after it is released, and the maturation process stops at this stage and continues only after fertilization. In *Cumingia tellinoides, Mytilus edulis, Argopecten* (= *Aequipecten*)

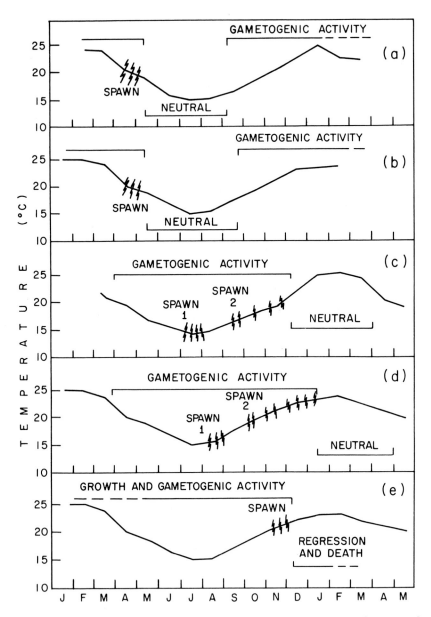

Fig. 37. Annual surface water temperature (data from Kwinana Refinery, Cockburn Sound, Fremantle area, Western Australia) and the major features of the reproductive cycles of five species of marine mussels. Temperature curves for each species represent the temperature values during the period in which gonad samples of that species were taken. (a) *Septifer bilocularis*, 1963–1964. (b) *Brachidontes* cf. *variabilis*, 1961–1962. (c) *Mytilus edulis planulatus*, 1960–1961. (d) *Xenostrobus pulex*, 1961–1962. (e) *Amygdalum glaberrimum*, 1961–1962. (From Wilson and Hodgkin, 1967.)

218

irradians, Mya arenaria, and *Mercenaria mercenaria,* the eggs shed into the seawater proceed spontaneously to metaphase of the first maturation division (Raven, 1958; Costello, *et al.,* 1957; Stickney, 1963; Sastry, 1966a). In *A. irradians,* oocytes stripped from the ovary or forced to shed with intact germinal vesicles undergo maturation and complete dissolution of the nuclear membrane *in vitro* and can be fertilized. However, oocytes having intact germinal vesicles when fertilized show surface protrusions and fail to undergo normal cleavage.

In other species, *Spisula solidissima, Crassostrea virginica, Ensis directus,* and *Barnea candida,* oocytes are shed with intact germinal vesicles. Maturation in these species does not begin spontaneously in unfertilized eggs under normal conditions (Allen, 1953; Raven, 1958; Costello *et al.,* 1957). In *S. solidissima,* the oocyte is shed with a large germinal vesicle containing a prominent nucleolus. The germinal vesicle breaks down immediately after insemination, and two polar bodies are quickly formed by two maturation divisions. According to Wada (1968) oocytes stripped from an ovary with intact germinal vesicles in members of Mactridae and Pholadidae are fertilizable. However, as normally released, the eggs show germinal vesicle breakdown. Iwata (1951c) has reported that the eggs of *Mactra veneriformis* are not fertilizable unless they have intact germinal vesicles.

In species in which the germinal vesicle remains intact, maturation may be provoked by ultraviolet irradiation (Pasteels, 1931). Oocytes stripped from the ovary have intact germinal vesicles in *Pinctada martensii* (Wada and Wada, 1953) and in *Mercenaria mercenaria* and *Tapes semidecussata* (Loosanoff and Davis, 1963). When subsequently exposed to seawater with ammonium hydroxide, the germinal vesicles break down and become fertilizable with spermatozoa. Allen (1953) found that the breakdown of the germinal vesicle could be initiated by ultraviolet irradiation, potassium, sodium, ammonium, hypertonicity, hypotonicity, and protamine with a minimum amount of calcium. It is indicated that the maintenance of oocytes in the germinal vesicle stage might be due to sodium and potassium antagonism (Raven, 1958). Raven (1958) also suggested that germinal vesicle breakdown was due to the release of calcium from protein binding in the cortex, followed by calcium activation of a proteolytic enzyme.

Sawada (1952, 1954a,b) has reported that germinal vesicle breakdown in *Mactra veneriformis* is inhibited by body fluids. It is suggested that the maturation divisions in the ovary are prevented by the presence of certain polysaccharides in the body fluid and egg cortex. The removal of inhibition evokes maturation, presumably through increased permeability to ions (Raven, 1958).

5.4.1.2 FERTILIZATION

According to Monroy (1965a) fertilization involves a series of events: contact between the spermatozoan and egg, penetration of the spermatozoan into the egg, activation of the egg, and fusion of male and female pronuclei, which results in formation of the zygote nucleus. The physiology of fertilization has been studied extensively in echinoderms, but not in pelecypods.

As in other animal groups, substances produced by the eggs and sperms of pelecypods may play a significant role in fertilization. Egg water, seawater in which the eggs have been left standing, agglutinates homologous spermatozoa and activates their motility. The active substance, Lillie's "fertilizin," has been considered to mediate egg and sperm interaction and also to be a key factor in fertilization. Fertilizin, or gynogamones, activates sperm and induces spawning. These egg substances have been found in *Solen* and *Pecten* (Tyler, 1949) and *Spisula solidissima* (Metz and Donovan, 1949). Sperms contain antifertilizin or androgamones which neutralize sperm agglutination and sometimes inactivate sperm of the same and other species. Sperm extracts also have egg membrane lysins (Monroy, 1965a).

In many pelecypods, sperms may enter at any point on the egg surface. At the time of fertilization striking changes take place in the acrosome region of the spermatozoan. On contact with the vitelline membrane, the spermatozoan undergoes deterioration of the acrosomal head region followed by projection of a stalklike filament into the egg. Dan and Wada (1955) studied the acrosome reaction of *Mytilus* spermatozoa. The sperm reacts upon contact with an unfertilized egg by extrusion of an acrosome filament. The acrosome reaction has also been observed in *Lithophaga, Spondylus, Crassostrea, Trapezium, Chama, Mactra,* and *Zirphea* (Dan and Wada, 1955). For further details on the acrosomal reaction consult Dan (1956, 1967), Dan *et al.* (1975), and Tilney and Mooseker (1976).

The acrosomal components function in the attachment and fusion of gametes in sperm penetration of the egg membrane and in activation of the egg (Franklin, 1970). It is known that a lytic factor contained in the acrosomal region enables the spermatozoan to penetrate the egg membrane. Sperm extracts of *Mytilus edulis* dissolve egg membranes, and fertilized eggs permitted to cleave in these extracts produce blastomeres that are only loosely joined (Berg, 1949, 1950). Wada *et al.* (1956), through a series of reciprocal experiments with sperm and sperm extracts, obtained evidence of the existence of a lysin in *M. edulis* and related the acrosomal reaction to the release of an egg membrane lysin. A single lytic factor extracted from *M. edulis* sperm attacks the major components of the vitelline envelope, which is the membrane responsible for the close contact of blastomeres during cleavage (Dan, 1962). The egg membrane lysin was not observed in spermatozoa of *Spisula solidissima* (Rebhun, 1962).

The contact established between the egg membrane and the spermatozoan initiates the fertilization reaction on the egg surface (Monroy, 1965a). The morphological and physiological aspects of the fertilization reaction in pelecypods have not been studied extensively. After sperm attachment, several minutes are usually required for the sperm head to disappear inside the egg. Following fertilization, the space between the vitelline membrane and the plasma membrane becomes a little wider (Allen, 1953; Wada, 1947).

Upon fertilization, changes occur rapidly on the egg surface, which make the eggs refractory to additional spermatozoa. However, fertilization can be reversed in *Spisula solidissima* by egg water, acidified seawater, or 0.3–0.5% ether in seawater (Allen, 1953). Ito and Leuchtenberger (1955) have suggested that fertilization in *S. solidissima* is dependent on a haploid amount of DNA in the sperm. Sperms with subhaploid DNA contents may penetrate eggs but do not activate them.

The nature and cause of the cortical reaction have not been studied in detail for molluscs (Raven, 1958). Sawada and Murakami (1959) observed that fertilization in *Mactra veneriformis* caused the disappearance of cortical granules, although often not as completely as in sea urchins. Cortical granules were observed in fertilized and cleaving eggs of *Mytilus edulis* (Reverberi and Mancuso, 1961). Pasteels and de Harven (1962) report that the expulsion of cortical granules in the egg of *Barnea candida* is independent of fertilization. According to Humphreys (1967) the cortical granules are not disrupted in large numbers on fertilization as in echinoderms. As to the possible role of cortical granules, the vitelline envelope material is gradually releaed into the environment of the developing eggs, and small numbers of cortical granules might discharge their content as necessary to replenish the substances of the vitelline membrane. A detailed discussion of the cortical reaction and its role in activation of the egg is given by Raven (1958, 1964), Monroy (1965a), Humphreys (1967), and Franklin (1970). Longo (1976) discussed the ultrastructural aspects of fertilization in eggs of *Spisula solidissima*.

Metabolic and biochemical aspects at the time of fertilization have not been studied extensively in pelecypods. For further discussion of the physiology of fertilization consult the reviews by Runnström *et al.* (1959), Raven (1958, 1964), Galtsoff (1964), Monroy (1965a,b), and Metz and Monroy (1967).

In pelecypods little detailed information exists on the events following sperm penetration into the egg cytoplasm that lead to fusion of the pronuclei of the spermatozoan and the egg. These details might be the same as those observed for other molluscs (Raven, 1958, 1964). A certain time after fertilization, the first polar body is extruded, and later the second polar body, both at the animal pole (Fig. 38). The time from fertilization to the expulsion of the polar bodies is apparently affected by the condition of the egg nucleus (Wada, 1968) and also

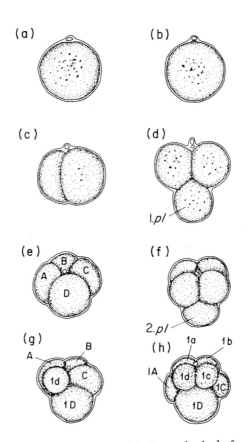

FIG. 38. Cleavage in *Pinctada maxima*. (a) First polar body formation. (b) Second polar body formation. (c) Two-cell stage. (d) Formation of first polar lobe, 1.pl, (trefoil stage). (e) Four-cell stage. (f) Formation of second polar lobe, 2.pl. (g) Cleavage of D cell. (h) Eight-cell stage. (From Wada, 1968.)

varies among species and with temperature (Costello *et al.*, 1957). At the time of polar body formation both the animal and vegetable poles undergo a change in form. The sperm lies in the egg cytoplasm near the periphery without much change in position until expulsion of the second polar body. When the maturation divisions are completed, the egg and spermatozoan pronuclei fuse near the center of the egg to result in the zygote (Wada, 1968).

5.4.1.3 THE ZYGOTE

Events taking place in the zygote before the beginning of cleavage have not been examined in detail for pelecypods, but they may not be very different from those observed for other molluscs (Raven, 1958). From ovulation until first cleavage, a shifting of various substances occurs through which they are accumulated or concentrated in certain places in the egg cell. In some pelecypods a polar lobe is formed before the first cleavage begins. Humphreys (1964) has indicated that, although subtle and consistent changes occur, the ultrastructure of the fertilized egg is similar to that of the unfertilized egg.

5.4.1.4 CLEAVAGE

Pelecypods produce an egg with mosaic cleavage. Cleavage is total and follows a spiral pattern. The first cleavage divides the egg meridionally into the AB blastomere at the animal pole and the CD blastomere at the opposite pole (Fig. 38). The second cleavage is also meridional at right angles to the first and results in equal sized A, B, and C blastomeres plus a large D blastomere. These four blastomeres are also called quadrants. Inequality among quadrants occurs as a result of formation of the polar lobe before cleavage. The vegetative side bulges out as the second cleavage furrow is developed and has the appearance of a three-cell embryo. This is called the trefoil stage. Polar lobe formation has been observed in *Pinctada, Pteria, Pecten, Mytilus*, and other genera (Raven, 1958; Wada, 1968).

The third cleavage divides each quadrant at the equatorial plane. The cells in the resulting first quartet of micromeres (1a–1d) at the animal pole are distinctly smaller than the macromeres of the vegetal pole (1A–1D). At the fourth cleavage, a second quartet of micromeres (2a–2d) is formed from the macromeres, and their remainder is designated 2A–2D. The first micromeres also divide into $1a^1$–$1d^1$ and $1a^2$–$1d^2$. The third quartet of micromeres (3a–3d) is formed at the fifth cleavage of the macromeres. The earlier-formed micromeres may be formed in the same direction to result in $1a^{11}$ and $1a^{12}$. The sixth cleavage results in the formation of a fourth quartet of micromeres

(4a–4d), and the remaining macromeres are designated 4A–4D. A regular succession of laeotropic and dexiotropic divisions occurs. According to Raven (1958) at about this time the regular character of spiral cleavage is lost and division is mostly bilaterally symmetric, beginning in the cells of the D quadrant.

Descriptions of early cleavage and the designation of each blastomere follow the terminology developed by Conklin (1897) in his work with the gastropod *Crepidula*. Detailed investigations on cell lineage in pelecypods are few. The knowledge of cell lineage for pelecypods is from the work on *Dreissensia* (Meisenheimer, 1901), *Unio* (Lillie, 1895), *Crassostrea gigas* (Fujita, 1929), and *Sphaerium striatinum* (Woods, 1931). For further descriptive details of cleavage patterns and cell lineage, consult MacBride (1914), Raven (1958), Galtsoff (1964), and Wada (1968). Cellular differentiation is certainly dependent on or correlated with the metabolism, but topics pertaining to the chemistry of development have been studied in few pelecypods. The lipid content of *Mytilus edulis* eggs is high, and it decreases during embronic development, suggesting that embryogenesis takes place largely by utilizing the lipid reserves of eggs (Bayne *et al.*, 1975). The early experimental work on the chemical embryology of pelecypods is reviewed by Raven (1958, 1972) and Reverberi (1971). Kidder (1976) reviewed the RNA synthesis patterns in some pelecypods and concluded that there was no common pattern characteristic of mosaic development. The pattern of gene expression was the same as in sea urchins. Costello and Henley (1976) provided a perspective of spiralian development.

5.4.1.5 BLASTULATION

Successive cleavages of the egg give rise to a blastula. In species with yolk-poor eggs, an unequal coeloblastula with a large cleavage cavity is formed. This is the case in *Cyclas, Sphaerium, Unio, Anodonta,* and *Dreissensia* (Raven, 1958). In other species with yolk-rich eggs, a solid stereoblastula is formed, as in the case of *Crassostrea, Teredo,* and *Adula californiensis* (Galtsoff, 1964; Lough and Gonor, 1971). The blastocoel forms early in eggs with less yolk, but it is never extensive. In the blastula stage, cilia are developed in certain regions of the embryo. Apical cilia are formed on the apical cells at the animal pole, and clumps of cilia are formed on the primary trochoblasts at four points around the equatorial region of the blastula. With further development, certain cells in the region of the primary trochoblasts produce a complete ring of cilia to form the prototroch. In most pelecypods, the apical cilia are prominent and remain so until the veliger stage, but in *Crassostrea virginica* they are absent at the beginning and in *Pinctada* they disappear early in the veliger stage (Wada, 1968). As the cilia develop, the blastula begins swimming within the vitelline membrane which remains intact.

Early development of *Crassostrea gigas* embryos to the swimming stage (late blastula–early gastrula) is under the control of information transcribed during oogenesis and stored in the egg. At the swimming stage, the nuclei become progressively active in RNA synthesis and produce at least one class of transcript. Transcript at the swimming stage appears to be necessary for further development (McLean, 1976). This pattern of gene expression correlates with results observed for *Spisula solidissima* and *Mulinia lateralis* (Firtel and Monroy, 1970; Kidder, 1972a,b). Gabrielli and Baglioni (1975) examined the maternal RNA and histone synthesis in embryos of *S. solidissima*.

5.4.1.6 GASTRULATION

Gastrulation takes place by epiboly, as well as by emboly (Raven, 1958). In species with a coeloblastula (e.g., freshwater bivalves, *Cyclas, Sphaerium, Unio, Anodonta,* and *Dreissensia*), invagination of the archenteron takes place first by macromeres, then by cells of the fifth quartet and the fourth quartet, and finally by all yolk-rich cells. The margin of the blastopore is formed by the stomatoblasts of the second quartet on the radii and the third quartet interradially (Raven, 1958). Many marine bivalves do not have a blastocoel. In species with a stereoblastula (e.g., *Teredo*), gastrulation takes place by epiboly. In *Mytilus* and *Ostrea* gastrulation begins with epibolic extension of the micromeres and then by invagination (Raven, 1958; Wada, 1968). Generally, the blastopore is wide at the beginning, but it is reduced later to a longitudinal slit and finally closes. In *Mytilus, Cyclas, Sphaerium, Anodonta,* and *Driessensia* the blastopore is completely closed, and the archenteron forms a closed sac and opens much later to the outside margin (Raven, 1958). In the yolk-rich eggs of *Teredo*, however, the archenteron is comprised of a mass of endoderm filled with yolk. The blastopore, which is originally located in the center of the vegetal hemisphere, becomes displaced anteriorly so that it comes to lie on the ventral side. Displacement of the blastopore occurs because of considerable growth of the dorsal region, especially the derivatives of the second blastomere, which form the somatic plate. Early formation of the shell gland may also assist in the shifting of the blastopore to the ventral side (Raven, 1958; Allen, 1961). Accompanying development of the archenteron and contraction of the blastopore, the stomatoblasts lining the periphery of the blastopore gradually invaginate to make up the stomodeum. The future mouth forms at the site of the blastopore.

The first quartet of micromeres gives rise to the ectoderm of the head region. Sometimes this region can be recognized in the advanced gastrula stage by a flattening of the cells, forming the head vesicle. Progeny of the second and third quartet give rise to the trunk region. The descendants of the first stomatoblast (2d) divide intensively to form the somatic plate. The anterior part of the somatic plate forms the shell gland, while the posterior part extends toward the ventral side to form the ventral plate, situated behind the mouth, from which the foot develops.

In forms in which gastrulation takes place by invagination, the endomeres change directly into cells bounding the archenteron. Mesoderm originates from the mesodermal teloblasts derived from the daughter cells of the 4d cell. The teloblasts situated on the left and right behind the archenteron bud off a small number of cells in front and produce the two mesodermal bands. For further details on gastrulation consult Raven (1958).

5.4.1.7 TROCHOPHORE

The gastrula develops typically into a top-shaped trochophore larva (Fig. 39). Development during this period mostly involves the rapid expansion of apical ectoderm, differentiation of cilia, and formation of

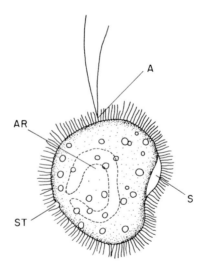

Fig. 39. Trochophore larva of *Mytilus edulis* at 24 hr. A, Apical tuft; AR, archenteron; S, shell; ST, stomodeum. (After Rattenbury and Berg, 1954; modified.)

the stomodeum and the lumen of the gut. The trochophore is divided by a band of cilia, the prototroch, into an upper prototrochal and a lower posttrochal region. The prototrochal portion bears large tufts of apical flagella. The prototroch of the trochophore larva eventually develops to form the organ of locomotion, the velum. The ectoderm of the apical region proliferates to give rise to the rudiments of the cephalic ganglia. The shell gland is formed dorsally as thickened ectoderm which first invaginates and later evaginates and begins secreting the larval shell.

The mouth invaginates on the ventral side just behind the prototroch in the prototrochal region. It leads into the stomodeum, which makes a connection with the archenteron (midgut) and eventually differentiates into the foregut (esophagus). The midgut differentiates into the stomach, intestine, and digestive diverticula. The digestive gland arises from the left and right anterior part of the midgut and the junction of the esophagus as a pair of foliated organs with a lumen. The crystalline style is secreted later in the veliger stage from a blind tube formed in the posterior part of the stomach. The posterior end of the larva may be marked by two protruding cells where the intestine makes connection with the surface ectoderm and the anus is formed.

During trochophore development, a pair of protonephridia differentiate ventrally in the body on the two sides of the foot. The protonephridia are straight organs composed of two to three cells (Wada, 1968). The main components of the larval musculature, velum, mantle, digestive tract, and foot are formed of spindle-shaped mesenchymal cells. The larval musculature degenerates at the time of metamorphosis. In pelecypods, the foot undergoes practically no further development during the trochophore stage.

Upon release from the vitelline membrane, the trochophore initially swims by revolving motions and then later moves forward jerkily with the apical tuft of flagella. In *Yoldia* and *Nucula*, the prototroch is strongly developed and consists of three rows of large vesicular cells which surround the posttrochal part of the body as a test (Raven, 1958).

Descriptive studies of embryonic development have been made on *Nucula delphinodonta, Yolida limatula,* and *Pecten tenuicostatus* (Drew, 1897, 1899, 1900, 1901, 1906), *Pecten opercularis* (Fullarton, 1890), *Tapes variegata* (Yoshida, 1953), *Bankia setacea* (Quayle, 1959), *Cerastoderma* (= *Cardium*) *edule* (Creek, 1960), *Pandora inaequivalvis* (Allen, 1961), *Argopecten* (= *Aequipecten*) *irradians concentricus* (Sastry, 1965), and *Adula californiensis* (Lough and Gonor, 1971). Detailed discussion of embryonic development in pelecypods may be found in MacBride (1914), Raven (1958, 1964), Galtsoff (1964), Wada (1968), and Reverberi (1971). The induction phenomenon plays an important role in embryogenesis, but experimental studies on pelecypods are very few (Raven, 1958; Collier, 1965).

5.4.1.8 Factors Influencing Successful Development

Temperature, salinity, chemicals, and other factors have been reported to affect the successful development of embryos to the veliger stage (Raven, 1958; Loosanoff and Davis, 1963; Collier, 1965). Normal development of embryos to the veliger stage seems to occur within a range of environmental conditions characteristic of a species. As conditions fall beyond the tolerance range, development is followed by either death or abnormal morphogenesis.

5.4.1.8.1 *Temperature.* Fertilization appears to occur over a wider range of temperatures than those that permit early cleavage. In *Mytilus edulis,* gametes released at 17°–18°C could be fertilized at temperatures between 5° and 22°C (Bayne, 1965). In *Adula californiensis,* fertilization is also reported to occur over a wider range of temperatures than those permitting normal development (Lough and Gonor, 1971). Development to metamorphosis occurs in *Mercenaria mercenaria* between 18° and 33°C (Loosanoff and Davis, 1963). Embryos transferred to 15°C 3 hr after fertilization fail to develop to the veliger stage. However, embryos transferred to 15°C 6 hr after fertilization develop to the veliger larval stage, but a high percentage of these embryos either die or develop abnormally. At 33°C a high percentage of mortality or abnormal development occurs. Apparently early cleavages are restricted to a narrow range of temperatures, but as development progresses some embryos can tolerate and develop at a wider range of temperatures, and these develop normally to the veliger stage.

Within the tolerance range, the rate of development is dependent on temperature (Fig. 40). The rate of embryonic development in *Mytilus edulis* increases with an increase in temperature (Bayne, 1965). The linear relationship between rate of development and temperature indicates that a constant proportion of time is spent in each stage. The Q_{10} for embryonic development of *M. edulis* is 2.63 between 10° and 18°C (Bayne, 1965). A similar linear relationship between rate of development and increase in temperature was found in *Adula californiensis* (Lough and Gonor, 1971). In *Mya arenaria,* the embryos can develop between 8.5° and 28°C, but the percentage of survival is greater at 15° and 18°C (Stickney, 1964). Kennedy *et al.* (1974a,b) determined the temperature–time relationships for embryonic development of *Mercenaria mercenaria* and *Mulinia lateralis.* For further dis-

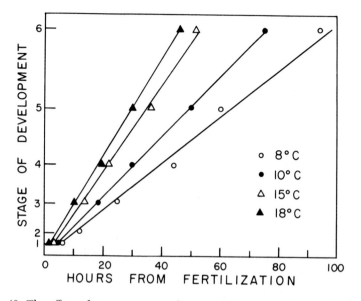

FIG. 40. The effect of temperature on the rate of cleavage and early development of *Mytilus edulis* embryos. Stage 1, Appearance of the first polar lobe; stage 2, macromeres surrounded by micromeres; stage 3, young trochophore; stage 4, young veliger; stage 5, first appearance of shell as a univalve pellicle on the dorsal surface of the shell; stage 6, the straight-hinge larva. (From Bayne, 1965; modified.)

cussion of temperature effects on embryonic development consult the reviews by Loosanoff and Davis (1963) and Kinne (1970).

5.4.1.8.2 *Salinity.* The salinity requirements for normal embryonic development of different species and their ability to withstand rapid and gradual changes may vary. Fertilization takes place over a wide range of salinities in *Mutilus edulis* (Bayne, 1965) and *Adula californiensis* (Lough and Gonor, 1971). Salinities below the tolerance range seem to result in retarded development and abnormal cleavage. A reduction in the salinity tolerance range following fertilization has been observed also for *Mytilus californianus* (Young, 1941) and *Spisula solidissima* (Schechter, 1956). In the latter case it was observed that the polar lobe phase was the stage most sensitive to reduced salinities.

Within the salinity range permitting embryonic development, there appears to be an optimum that produces a high percentage of normal larvae. Eggs of *Mercenaria mercenaria* from Long Island Sound have an optimum for development at about 27.5‰ but, at salinities below 17.5‰ and at 35‰, normal development does not occur (Davis, 1958;

A. N. SASTRY

Loosanoff and Davis, 1963). In the coot clam, *Mulinia lateralis*, from Long Island Sound, embryonic development to the veliger stage occurs over a wide range of salinities between 15 and 35‰ at 25°C. Salinities between 22.5 and 30‰ were optimal for the development of embryos to veligers (Calabrese, 1969b). An increase or decrease in salinity from the optimum seems to result in a lower percentage of normal development. The embryos fail to develop at 12.5‰, and only 1.2% develop normally at 37.5‰. Stickney (1964) also determined the influence of different salinities on the embryonic development of populations of *Mya arenaria* from Maine, Woods Hole, and Chesapeake Bay. He found that the percentage of embryos that survived to the veliger stage was the same in salinities of 10, 16, and 32‰, but transformation to the veliger was better in intermediate salinities for the Chesapeake Bay population. The embryos of populations from Maine and Woods Hole survived better at higher salinities. Thus the optimum, and the upper and lower limit for salinity tolerance, may vary for populations of the same species from different geographic locations.

5.4.1.8.3 *Temperature and Salinity*. The effects of temperature–salinity interaction on embryonic development have been studied in *Mercenaria mercenaria* (Davis and Calabrese, 1964) and *Mulinia lateralis* (Calabrese, 1969b). In *Mulinia lateralis*, 83.3–95% of embryos developed to the straight-hinge veliger stage at combinations of 17.5°–27.5°C temperature and 20–30‰ salinity (Table XI). At 32.5°C, development to veliger stage was poorer at all salinities. No normal development

TABLE XI

PERCENTAGE DEVELOPMENT OF *Mulinia lateralis* EMBRYOS TO STRAIGHT-HINGE LARVAE AT DIFFERENT COMBINATIONS OF SALINITY AND TEMPERATURE [a]

Salinity (‰)	Mean values from three experimental runs at:					
	32.5°C	27.5°C [b]	22.5°C	17.5°C	12.5°C [b]	7.5°C
35.0	1.0	41.3	58.5	42.8	2.6	0.0
30.0	20.3	88.5	93.9	89.1	83.4	0.0
25.0	31.6	84.8	95.9	91.8	69.2	0.0
20.0	27.3	84.5	91.4	83.3	41.7	0.0
15.0	1.5	6.4	12.8	0.7	0.0	0.0
10.0	0.0	0.0	0.0	0.0	0.0	0.0

[a] From Calabrese (1969b).

[b] Results of two experiments—the temperature control unit went out of order in the third experiment.

occurred at any salinity at 7.5°C nor at any of the experimental temperatures at 10‰ salinity.

Lough and Gonor (1971) determined the rate of development of embryos of *Adula californiensis* at various temperatures and salinities (Fig. 41). Normal development occurred between 26.3 and 33.2‰ salinity, the rate of development being linear with temperature. The rate of development at comparable temperatures was slower at lower salinities. In salinities below 26.3‰, there were abnormalities, asynchronous cleavages, irregular elongation of polar lobes, and irregular cell formation. Brenko (1974) determined the temperature and salinity requirements for the embryonic development of *Mytilus galloprovincialis*. For a detailed discussion on the effect of temperature and salinity consult the review by Kinne (1964).

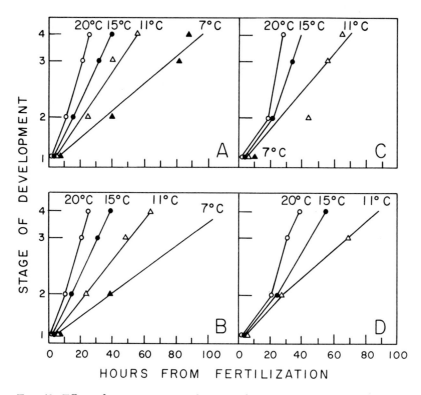

Fig. 41. Effect of temperature on the rate of embryonic development of *Adula californiensis* at salinities of (A) 33.2‰, (B) 26.3‰, (C) 22.6‰, and (D) 20.4‰. Stage 1, Second cleavage; stage 2, early trochophore; stage 3, late trochophore; stage 4, veliger larva. (From Lough and Gonor, 1971; modified.)

5.4.1.8.4 *Other Factors.* The pH of seawater is also reported to affect the successful development of some pelecypod embryos (Calabrese and Davis, 1966, 1970; Calabrese, 1969b). In *Mulinia lateralis,* embryonic development is not significantly affected within a pH range between 7.25 and 8.25. The pH for maximum rate of development of embryos was found to be 7.75. Successful embryonic development decreased precipitously below pH 7.0 and above pH 8.50.

Chemicals, toxicants, and turbidity-producing materials may kill zygotes or produce abnormalities in embryos if their concentrations in the seawater rise beyond the tolerance range (Raven, 1958, 1964; Davis, 1960, 1961; Davis and Hidu, 1969a,b; Calabrese and Davis, 1967; Hidu, 1965; Loosanoff, 1971; Granmo, 1972; Calabrese and Nelson, 1973; Calabrese and Rhodes, 1974; Brenko et al., 1977).

Metabolites released by microorganisms, especially dinoflagellates, may adversely affect the development of embryos. On the other hand, metabolites released by dinoflagellates or substances removed from the medium by the algal cells might be necessary for normal embryonic development of *Mercenaria mercenaria* (Loosanoff and Davis, 1963). In contrast, toxic metabolites of algae that may persist in seawater several weeks after the algal bloom could adversely affect normal development. In addition, eggs obtained from adults in poor physical condition, or eggs retained for a prolonged period in the seawater in which adults have been kept, also seem to develop abnormally. It has been suggested that the adults might deplete some substances from seawater necessary for shell formation in the larvae (Loosanoff and Davis, 1963). High concentrations of eggs can also cause abnormalities, perhaps by depleting oxygen in the seawater (Loosanoff and Davis, 1963). High concentrations of eggs and sperms are reported to cause abnormal development of *Pecten maximus* (Gruffydd and Beaumont, 1970). For additional discussion of various factors influencing embryonic development in pelecypods, consult the reviews by Loosanoff and Davis (1963) and Calabrese and Davis (1970).

5.4.2 Larvae

5.4.2.1 TYPES

Three ecologically and morphologically well-defined developmental types have been recognized in pelecypods by Thorson (1946, 1950) and Ockelmann (1962). Planktotrophic larval development is the most common type. Species exhibiting this type produce a large number of small eggs which are released into the surrounding water and typically develop to the veliger larval stage. The diameter of the yolk mass of ripe eggs in species with planktotrophic larvae ranges between 40 and 85 μm. In almost all species of this group, the diameter of prodissoconchs ranges between 75 and 150 μm. Metamorphosis of these larvae may take place when the larvae are less than 200 μm to more than 600 μm (Ockelmann, 1962). The duration of larval development may extend up to 3 months. The larvae invariably feed on phytoplankton up to metamorphosis.

Lecithotrophic development with a short pelagic larval stage occurs in a number of other species. These larvae reach metamorphosis utilizing the nutrient matter from the egg. In all these species, the veligers have a short pelagic life lasting from a few hours to a few days. The larvae are independent of planktonic food, although feeding before metamorphosis may occur as in the case of Pandora inaequivalvis (Allen, 1961). In these species, the velum may not develop elaborately, and planktonic food is apparently of secondary importance. The larvae grow very little between hatching and settling. Lecithotrophic larvae with a short pelagic life have been reported in Nucula proxima and Yoldia limatula (Drew, 1900, 1901), N. turgida and N. nucleus (Lebour, 1938a), Nuculana pernula, Thyasira flexuosa, T. sarsi, Teredo pedicellata, and others (Ockelmann, 1958, 1962), Ostrea chilensis (Soot-Ryen, 1959; Walne, 1963), and P. inaequivalvis (Allen, 1961).

Direct development without any larval stage also occurs in a number of species, and this is frequently accompanied by brood protection (see Sections 5.4.4.2 and 5.4.4.3). In Liocyma fluctuosa, development is direct, but there is no brood protection (Thorson, 1936; Ockelmann, 1964). Direct development with some type of brood protection occurs in Nucula delphinodonta (Drew, 1901), Musculus discors and M. niger (Thorson, 1935), some species of Astarte (Ockelmann, 1962) and Loripes lacteus (Myazaki, 1938), Transenella tantilla (Hansen, 1953), Cardita ventricosa, C. bailyi, and C. barbarensis (Jones, 1963), and Lasaea rubra (Oldfield, 1964). In Gemma gemma, the young are released as miniature adults (Sellmer, 1967). In the viviparous species Cerastoderma (=Cardium) elegantulum, the young attain a shell length of 1.2 mm when released (Matveeva, 1953).

Species with direct development produce large eggs. The yolk mass of ripe eggs commonly measures 150–200 μm or more in diameter, and such eggs are frequently surrounded by conspicuous membranes. In most species the embryonic shell at hatching measures at least 230–500 μm or larger (Ockelmann, 1958). Direct development implies the absence of a pelagic larval stage. Rather, embryonic development is immediately followed by a free-living bottom stage. Development of these species is lecithotrophic.

5.4.2.2 Veliger Development

The trochophore develops to a straight-hinge veliger larval stage. The veliger stage has a more complex morphology than the trochophore, with the various organs attaining a further stage of development (Fig. 42). Veliger development, which is similar in most species, is discussed in detail by Raven (1958) and Wada (1968). Recent literature on the development and anatomy of veliger larval stages includes work with *Cerastoderma (= Cardium) edule* (Creek, 1960), *Pandora inaequvalvis* (Allen, 1961), *Venus striatula* (Ansell, 1962), *Argopecten (= Aequipecten) irradians concentricus* (Sastry, 1965; Castagna and Duggan, 1971), *Mytilus edulis* (Brenko, 1973; Bayne, 1976c), *Placopecten magellanicus* (Culliney, 1974), *Modiolus modiolus* (De Schweinitz and Lutz, 1976), *Mya arenaria* and *Hiatella* sp. (Savage and Goldberg, 1976), *Martesia striata* (Boyle and Turner, 1976), and *Xylophaga atlantica* (Culliney and Turner, 1976).

The prototroch of the trochophore develops by outgrowth into a

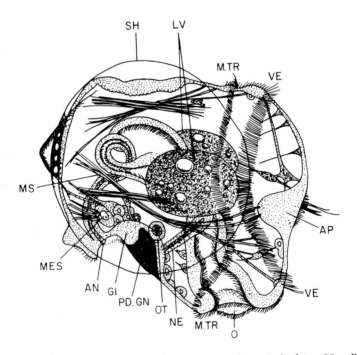

Fig. 42. Late veliger larva of *Teredo*. AN, Anus; AP, apical plate; GI, gill rudiment; LV, liver; MES, mesoderm; MS, larval muscle; M.TR, metatrochal ring; NE, nephridium; O, mouth; OT, otocyst; PD.GN, rudiment of pedal ganglion; SH, shell; VE, velum. (After Hatschek, 1881; modified.)

bilobed velum with cilia on the outer margin. This serves as an effective locomotory organ for the veliger larva.

In the region of the apical tuft or beside it, the apical plate is formed in the prototrochal region. On the right and left of the apical plate lie the cephalic plates from which the cerebral ganglia, eyes, and tentacles develop. The pedal ganglia appear in the ventral region between the mouth and anus where the foot develops later. The visceral ganglia develop near the pedal ganglia, and parietal ganglia close to the cerebral ganglia. In higher pelecypods, the parietal ganglia fuse with the cerebral ganglia (Raven, 1958; Wada, 1968). A pair of statocysts develops in the vicinity of the pedal ganglia in the late veliger stage. A pair of pigmented eyespots also develops below the velum.

The larval musculature in *Cerastoderma edule* and *Venus striatula* has been described in detail by Creek (1960) and Ansell (1962). The anterior adductor muscle develops near the anterior part of the shell. The posterior adductor muscle develops in the late veliger stage at a point anterior to the anus. Both adductors develop during the larval period in the case of monomyarian species, but later the anterior adductor degenerates and it is only the posterior adductor which develops (Sastry, 1965; Wada, 1968). The larval musculature also includes velar retractor muscles which serve to contract the whole body of the larva so that it can be withdrawn into the shell. These muscles extend from the dorsal body wall in the region of the hinge line and are attached to the velum and the digestive tract. For further details on larval musculature consult the descriptions given by Creek (1960), Raven (1958), and Wada (1968).

The digestive system becomes functional during the veliger stage with the opening of the anus. The digestive diverticula appear in this stage as paired rudiments. The intestine begins to acquire loops as development of the veliger progresses. The crystalline style sac develops as a blind sac at the posterior end of the stomach.

The foot forms on the ventral side of the body between the mouth and anus as the late veliger begins to settle, just before metamorphosis. An ectodermal invagination at the base of the foot gives rise to the pedal and byssal glands. The mucus threads secreted by the byssal gland allow the larva to attach to the substratum after metamorphosis. In some species, the byssal gland remains functional in the adult, whereas in others it degenerates as development continues. A cement gland, similar in nature to the byssal gland, develops in species that attach permanently to the substratum after metamorphosis.

The mantle develops during the trochophore stage as an expansion of ectoderm on the posterior surface of the larva and develops later as a lateral fold toward the anterior region. In the veliger stage the ventral edge of the mantle is thickened and forms two folds in *Venerupis pullastra* (Quayle, 1952) and *Venus striatula* (Ansell, 1962). The shell is deposited by the specialized portion of the mantle folds (Beedham, 1958; Ansell, 1961b, 1962). Humphreys (1969) made an electron microscopic study of larvae of *Mytilus edulis* during up to 70 hr of development to follow the process of larval calcification.

According to Raven (1958) and Wada (1968) a larval protonephridium of ectodermal origin develops near the tip of the pair of long, narrow mesodermal bands. The adult kidney is formed adjacent to these. Creek (1960) and Ansell (1962) found no larval excretory organs in *Cerastoderma edule* and *Venus striatula*, respectively.

The ctenidium appears at the time of metamorphosis as a row of papillae. The inner ctenidial lobes appear earlier than the outer pair. The ctenidia do not perform the function of food collection until sometime after metamorphosis.

According to Wada (1968) the pericardium and its accessory organs are mesodermal in origin. The cells of the anterior dorsal part of the left and right mesodermal bands form two coelomic cavities which constitute the rudiment of the pericardium. The coelomic cavaities increase in size on the two sides of the intestine and eventually fuse to form the pericardium. The ventricle is formed between the coelomic cavity and the intestine. An invagination appears in the pair of cavities to form the auricles, and they make a connection with the ventricle. The rudiment of the kidney is found adjacent to the pericardium as a filiform tissue formed by mesodermal cells and contains a lumen. The lumen of the kidney opens into the pericardium, while the other end extends toward the body wall and opens to the outside (Wada, 1968).

5.4.2.3 SETTLING AND METAMORPHOSIS

5.4.2.3.1 *Settling.* Toward the close of veliger development, the larvae tend to keep to the bottom, having passed through the swimming stage to a crawling form called the "pediveliger" prior to metamorphosis (Carriker, 1961). The name "veliconcha" has been proposed by Werner (1939) for the fully developed veliger stage immediately preceding metamorphosis. Swimming and crawling stages have been reported for a number of pelecypods (Grave, 1927b; Lebour, 1938a; Belding, 1930; Werner ,1939; Turner and George, 1955; Allen, 1961; Ansell, 1962; Bayne, 1965; Sastry, 1965). The morphology and anatomy of the pediveliger stage has been described for *Pandora inaequivalvis* (Allen, 1961), *Venus striatula* (Ansell, 1962), *Mytilus edulis* (Bayne, 1965), and *Argopecten (= Aequipecten) irradians concentricus* (Sastry, 1965).

The most prominent features of the pediveliger stage include the rounded shell, velum, conspicuous foot, and two or three primary gill filaments. According to Bayne (1965) the velum is reduced in size with the delay in metamorphosis as the pediveliger continues to grow (Fig. 43).

The pediveliger stage is usually associated with exploratory behavior when the larva tests the substratum. The presence or absence of a suitable substratum apparently has a considerable influence on the duration of the pediveliger stage, with metamorphosis being postponed until suitable conditions are found. Settlement is considered a critical period for many pelecypods. At least in laboratory cultures high mortality usually occurs (Allen, 1961; Sastry, 1965).

It has been well established that larvae of marine invertebrates, including some pelecypods, choose a suitable substratum for settling and metamorphosis. This phenomenon has been reviewed by Thorson (1946, 1964) and Wilson (1960) (see also Section 6.4.2). The larvae delay metamorphosis when a suitable substratum is not encountered. The delay in metamorphosis permits larvae to maintain a certain level of organization, subsequently to select a suitable site for further develop-

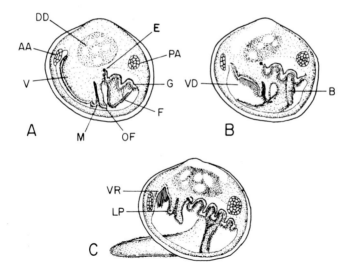

Fig. 43. Changes in the gross morphology of *Mytilus edulis* larvae during a delay in metamorphosis. (A) Pediveliger stage 1. (B) Pediveliger stage 2. (C) Pediveliger stage 3. Compare with Fig. 51. AA, Anterior adductor muscle; B, byssus duct; DD, digestive diverticulum; E, eyespot; F, foot; G, primary gill filaments; LP, labial palps; M, mouth; OF, oral flap; PA, posterior adductor muscle; V, velum; VD, degenerating velum; VR, reduced velum. (After Bayne, 1965.)

ment. Metamorphosis to the postlarval stage may occur after a shorter or a longer pediveliger stage.

5.4.2.3.2 *Metamorphosis.* The changes occurring in the larva between attachment and the formation of the postlarval, or dissoconch, shell are considered to signify metamorphosis in pelecypods (Bayne, 1965; Sastry, 1965). The swimming and crawling stage may last for a few hours or several days before attachment to the substratum occurs by byssal secretion. In sedentary species, such as oysters, the byssal gland secretes a cementing fluid which provides a permanent attachment to the substratum. In others, the byssal gland secretes mucus threads for temporary attachment to the bottom. Metamorphosis begins immediately after attachment to the substratum; certain larval organs disappear, and reorientation of the organs takes place, resulting in the adult form. Metamorphosis has been described for the larvae of *Venerupis pullastra* (Quayle, 1952), *Pandora inaequivalvis* (Allen, 1961), *Cerastoderma* (= *Cardium*) *edule* (Creek, 1960), *Venus striatula* (Ansell, 1962), *Mytilus edulis* (Bayne, 1965), and *Argopecten* (= *Aequipecten*) *irradians concentricus* (Sastry, 1965) but not in as much detail as for *Ostrea edulis* (Cole, 1937, 1938). The changes in anatomical organization during the metamorphosis of pelecypod larvae are similar in many respects.

During metamorphosis the mouth shifts in a clockwise direction and occupies an area anteriodorsally near the hinge. The anus also shifts in the same direction from a posteriordorsal orientation to a posterio-ventral position. The velar retractor muscles disappear and are replaced by the pallial muscles. The velum undergoes extensive changes during the process of metamorphosis and gradually disintegrates or is cast off in fragments at a certain moment. Cole (1938) showed that the velum collapsed immediately after attachment in *Ostrea edulis* and that the bulk of the structure was subsequently either cast off or disintegrated. In *Argopecten* (= *Aequipecten*) *irradians concentricus,* the velum disintegrates gradually (Sastry, 1965). According to Bayne (1965) the velum in *Mytilus edulis* gradually disintegrates when metamorphosis is delayed. In view of these findings, it appears that the velum is cast off soon after settlement, if the larva encounters a suitable substratum for attachment, or it degenerates when the exploratory period for a suitable attachment site is prolonged.

The apical plate of the velum becomes detached from the rest of the body and sinks to a position dorsal to the esophagus to fuse with the upper lip of the mouth. The upper lip extends to contribute to formation of the labial palps (Quayle, 1952). The foot undergoes progressive changes and acquires its definitive (adult) shape or, as in the case of oysters, is resorbed after disintegration. In the case of monomyarians, it is at metamorphosis that the anterior adductor disintegrates and the posterior one moves counterclockwise in the same direction as the mouth. The postlarval, or dissoconch, shell is secreted as metamorphosis is completed (Sastry, 1965). In the protobranchs *Nucula delphinodonta* and *Yoldia limatula*, metamorphosis takes place within the prototroch or test. At metamorphosis, the prototroch, apical plate, and even part of the stomodeum are suddenly thrown off within a few minutes (Drew, 1899). For further details on the anatomical changes during the process of metamorphosis consult the reviews by Raven (1958) and Galtsoff (1964).

5.4.3 Factors Influencing Growth, Settling, and Metamorphosis

5.4.3.1 GROWTH

With the development of laboratory culture methods, the veliger larvae of a number of species have been reared to metamorphosis (Loosanoff and Davis, 1963; Walne, 1964). The dimensions of veliger larval shells during development to metamorphosis have been reported for a number of species by Loosanoff *et al.* (1966). The sizes of both the early veliger larva and the later veliger at the time of metamorphosis are summarized in Table XII. However, the influence of environmental factors on survival and growth has been studied for only a few species. Included are such commercially important pelecypods as *Crassostrea virginica, Ostrea edulis,* and *Mercenaria mercenaria.* Loosanoff and Davis (1963), Walne (1964), and Loosanoff (1971) have reviewed the factors affecting the survival and growth of larvae of pelecypods.

5.4.3.1.1 *Temperature.* The temperature tolerance range for the larval stages is reported to be wider than during early cleavage in *Mercenaria mercenaria* (Loosanoff and Davis, 1963) and *Mulinia lateralis* (Calabrese, 1969b). In *Mulinia lateralis,* the larvae survived at 7.5°C, although growth was negligible; between 7.5° and 27.5°C, the growth rate increased with increase in temperature; and at 30° and 32.5°C, the growth rate decreased. The optimum temperature for growth was 27.5°C.

Loosanoff *et al* (1951) cultured the larvae of *Mercenaria mercenaria* at five different temperatures and determined the increase in length of shell up to metamorphosis (Fig. 44). The larval growth rate was higher at high temperatures than at lower temperatures. Larvae maintained at 30°C began metamorphosis as early as 7 days, while those

TABLE XII

Size of Egg and Early Veliger, Size at Metamorphosis, and Duration of Larval Development of Certain Pelecypods

Species	Temperature (°C)	Egg diameter (μm)	Size of early veliger (μm)	Size at the time of metamorphosis (μm)	Duration of larval development (days)	Reference
Argopecten irradians	20–23	55–65	80	175	14	Loosanoff and Davis (1963)
Argopecten irradians concentricus	24 ± 1.0	62	101	190	15–19	Sastry (1965)
Anomia simplex	?	42–45	58	195–210	?	Loosanoff and Davis (1963)
Arca transversa	20	52	70	240–260	27–37	Loosanoff and Davis (1963)
Bankia fimbriatula	28	45	70–90	250	40	Turner and Johnson (1971)
Bankia setacea	?	?	120–150	350–385	17–31	Quayle (1959)
Barnea truncata	19–26	40–45	55–95	250–315	35	Chanley (1965)
Cerastoderma edule	10–15	50–60	80	270	24	Creek (1960)
Donax variabilis	20–25	?	70	275–340	?	Chanley (1969)
Ensis directus	24	64–73	80	210	10	Loosanoff and Davis (1963)
Laevicardium mortoni	20	60–65	85	245	8–10	Loosanoff and Davis (1963)
Martesia striata	26	46.8	60–70	236	48–53	Boyle and Turner (1976)
Mercenaria mercenaria	18	70–73	86	175–240	16	Loosanoff and Davis (1963)

Species						Reference
Montacuta percompressa	?	?	120–150	350–385	17–31	Chanley and Chanley (1970)
Mulinia lateralis	?	?	60	210–230	6	Loosanoff and Davis (1963)
Mya arenaria	20	68–73	85–95	190	28	Loosanoff and Davis (1963)
Mytilus edulis	?	?	96	215	?	Loosanoff and Davis (1963)
Noeteredo knoxi	28	45	70–90	?	?	Turner and Johnson (1971)
Noetia ponderosa	19–26	65	80–85	200–300	28–42	Chanley (1966)
Pandora inaequivalvis	18 ± 1.0	105–125	150	175	3	Allen (1961)
Petricola pholadiformis	Room temp.	51–55	79	186	13	Loosanoff and Davis (1963)
Pinctada maxima	27–31	59–60	101	130	14–20	Minaur (1969)
Pitar (Calliocardia) morrhauna	?	49–60	78	180	?	Loosanoff and Davis (1963)
Spisula solidissima	22	56.5	79	230–250	19	Loosanoff and Davis (1963)
Tapes semidicussata	Room temp.	60–75	95	175	14	Loosanoff and Davis (1963)
Teredo navalis	20	50–60	85–95	190	28	Loosanoff and Davis (1963)
Volsella demissus	22	?	110–115	220	12–14	Loosanoff and Davis (1963)

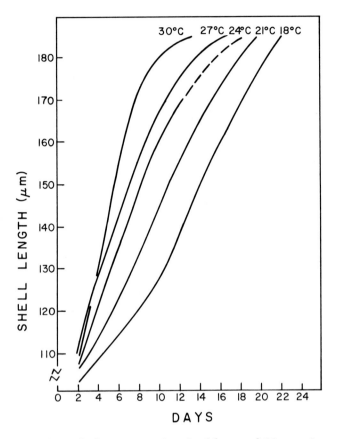

Fig. 44. General trend of increase in length of larvae of *Mercenaria mercenaria* grown at different temperatures, based on measurements made in four experiments. (From Loosanoff *et al.*, 1951; modified.)

at 18°C reached the same stage only after 16 days. Bayne (1965) determined the effect of temperatures between 5° and 22°C on the growth rate of larvae of *Mytilus edulis* fed on *Isochrysis galbana* and *Monochrysis lutheri*. Larvae cultured at 5°C remained swimming for more than 30 days, but they did not feed and there was no growth after the first 3 days (Fig. 45). The growth rate of the larvae increased with an increase in temperature between 10° and 21°C, but growth was relatively independent of temperature between 14° and 20°C. These larvae also showed a decline in growth rate with increase in shell length.

Bayne (1965) compared the growth rate of larvae of *Mytilus edulis* obtained from adults collected from Talyfoel, North Wales, and

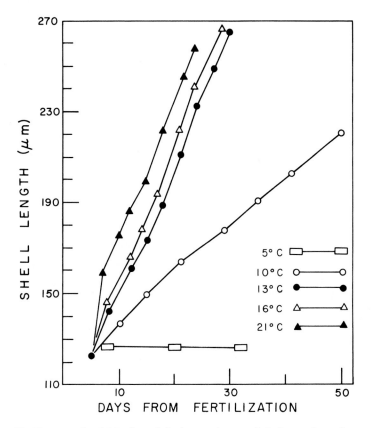

FIG. 45. The growth of *Mytilus edulis* larvae from Talyfoel, North Wales, at different temperatures; 25 cells *Isochrysis* per μl, 31–33‰ salinity. (From Bayne, 1965; modified.)

Helsingør, Denmark. The North Wales population occurs intertidally in water temperatures between 8° and 18°C and salinities between 29 and 30‰, while the Danish population is subtidal where temperatures range between 8° and 17°C and salinities between 13 and 23‰. The instantaneous relative growth rate k, used to compare the growth rate of larvae under different conditions, is calculated by the formula.

$$k = \frac{\log_e L_2 - \log_e L_1}{t}$$

where L_1 is the initial mean length of the larvae and L_2 is the mean length at time t. The growth rate of larvae of both populations

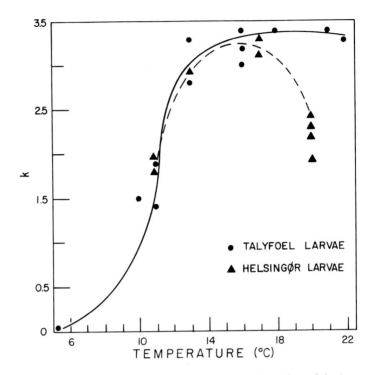

Fɪɢ. 46. The effect of temperature on growth rates of *Mytilus edulis* larvae from Talyfoel, North Wales, and Helsingør, Denmark. Each value of k is calculated for $L_1 = 130$–140 μm and $t = 20$ days. (From Bayne, 1965; modified.)

$(K = 100k)$ was similar between 11° and 17°C (Fig. 46). Above 17°C, the growth rate of the Talyfoel population did not change markedly, whereas it decreased for the larvae of the Helsingør population. Maximum larval growth rates occurred near the maximum temperatures of their respective habitats. Stickney (1964) reported that larvae of a *Mya arenaria* population from Maine grew faster than those from Chesapeake Bay when cultured at 14°–15°C. Loosanoff (1959) found that the rate of larval growth of the southern clam *Mercenaria campechiensis* was the same as that of the northern clam *M. mercenaria* if maintained under identical culture conditions.

In culturing pelecypod larvae, a number of workers have observed considerable variations in growth rate and time needed for metamorphosis among larvae reared under uniform conditions (Loosanoff *et al.*, 1951; Loosanoff and Davis, 1963; Ansell, 1962; Sastry, 1965). During the early veliger stage in *Mercenaria mercenaria,* variation in size is com-

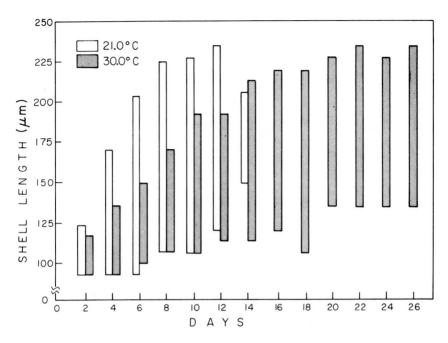

Fig. 47. Extent of variations in length of larvae of *Mercenaria mercenaria* on different days of cultivation. (From Loosanoff *et al.*, 1951; modified.)

paratively small, but with the progress of development, the difference between minimum and maximum sizes increases (Fig. 47). Chanley (1955) discussed causes for variation in the growth rate of *M. mercenaria* larvae and suggested that slight differences between young larvae were due to a combination of inherited characteristics and differences in the physiological condition of the eggs.

5.4.3.1.2 *Salinity.* In a number of pelecypods larvae are reported to tolerate a wider range of salinities than the early cleavage stages. Davis (1958) determined the effect of various salinities on the growth of *Mercenaria mercenaria* larvae (Fig. 48). They grew to metamorphosis at 17.5‰, but at lower salinities there was slight growth and 100% mortality occurred before metamorphosis. Growth of larvae was better at 20‰ and higher. Larvae cultured at 15‰ grew slightly, but none of them metamorphosed. At 12.5‰ the larvae did not grow, and all were dead by the tenth day. Calabrese (1969b) showed the effect of different salinities on larval growth in *Mulinia lateralis.* Larval growth was optimum at 27.5‰, but at lower salinities, namely, 7.5 and 10‰, growth was limited. At salinities above 27.5‰ growth decreased slowly. Stickney

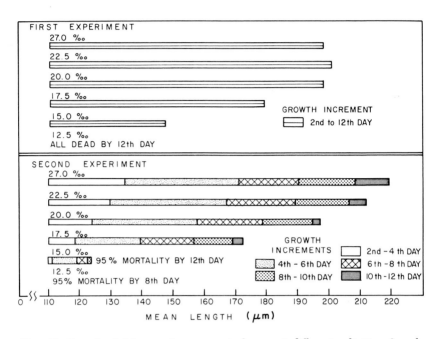

FIG. 48. Growth of *Mercenaria mercenaria* larvae at different salinities. Samples were taken, and measurements were made only at the beginning and termination of the first experiment. In the second experiment salinities were more carefully controlled, and samples from each of the duplicate cultures at each salinity were taken every second day. One hundred larvae from each sample were measured. (From Davis, 1958; modified.)

(1964) found that the straight-hinge larvae of *Mya arenaria* from Chesapeake Bay grew best at 16‰ and successfully metamorphosed. At 32‰, the larvae grew to metamorphosis, but mortality was higher. At 10‰ salinity, the larvae grew, but no metamorphosis occurred. In contrast, the larvae of *Mya arenaria* from Woods Hole, Massachusetts, showed no significant differences in growth at salinities between 10 and 32‰. However, larval growth was better at the higher salinities.

5.4.3.1.3 *Temperature and Salinity.* The effect of the interaction of temperature and salinity on the growth of veliger larvae of *Mulinia lateralis* was determined by Calabrese (1969b). The percentage increase in length of larvae was most rapid at 22.5° and 27.5°C and at salinities between 20 and 35‰ (Table XIII). The larvae survived at different salinities at 32.5°, 17.5°, 12.5°, and 7.5°C, but growth was slower. Brenko and Calabrese (1969) made a similar study on the effect of various temperatures and salinities on the growth and survival of larvae

TABLE XIII

PERCENTAGE INCREASE IN MEAN LENGTH OF *Mulinia lateralis* LARVAE AT DIFFERENT COMBINATIONS OF SALINITY AND TEMPERATURE [a]

Salinity (‰)	Temperature (°C)					
	32.5	27.5	22.5	17.5	12.5	7.5
35.0	63.5	83.6	70.2	25.4	2.7	1.2
30.0	65.0	99.4	90.5	43.1	3.2	0.7
25.0	45.6	81.0	82.3	43.2	4.9	0.7
20.0	61.7	84.6	92.9	61.7	13.3	1.1
15.0	38.8	57.5	68.2	38.6	3.5	1.2
10.0	0.0	40.9	44.9	15.8	0.9	0.5

[a] From Calabrese (1969b).

of *Mytilus edulis*. The growth of *Mytilus edulis* larvae was rapid at 15° and 20°C at salinities between 25 and 35‰. Larval growth decreased at both 25° and 10°C; the decline was most pronounced at 40 and 20‰ and below. Lough and Gonor (1973) determined the combined effects of temperature and salinity on the growth and survival of 3- and 15-day-old larvae of *Adula californianus*. Three-day-old larvae were more sensitive to reduced salinity than 15-day-old larvae. However, 15-day-old larvae showed a narrower temperature tolerance than 3-day-old larvae.

Bayne (1965) showed that larvae of *Mytilus edulis* from Talyfoel grew rapidly at 30–32‰, but growth was slower at 24‰ and no growth occurred at 9‰. The larvae of *M. edulis* from Helsingør were cultured between 10.0 and 30.5‰ at 13° and 16°C. The optimum growth of larvae at 16°C was between 20 and 30‰ and at 13°C between 17.5 and 26‰. Comparison of optimum salinities for Talyfoel and Heslingør larvae indicated a greater tolerance for lower salinity among the Danish animals. Growth in the Helsingør larvae occurred even at 14–15‰ at both 13° and 16°C, about 8–10‰ below the lower limit for the Talyfoel larvae.

5.4.3.1.4 *Food.* The influence of food on survival and growth of larvae of commercially important species, *Crassostrea virginica, Ostrea edulis, Mercenaria mercenaria,* and *Mytilus edulis* has been studied extensively, but similar investigations on noncommercial pelecypods are few (Davis, 1953; Davis and Guillard, 1958; Davis and Calabrese, 1964; Loosanoff et al., 1951, 1953; Walne, 1956, 1965; Bayne, 1965; Loosanoff, 1971; Ukeles, 1971, 1975; Masson, 1977). The larvae of pelecypods show differences in qualitative and quantitative food requirements for growth and survival (Loosanoff and Davis, 1963). Of 10 genera of microorganisms tested as food for larvae of *Mercenaria mercenaria, Monochrysis lutheri* and *Isochrysis galbana* proved to be the best for sustaining growth (Davis and Guillard, 1958). Larvae fed on mixtures of food organisms grew significantly larger than those receiving a single food organism (Fig. 49). Bayne (1965) also showed that larvae of *Mytilus edulis* grew equally well when fed on *Monochrysis* or *Isochrysis,* but equal mixtures by volume of both food organisms supported faster growth than similar volumes of either species alone.

Loosanoff *et al.* (1951) and Davis (1953) showed that the larvae of *Mercenaria mercenaria* were capable of qualitative food selection from mixtures of food organisms provided to them. According to Loosanoff

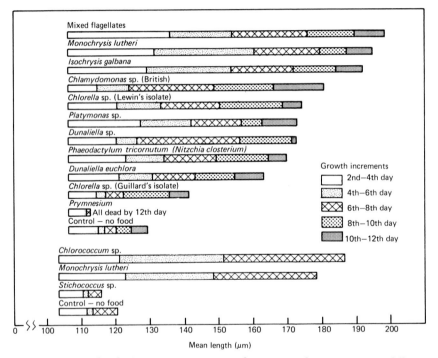

Fig. 49. Growth of *Mercenaria mercenaria* larvae in cultures receiving different microorganisms at the rate of 10×10 mm^3 of packed cells per milliliter per day. Concentrations of clam larvae averaged 14 per ml in one experiment (top) and approximately 10 per ml in the second experiment (bottom). Plots were based on mean length of 100 larvae from each of duplicate cultures. Mixed flagellates consisted of equal quantities (by packed cell volume) of *Isochrysis, Monochrysis, Dunaliella euchlora,* and *Platymonas.* (From Davis and Guillard, 1958; modified.)

and Davis (1963) the larvae of some species (e.g., *Mercenaria mercenaria* and *Mytilus edulis*), utilize a wide variety of food organisms, while others feed on only a few species. There is also an indication that larvae of some species (e.g. *Mercenaria mercenaria*) in all developmental stages can utilize *Chlorella*, while the early larval stages of other species (e.g., *Crassostrea virginica*) can not. This suggests that qualitative requirements for food organisms can change with the growth of larvae in some species (Davis, 1953; Davis and Guillard, 1958; Loosanoff and Davis, 1963). The presence and thickness of the cell wall and the degree of toxicity of metabolites are important factors in determining the suitability of photosynthetic microorganisms as food for pelecypod larvae. Some microorganisms, such as *Prymnesium parvum, Stichococcus* sp., and high concentrations of *Chlorella* kill larvae of *Mercenaria mercenaria* by producing toxic metabolites. Bayne (1965) found that *Mytilus edulis* larvae did not grow when provided with *Phaeodactylum tricornutum* as food. The nutrition of bivalve larvae has been reviewed further by Ukeles (1975).

Loosanoff *et al.* (1951) showed that toxic metabolites from *Chlorella* sp. (Lewin's isolate) killed larvae. Davis and Chanley (1956) showed that several toxic substances reduced the growth of *Mercenaria mercenaria* larvae. Larvae of *Mytilus edulis* fed on *Nannochloris atomus* and *Chorella* grew slower than those receiving *Monochysis* as food, either because of secretion of growth-inhibiting metabolites by the algae present in greatest concentrations in the old cultures, or because of bacterial blooms that accompanied the algae (Bayne, 1965).

Davis and Guillard (1958) found that the growth rate of *Mercenaria mercenaria* larvae was not affected when 50,000–400,000 cells/ml of *Isochrysis* and 125,000–500,000 cells/ml of *Monochrysis* were fed through the eighth day of development (Fig. 50). In contrast, 200,000 cells/ml of *Chlorella* caused mortality of the larvae. It is suggested that the external metabolites of *Monochrysis* and *Isochrysis* are much less toxic to *Mercenaria mercenaria* larvae than those of *Chlorella*. Bayne (1965) determined the effect of different food concentrations on the growth of *Mytilus edulis* larvae and found that, at the same temperature (16°C), the larvae receiving higher concentrations (100,000 cells/ml) reached the pediveliger stage after 16 days, while those receiving lower concentrations (25,000 cells/ml) took 30 days. A similar increase in larval growth rate with an increase in food concentration was observed at 11°C. The larvae of a number of pelecypods survived for prolonged periods even when deprived of food organisms. Bayne (1965) reported that starved *Mytilus edulis* larvae began feeding immediately when food was added and that growth followed feeding.

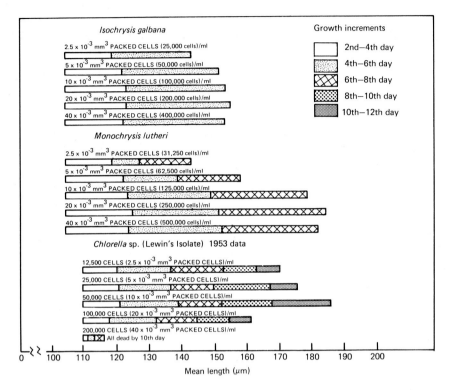

FIG. 50. Growth of *Mercenaria mercenaria* larvae in cultures receiving five different concentrations of *Isochrysis, Monochrysis,* and *Chlorella* sp. (Lewin's isolate). The concentration of larvae averaged 10 per ml. Plots were based on mean length of 100 larvae from each of duplicate cultures. (From Davis and Guillard, 1958; modified.)

The growth rate of *Mercenaria mercenaria* larvae is critically affected by the type of food organisms provided to them (Davis and Calabrese, 1964). The larvae of *M. mercenaria* can use *Chlorella* sp. and grow at temperatures of 25°–30°C, but not at 15°C. At 15°C, the larvae can ingest but not digest and assimilate *Chlorella* sp., although they can utilize naked flagellates, *Monochrysis* and *Isochrysis*. It has been suggested that the enzymes required by the larvae for digesting naked flagellates are active at lower temperatures than those necessary for the digestion of food organisms with more resistant cell walls (Loosanoff and Davis, 1963).

Loosanoff and Davis (1963) and Walne (1964, 1965) discussed in detail the influence of food on growth and survival of the larvae of some pelecypods.

5.4.3.1.5 *Other Factors.* The larvae of *Mytilus edulis* resulting from gametes of stressed adults grow slower than those from unstressed animals (Bayne, 1972). Previous environmental experiences are manifested in early development up to prodissoconch larvae and later at the time of metamorphosis (Bayne *et al.*, 1975). The effect of pH on survival and growth of larvae has been studied in *Mercenaria mercenaria* (Davis and Calabrese, 1964) and in *Mulinia lateralis* (Calabrese, 1970). The larvae of *Mulinia lateralis* grow satisfactorily at a pH range of 7.0–8.25, but a precipitous decline in growth occurs beyond this range. The growth of larvae within the satisfactory pH range does not vary significantly, and they seem to survive a wider range of pH than is suitable for growth.

Crowding also affects the growth of pelecypod larvae. The optimal concentration for larval growth appears to vary with the species and the quality and quantity of food (Loosanoff and Davis, 1963). The optimum population density for larval growth is often greater than the minimum density. The mean growth rate of larvae of *Mercenaria mercenaria* shows an inverse relationship to the population density within the density range tested (Loosanoff *et al.*, 1955). The slow growth in crowded cultures has been attributed to the deleterious effects of excretory products; a proportional increase in the quantity of food does not remedy the slow growth among crowded larvae. Substances other than those already discussed have been reported to increase the growth rate of *M. mercenaria* larvae when added in low concentration to cultures (Davis, 1960, 1961; Loosanoff and Davis, 1963; Calabrese and Davis, 1970). These include turbidity-producing materials, certain antibiotics, pesticides, insecticides, and inert substances such as silt and kaolin. It is suspected that the observed increase in larval growth rate may be due to the removal of toxins from seawater in larval cultures by absorption or formation of complexes with large organic molecules of antibiotics and pesticides.

5.4.3.2 SETTLING

The early veliger larvae of many pelecypods swim upward as a positive response to light and aggregate at an illuminated surface (Allen, 1961; Ansell, 1962; Sastry, 1965; Minaur, 1969). Bayne (1964a) made a detailed study on the response of *Mytilus edulis* larvae to light and gravity. Changes occurring between the early veliger stage and the pediveliger are apparently a function of size as well as behavior and morphology. As discussed in Sections 5.4.1.8 and 5.4.3.1, development to the pediveliger stage is affected by environmental factors as well as by inherent differences in individual larvae. *Mytilus edulis* larvae with eyespots show photopositive and geonegative responses. Many marine planktotrophic larvae are photopositive when young and then become photonegative when approaching metamorphosis. This behavior brings the larvae up to plankton-rich surface water and may also serve as a mechanism for planktonic dispersal. The larvae also swim upward in response to increased hydrostatic pressure, and this apparently allows young larvae to remain at the surface (Bayne, 1963). The light and gravity responses are variably affected by changes in temperature and light intensity. Turner and George (1955) observed the response of *Mercenaria mercenaria* larvae to a salinity gradient and found that they swam upward across a sharp gradient with no loss in velocity until they had passed the boundary between 20 and 15‰. At 15‰ salinity, not only the velocity of the larvae decreased, but they no longer swam upward, only in a circular pattern.

As the larvae approach settlement, they become photonegative and geopostive (Bayne, 1964a). The response to hydrostatic pressure also weakens as the larvae grow older, which allows the later stages to come in contact with the bottom (Bayne, 1963). The pediveliger larvae of a number of species reportedly are able to keep to the bottom and pass through a period of swimming and crawling where they test the substratum for a suitable site for attachment (Isham and Tierney, 1953; Turner and George, 1955; Allen, 1961; Ansell, 1962; Sastry, 1965; Bayne, 1965). Bayne (1965) has shown that the pediveliger larvae of *Mytilus edulis* respond to light negatively and are also positively geotactic, which appears to explain the tendency of pediveligers to concentrate on the bottom. After larvae return to the bottom, there is a testing period characterized by momentary settlement. This has been described for larvae of a number of invertebrate species (Thorson, 1946). The presence or absence of a suitable substratum has a considerable influence on the duration of this final motile stage. The larvae can postpone attachment if the substratum is unsuitable. Bayne (1965) has described the behavior of pediveliger larvae during the exploratory period prior to attachment. The larvae of *M. edulis* extend their foot frequently during swimming. If a suitable substratum is encountered, the velum is withdrawn and the larvae start to crawl. Crawling continues for a period after which the larvae either attach by the byssus or withdraw the foot and swim again. If attachment occurs, the larvae gradually cease to crawl, the foot is extended outside the shell, and a single byssus thread is secreted. The pediveliger larvae of many pelecypods are able to delay metamorphosis if the conditions of the environment are unsuitable for completion of this important process (Thorson, 1946; Wilson, 1960). Bayne (1965) studied the factors influencing the events prior to and during the delay of metamorphosis of *M. edulis* larvae (Fig. 51).

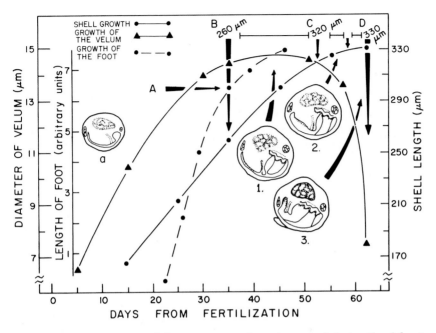

FIG. 51. A representation of the events occurring prior to and during the delay in metamorphosis of *Mytilus edulis* larvae at 13°C. A, Foot becomes functional; B, larva becomes capable of metamorphosis; C, velum begins to degenerate; D, maximum size of larva; a, appearance (diagrammatic) of veliconcha of length 190 μm; 1, appearance of pediveliger stage 1; 2, pediveliger stage 2; 3, pediveliger stage 3. Compare with Fig. 43. (From Bayne, 1965; modified.)

Observations on the influence of the substratum on the settling of pelecypod larvae have been reported in the literature. de Block and Geelen (1958) found that in the field *Mytilus edulis* larvae attached more readily to filamentous substrates than to others. A preference of *M. edulis* larvae to attach to filamentous algae, *Ceramium rubrum* and *Polysiphonia* sp., and to embroidery silk was subsequently observed in the laboratory (Bayne, 1965). When filamentous substrates were withheld from the larvae, the pediveliger stage was prolonged. The larvae continued alternate swimming and crawling, and the majority eventually died. The larvae of *Argopecten (= Aequipecten) irradians* are reported to attach to eel grass and also to a variety of solid surfaces (Belding, 1910; Sastry, 1961, 1965). Allen (1961) found that scattering a few sand grains on the bottom of culture containers encouraged settlement of larvae of *Pandora inaequivalvis* and also reduced mortality at metamorphosis. With the exception of Bayne's studies (1963, 1964a,b,

1965) on the larvae of *M. edulis,* the factors influencing the settling behavior of pelecypods have not been examined in detail (see also Section 6.4).

5.4.3.3 METAMORPHOSIS

Larval metamorphosis begins within a few hours to several days after the pediveliger stage is reached. The growth of larvae preceding the pediveliger stage is affected by a number of factors, but it apparently does not influence the time at which the larvae begin metamorphosis. Attachment of the pediveliger and the beginning of metamorphosis occur as a reaction to definite ecological stimuli, although a certain level of organization is a requisite (Ansell, 1962; Loosanoff and Davis, 1963; Bayne, 1965). The time required from fertilization to metamorphosis of *Mercenaria mercenaria* larvae cultured at different temperatures has been determined by Loosanoff *et al.* (1951). Larvae cultured at 30°C metamorphosed between days 7 and 14 after fertilization, at 24°C between days 11 and 22, and at 18°C between days 16 and 30. Loosanoff (1959) has shown that, although larvae grown at different temperatures require different lengths of time to initiate metamorphosis, the size at which larvae begin metamorphosis is not significantly different. Apparently the larvae have to reach a certain size before metamorphosis can occur. The larvae of *M. mercenaria* usually begin metamorphosis between 210 and 220 μm in size, but larvae as small as 175 μm have also been observed to begin metamorphosis (Loosanoff and Davis, 1963). The initiation of metamorphosis is also believed not to be related to the size of the pediveliger larvae, since smaller larvae were observed to begin metamorphosis while larger larvae were still swimming.

If metamorphosis of the pediveliger is delayed, the velum gradually degenerates, the foot grows, and gill filaments form (Bayne, 1965; see also Fig. 43). Early pediveligers with a fully developed velum are capable of swimming and feeding, but advanced stages with a degenerating velum stop feeding, swim weakly, and often crawl on the bottom with the foot. Temperature and salinity affect the ability to delay metamorphosis, but nutritional level does not. Growth gradually declines to zero in pediveligers which delay metamorphosis but, if a suitable substratum is offered, the larvae metamorphose and normal growth is then resumed. In *Mytilus edulis,* if a suitable site is not available for settlement, the larvae maintain a level of organization for a considerable period and are capable of attachment, metamorphosis, and further development soon after a suitable substratum is encountered. As metamorphosis is delayed, the threshold stimuli necessary to induce it decrease and the advanced stages become less discriminatory in substratum preference. Ecologically, the ability of larval marine inverte-brates to delay metamorphosis is significant in that it increases the chances for selection of a suitable substratum for attachment, hence contributes to the success of the population (Thorson, 1946; Wilson, 1960).

5.4.4 Juvenile Development

5.4.4.1 POSTLARVAL DEVELOPMENT

The Postlarval development in pelecypods is characterized principally by differentiation and growth of various organs from larval primordia. Aspects of postlarval organogenesis are described for *Venerupis pullastra* (Quayle, 1952), *Cerastoderma (= Cardium) edule* (Creek, 1960), *Pandora inaequivalvis* (Allen, 1961), *Venus striatula* (Ansell, 1962), and *Argo-pecten (= Aequipecten) irradians* (Belding, 1910; Sastry, 1965). A de-tailed discussion of organogenesis in pelecypods is given by Raven (1958). Immediately after metamorphosis, the postlarval shell (disso-conch) is secreted, and it is clearly demarcated from the larval (prodis-sonch) shell by its structure and composition. The postlarval shell attains the shape of the adult shell by differential growth. Sastry (1965) described the changes in shell morphology of postlarval stages of *Argo-pecten (=Aequipecten) irradians concentricus*. The postlarval and adult shell is secreted by the mantle edge. Immediately after metamorphosis, the mantle consists of two folds. Later the outer fold becomes divided into the outer and middle marginal folds of the adult (Ansell, 1962). The outer surface of the outer fold of the mantle secretes the calcareous layer, and the inner surface the periostracum. According to Ansell (1962) the middle fold of the mantle edge of the adult is later duplicated and takes part in the formation of siphons in *V. striatula*. An excellent dis-cussion of shell formation and regeneration in molluscs is given by Wilbur (1964) and Clark (1976).

The formation of gills in pelecypods involves varying degrees of complexity. The simplest type of gill formation is found in protobranchs (e.g., *Nucula delphinodonta* and *Yoldia limatula*), where the gills arise from a thickening of the posterior mantle edge and grow in an anterior direction. Gill formation in these species is discussed in detail by Drew (1899, 1901) and Raven (1958). In other pelecypods, gill formation differs in two fundamental ways. In the first type, rows of papillae are formed on either side of the foot in an anterior-posterior sequence. The papillae grow out to form the gill filaments which unite secondarily with the inner gills. The outer gills arise later from a similar row of papillae. This type of gill formation is found in *Mytilus, Arca, Pecten, Ostrea, Modiola, Anomia*, and other genera. In the second type, the gills are formed as a fold arising from behind and forward on both sides of the foot in which perforations occur secondarily. This type of gill formation occurs in *Teredo* and *Zirphea*.

The apical region of the velum becomes attached to the dorsal region of the mouth after metamorphosis. This contributes to the labial palps. A detailed account of labial palp formation in *Venerupis pullastra* is given by Quayle (1952).

The cerebral, pedal, and visceroparietal ganglia arise from independent pairs of ectodermal thickenings which split off from the epithelium and then connect by commissures and connectives. A brief description of the differentiation of the nervous system in *Venus striatula* is given by Ansell (1962). Further details on the differentiation of the nervous system in pelecypods can be found in Raven (1958).

The larval eyes, which persist in adults, are situated at the base of the first inner gill filaments immediately above the cerebral ganglion (Raven, 1958). In addition, some bivalves develop eyes of various types on the mantle edge. These attain a high degree of differentiation in *Pecten, Spondylus, Chlamys*, and *Amussium* (Dakin, 1928; Sastry, 1965). Statocysts, which develop during the veliger stage on the side of the foot primordium, are retained; they lie close to the pedal ganglia or cerebropedal connectives. The development of tentacles on the mantle edge of *Argopecten (= Aequipecten) irradians concentricus* is described by Sastry (1965).

Upon loss of the velum at metamorphosis, the digestive system, the organs of the visceral mass, and the foot shift in a clockwise direction, resulting in reorientation of the digestive system. The digestive gland enlarges, with the style sac now well developed. Development of the digestive system and associated structures is described for *Venus striatula* (Ansell, 1962) and *Cerastoderma (=Cardium) edule* (Creek, 1960).

At metamorphosis, the larval musculature degenerates, with the exception of the adductor muscles. Other adult musculature develops independently of larval musculature from scattered mesenchymal cells which arrange into strands and then become fibrillar and differentiate into muscle cells (Raven, 1958). The anterior adductor muscle, which appears first in the early veliger stage, comes to lie close to the mouth. The posterior adductor muscle, which develops in the late veliger stage, is formed by an aggregation of spindle-shaped mesenchymal cells lying behind the visceral ganglion and above the hindgut (Wada, 1968). Both anterior and posterior adductor muscles are developed in monomyarians during the larval period, but the anterior adductor degenerates later in the adults, while only the posterior adductor develops. The posterior and anterior retractor muscles develop from paired groups of myocytes. Finally, transverse musculature is formed in the foot. The development of heart, pericardium, and kidneys is described in detail for *V. pullastra* by Quayle (1952). In pelecypods, the stage at which the gonad primordium becomes recognizable is variable, depending on differentiation of the primordial germ cells. In *Cerastoderma (= Cardium) edule* and *Argopecten irradians concentricus*, the gonads differentiate in the late juvenile stage (Creek, 1960; Sastry, 1965). The origin and development of the primary gonads and germ cells are discussed in Section 5.3.4.

5.4.4.2 DIRECT DEVELOPMENT

In species with direct development, miniature adults are released, whether the embryos develop within the mantle, gill lamellae, or egg capsules (see Section 5.4.2.1). Development initially follows the typical course for most other lamellibranchs, and after gastrulation the stomodeum, shell gland, and foot rudiment are formed as in normal trochophores. The prototroch of the trochophore is reduced and occupies only a small area in the neighborhood of the mouth. The velum is absent in the developmental stages of *Lasaea rubra* (Oldfield, 1964) and *Transennella tantilla* (Hansen, 1953), two species with direct development. Except for this and a few other minor differences that accompany the development of large, yolky eggs, the remainder of the development is similar in many respects to that of other pelecypods (Oldfield, 1964).

5.4.4.3 BROOD PROTECTION

5.4.4.3.1 *Types of Broods.* Brood protection occurs in a number of species in combination with any of the three developmental types. The degree of parental care exhibited by different species is variable (Table XIV).

Among incubatory species, it is presumed that spermatozoa released into the water by the male are drawn into the inhalant water by the female and that the eggs are fertilized in the demibranchial or epibranchial chamber (Oldfield, 1964; Sellmer, 1967). The eggs of some specialized species, *Arca vivipara, Philobrya, Lasaea, Entovalva, Teredo,* and *Pseudokellia* (Lankester, 1900), *Kellia, Lepton,* and *Montacuta* (Pelseneer, 1935), *Galeomma* (Lebour, 1938b), *Transennella tantilla* (Hansen, 1953), and *Turtonia minuta* (Oldfield, 1964), are fertilized within the suprabranchial chamber or within the gills.

The fertilized eggs continue development in the pallial cavity of *Ostrea, Entovalva,* and some species of *Teredo* (Pelseneer, 1935). In

TABLE XIV

Brood Protection in Some Pelecyopods

Species	Notes on brood protection	Reference
Arca vivipara	Incubatory	Lankester (1900)
Astarte borealis	Eggs in mucous and adhesive membrane attached to substratum	Thorson (1946)
Astarte elliptica	Eggs in mucous and adhesive membrane attached to substratum	Ockelmann (1958)
Astarte sulcata	Very short or no pelagic development, adhesive eggs	Ockelmann (1958)
Cardita bailyi	Incubates young in interlamellar spaces of the inner and outer demibranches; released as miniature adults	Jones (1963)
Cardita barbarensis	Incubates young in interlamellar spaces of the inner and outer demibranches; released as miniature adults	Jones (1963)
Cardita ventricosa	Incubates young in interlamellar spaces between inner and outer demibranches; released as miniature adults; very young specimens attached to the parent by byssus	Jones (1963)
Cerastoderma (= Cardium) elegantulum	Eggs develop in brood pouches formed by folds of the ventral part of the mantle; young hatch as fully developed bottom stage	Matveeva (1953); Ockelmann (1958)
Cerastoderma (= Cardium) exigium	Eggs in gelatinous envelope attached to substratum	Thorson (1946)
Entovalva sp.	Incubates young in the pallial cavity	Lankester (1900)
Galeomma turtoni	Released as shell-less larvae, presumably as trochophores	Lebour (1938b)
Gemma gemma	Incubates embryos in the demibranches; released as miniature adults	Sellmer (1967)
Kellia suborbicularis	Eggs develop in the gills; released as veliger larvae	Lebour (1938b); Howard (1953); Oldfield (1964)
Lasaea rubra	Incubates embryos until they are released as miniature adults	Oldfield (1964)
Lepton parasiticum	Incubates young in mantle cavity; released as miniature adults	Pelseneer (1935)
Loripes lacteus	Eggs develop in a protective capsule	Miyazaki (1938)
Lyrodus massa	Embryos developed within gills, released as straight-hinge veligers	Turner and Johnson (1971)
Lyrodus pedicellatus	Embryos develop within gills, released as pediveligers	Turner and Johnson (1971)
Modiolaria discors	Egg strings in the nest of the parent; embryos develop without any pelagic larval stage	Ockelmann (1958)

TABLE XIV *(Continued)*

Species	Notes on brood protection	Reference
Modiolaria nigra	Spawn egg strings attached to the byssus of the parent	Thorson (1946); Ockelmann (1958)
Montacuta ferruginosa	Incubates embryos in demibranches; released as well-developed veliger larvae	Oldfield (1964)
Montacuta percompressa	Incubates in the suprabranchial chamber; released as veligers	Chanley and Chanley (1970)
Montacuta phascolionis	Embryos develop in the demibranches; release young as miniature adults	Deroux (1960, 1961)
Montacuta substriata	Embryos develop in the demibranches; released as late veligers	Oldfield (1964)
Mysella bidentata	Released as larvae	Lebour (1938b); Howard (1953)
Nucula delphinodonta	20–70 eggs develop in a gelatinous sac attached to posterior end of valves of the parent; nonpelagic development for about 3–4 weeks; release young as miniature adults	Drew (1901)
Ostrea sp.	Embryos develop in the pallial cavity; released as free-swimming larvae	Lankester (1900); see also this volume (Chapter 6)
Philobrya setosa	Incubatory	Howard (1953)
Pseudokellia	Incubates embryos in interlamellar spaces	Lankester (1900)
Teredo bartschi	Embryos develop within gills; released as pediveligers	Turner and Johnson (1971)
Teredo clappi	Embryos develop within gills; released as pediveligers	Turner and Johnson (1971)
Teredo diegensis	Embryos develop in a brood pouch	Hill and Kofoid (1927)
Teredo furcifera	Embryos develop within gills; released as pediveligers	Turner and Johnson (1971)
Teredo navalis	Embryos develop within gills; brood pouch contains about 500,000–1,000,000 embryos; released as straight-hinge veligers	Hill and Kofoid (1927); Turner and Johnson (1971)
Teredo parksi	Embryos develop in a brood pouch	Hill and Kofoid (1927)
Transennella tantilla	Embryos develop in a brood chamber situated between inner gill and body wall, young pass directly into a bottom stage	Hansen (1953)
Turtonia minuta	Eggs develop in a protective capsule outside the parent; capsules attached to byssus of the female parent; each capsule contains 8–19 embryos	Ockelmann (1964); Oldfield (1964)

(continued)

TABLE XIV *(Continued)*

Species	Notes on brood protection	Reference
Xylophaga africana	No pelagic stage assumed, juveniles attach by means of byssus to dorsal part of shell of the parent	Knudsen (1961)
Xylophaga atlantica	Although reported to brood by Turner (1954) on the basis of one specimen, no other brooding individuals have been found	Turner and Johnson (1971)
Xylophaga bruuni	No pelagic stage assumed; juveniles attached to posteriodorsal part of shell of the parent	Knudsen (1961)
Xylophaga concava	No pelagic stage assumed; juveniles attached to external surface of right side of mantle of the parent	Knudsen (1961)
Xylophaga lobata	No pelagic stage assumed; juveniles attached to posteriodorsal part of shell of the parent	Knudsen (1961)
Xylophaga panamensis	No pelagic stage assumed; juveniles attached to posteriodorsal part of shell of the parent	Knudsen (1961)
Xylophaga tubulata	No pelagic stage assumed; juveniles are found in a depression of the mantle tissue of the parent	Knudsen (1961)
Xylophaga wolffi	No pelagic stage assumed; juveniles attached to mantle edge of the parent	Knudsen (1961)

Montacuta percompressa, the embryos are incubated in a brood pouch formed by a posterior extension of the mantle (Chanley and Chanley, 1970).

Incubation may occur in the interlamellar spaces of both inner and outer demibranchs, or in only one. In *Transennella tantilla*(Hansen, 1953) and *Lasaea* and *Pseudokellia* (Oldfield, 1964) the embryos develop in a brood chamber situated between the inner gill and the body wall (Fig. 52). In *Cardita ventricosa, C. bailyi,* and *C. barbarensis* (Jones, 1963) and *Gemma gemma* (Sellmer, 1967) the embryos are incubated in the interlamellar spaces between the inner and outer demibranchs. In *Cardita* spp. maternal care may extend to protection of the juveniles, since the young are attached to the adult by byssus threads (Jones, 1963). Brood protection has been reported in some species of bathyal and abyssal *Xylophaga* by Knudsen (1961). The juveniles are attached by means of a single byssus thread to the siphon or the posterior part of the umbo of the parent (Fig. 52).

In certain species, external incubation seems to occur within gelati-

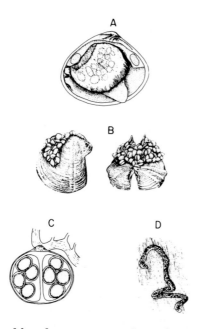

FIG. 52. Some types of brood protection in pelecypods. (A) *Transennella tantilla*. The brood is protected in the space between the inner gill and the body wall; all stages of development from newly laid eggs to young ready to leave the mother occur in the same individual. (After Hansen, 1953; modified.) (B) *Xylophaga panamensis*. About 50 young are attached to the posteriodorsal part of the shell of the mother. (After Knudsen, 1961; modified.) (C) *Turtonia minuta*. The eggs are protected by a capsule attached to the byssus of the female; the young emerge from the capsule as miniature adults. (After Oldfield, 1964; modified.) (D) *Musculus niger* spawns egg strings and attaches them to a substratum; the young hatch directly as bottom stages. (After Thorson, 1935; modified.)

nous egg masses or strings and in egg capsules (Fig. 52). *Musculus discor* and *M. niger* deposit egg strings from which the young hatch directly as bottom forms (Thorson, 1935; Miyazaki, 1938). Ockelmann (1958) indicated that eggs were produced with a mucous and adhesive membrane which may attach to the substratum in *Astarte borealis, A. montagui, A. crenata, A. elliptica; Macoma moesta, M. loveni,* and *M. torelli.* Larval development of the arctic bivalve *Thyasira gouldi* has been studied in detail by Blacknell and Ansell (1974). This species produces large, yolky eggs which sink to the bottom and attach to the sediments with their sticky membrane. The outer gelatinous coat of the egg develops a strong, protective capsule after fertilization. Development is direct and takes place within the capsule, and the young hatch as benthic juveniles. Hatching seems to take place by means of an enzyme secreted by the mantle. Development of all the stages takes place slowly, and growth after hatching is slow. The development period and time of hatching are influenced by salinity, but the time for hatching appears to be related to the amount of stored reserves remaining.

Embryos are incubated in protective egg capsules in *Nucula delphino-donta* (Drew, 1901), *Loripes lacteus* (Miyazaki, 1938), and *Turtonia minuta* (Oldfield, 1964; Ockelmann, 1964). Oldfield (1964) has found that, in adults of *T. minuta*, the mantle edge of the female undergoes seasonal changes corresponding to the reproductive cycle. Before the onset of spawning, the anterior ventral region of the mantle thickens, owing to the development of epithelial cells of the subepithelial gland. After the reproductive season, it is again reduced and appears similar to the mantle edge in the male. According to Oldfield (1964) the modified mantle edge secretes the capsule in which the fertilized eggs develop. From one to four egg capsules are attached to the byssus of the female.

The duration of incubation and the developmental stage at which the brood is released vary in different species. The incubation period can last from a few hours to the entire length of the larval period. In *Ostrea* (Pelseneer, 1935) and *Galeomma* (Lebour, 1938b) the embryos are released as early free-swimming larvae. In species with planktotrophic larval development, brood protection occurs in some species of *Ostrea, Mysella, Montacuta, Kellia,* and *Teredo,* and the larvae are released at about the veliger stage (Ockelmann, 1962). In *Montacuta percompressa,* the brood is released 5–7 days after fertilization as straight-hinge veligers (Chanley and Chanley, 1970). Miniature adults are released in *Musculus discor* and *M. niger* (Thorson, 1935), *Transennella tantilla* (Hansen, 1953), *Cardita* spp. (Jones, 1963), and others (Table XIV).

Oldfield (1964) studied four genera of the families Erycinidae and Montacutidae and found a series from the production of small eggs with little yolk in *Kellia* and *Montacuta* to the production of a few very large eggs in *Turtonia* and *Lasaea.* In these genera there appears to be a relationship between the size of the eggs produced and the duration that the brood is protected. *Kellia suborbicularis, Montacuta ferruginosa,* and *M. substriata* release bivalved veligers, whereas the life history of *Turtonia minuta* and *Lasaea rubra* is atypical in that the free-swimming veliger stage is suppressed and the young are released as miniature adults. The extended incubation period in *L. rubra* may be an adaptation to the prolonged exposure to air that individuals of this intertidal species experience during the spring tides.

5.4.4.3.2 *Relationship between Parent and the Brood.* The relationship between brooding young and the parent animals has not been examined

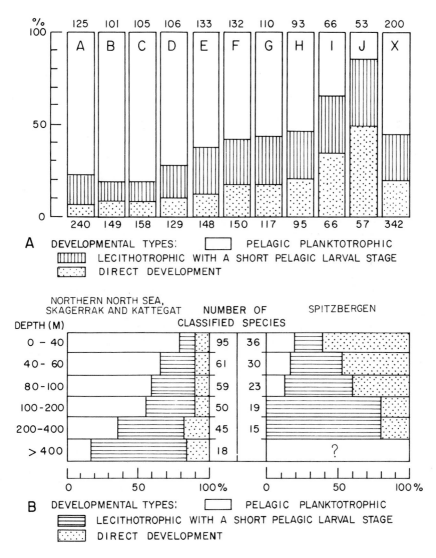

FIG. 53. (A) Percentage of bivalve species, classified according to main type of development, in different shelf regions from Gibraltar to Murmansk, and Spitsbergen on the European west coast. Figures above the graph represent the number of species classified; figures below the graph represent the number of species recorded from each region down to 300 m. Regions: A, Gibraltar to Arcachon; B, Brest to St. Malo; C, English and French Channel coasts; D, southern North Sea including Dogger Bank; E, northern North Sea, Kattegat, Skagerrak to Stavanger; F, north of Stavanger, Bergen and Trondhjem area; G, Loften; H, West Finmark; I, East Finmark, Murman coast; J, Spitsbergen; X, all regions combined. (B) Percentage of bivalve species, classified according to the main type of development, at different depths in a boreal (left) and in an arctic (right) region. (From Ockelmann, 1962; modified.)

in detail but, according to Pelseneer (1935), the embryos of *Anodonta, Ostrea edulis,* and *Montacuta ferruginosa* develop after removal from the parent. Recently fertilized eggs and embryos removed from the gill lamellae of *Teredo navalis* develop normally to metamorphosis when cultured in the laboratory (Loosanoff and Davis, 1963). Late-stage embryos of *Gemma gemma* are able to develop *in vitro* to maturity (Sellmer ,1967). In contrast, the aborted embryos of *Montacuta percompressa* do not develop normally and eventually die (Chanley and Chanley, 1970).

Apparently, brooding young of *Transennella tantilla* do not receive nutrients from the parent while they are in the brood chamber. The gills of the parent are not modified for nutrient transfer, and the brooding embryos are completely surrounded by an egg membrane that does not break until they are released (Hansen, 1953).

5.4.4.3.3 *Climatic Variation and Developmental Types.* Species with planktotrophic larvae predominate in most of the European region according to Ockelmann (1962), who observed pelecypods between Gibraltar and Murmansk on the European west coast. However, proceeding from south to north along the European west coast, species with lecithotrophic larvae in particular, as well as those with direct development, become more common. In the arctic region near Spitsbergen, Norway, few species have planktotrophic larvae, whereas almost half of the total number of species examined have direct development and one-third have lecithotrophic development (Fig. 53).

According to Ockelmann (1962) species with lecithotrophic larvae become more common with increasing depth in both arctic and boreal regions (Fig. 53). In the arctic region, species with direct development predominate only at depths less than 100 m, replacing boreal species with planktotrophic larval development. In both arctic and boreal regions, there are only a small number of species with direct development in deeper waters. Lecithotrophic larval development with a short pelagic life is most common in the deep sea. Direct development and brood protection occur in only a small number of species in the deep sea and appear to be rare in abyssal bivalves.

Acknowledgments

The original work of the author referred to in this chapter has been supported partly by grants from the National Science Foundation (GB-1356) and the Office of Sea Grants, NOAA. The author wishes to thank Miss L. Armstrong for preparation of the figures, Miss M. Kohl for bibliographic research, and Mrs. J. Glass for typing the manuscript. The author also thanks Drs. Don C. Miller and David M. Pratt for reading an earlier version of this manuscript.

References

Abraham, K. C. (1953). Observations on the biology of *Meretrix casta* (Chemitz). *J. Zool. Soc. India* 5, 163–190.

Ahmed, M., and Sparks, A. K. (1970). Chromosome number, structure and autosomal polymorphism in the marine mussels *Mytilus edulis* and *Mytilus californianus*. *Biol. Bull. (Woods Hole, Mass.)* 138, 1–13.

Aiello, E. L. (1957). The influence of the branchial nerve and 5-hydroxytryptamine on the ciliary activity of *Mytilus* gill. *Biol. Bull. (Woods Hole, Mass.)* 113, 325.

Aiello, E. L. (1960). Factors affecting ciliary activity on the gill of the mussel, *Mytilus edulis*. *Physiol. Zool.* 33, 120–135.

Aiello, E. L. (1962). Identification of the cilio-excitatory substance present in the gill of the mussel *Mytilus edulis*. *J. Cell. Comp. Physiol.* 60, 17–21.

Aiello, E. L. (1965). Pharmacology of cilio-excitation in the mussel. *Fed. Proc. Fed. Am. Soc. Exp. Biol.* 24, 758.

Aiello, E. L. (1970). Nervous and chemical stimulation of the gill cilia in bivalve molluscs. *Physiol. Zool.* 43, 60–70.

Alagarswami, K. (1966). Studies of some aspects of biology of the wedge-clam, *Donax faba* Gmelin from the Mandapam coast in the Gulf of Mannar. *J. Mar. Biol. Assoc. India* 8, 56–75.

Allen, J. A. (1961). The development of *Pandora inaequivalvis* (Linne). *J. Embryol. Exp. Morphol.* 9, 252–268.

Allen, J. A. (1962). Preliminary experiments on the feeding and excretion of bivalves using *Phaeodactylum* labelled with [32]P. *J. Mar. Biol. Assoc. U.K.* 42, 609–623.

Allen, J. A. (1963). Ecology and functional morphology of molluscs. *Oceanogr. Mar. Biol,* 1, 253–288.

Allen, J. A. (1969). Observations on size, composition and breeding of Northumberland populations of *Zirphaea crispata* (Pholadidae: Bivalvia). *Mar. Biol.* 3, 269–275.

Allen, J. A. (1970). Experiments on the uptake of radioactive phosphorus by bivalves and its subsequent distribution within the body. *Comp. Biochem. Physiol.* 36, 131–141.

Allen, J. F. (1962a). Gonad development and spawning of *Brachidontes recurvus* in Chesapeake Bay. *Nautilus* 75, 149–156.

Allen, J. F. (1962b). Gonad development and spawning of *Brachidontes recurvus* in Chesapeake Bay. *Nautilus* 76, 9–16.

Allen, R. D. (1953). Fertilization and artificial activation in the egg of the surf-clam, *Spisula solidissima*. *Biol. Bull. (Woods Hole, Mass.)* 105, 213–239.

Altman, G. (1959). Atmungsphyiologische untersuchungen *Mytilus galloprovincialis*. *Experientia* 15, 192.

Amirthalingam, C. (1928). On lunar periodicity in the reproduction of *Pecten opercularis* near Plymouth in 1927–1928. *J. Mar. Biol. Assoc. U.K.* 15, 605–641.

Anderson, W. A., and Personne, P. (1976). The molluscan spermatozoan: Dynamic aspects of its structure and function. *Am. Zool.* **16**, 293–313.

Ansell, A. D. (1961a). The development of the primary gonad in *Venus striatula* (Da Costa). *Proc. Malacol. Soc. London* **34**, 243–247.

Ansell, A. D. (1961b). Reproduction, growth and mortality of *Venus striatula* (Da Costa) in Kames Bay, Millport. *J. Mar. Biol. Assoc. U.K.* **41**, 191–215.

Ansell, A. D. (1962). The functional morphology of the larva and the postlarval development of *Venus striatula* (Da Costa). *J. Mar. Biol. Assoc. U.K.*, **42**, 419–443.

Ansell, A. D. (1969). Behavioral adaptations of intertidal molluscs from a tropical sandy beach. *J. Exp. Mar. Biol. Ecol.* **4**, 9–35.

Ansell, A. D. (1972). Distribution, growth and seasonal changes in the biochemical composition of the bivalve *Donax vittatus* (Da Costa) from Kames Bay, Millport. *J. Exp. Mar. Biol. Ecol.* **10**, 137–150.

Ansell, A. D. (1974a). Seasonal changes in the biochemical composition of the bivalve *Abra alba* from the Clyde Sea area. *Mar. Biol.* **25**, 13–20.

Ansell, A. D. (1974b). Seasonal changes in the biochemical composition of the bivalve *Chlamys septemradiata* from the Clyde Sea area. *Mar. Biol.* **25**, 85–99.

Ansell, A. D. (1974c). Seasonal changes in the biochemical composition of the bivalve *Nucula sulcata* from the Clyde Sea area. *Mar. Biol.* **25**, 101–108.

Ansell, A. D., and Lander, K. F. (1967). Studies on the hard-shell clam, *Venus mercenaria*, in British waters. III. Further observations on the seasonal biochemical cycle and on spawning. *J. Appl. Ecol.* **4**, 425–435.

Ansell, A. D., and Trevallion, A. (1967). Studies on *Tellina tenuis* Da Costa. I. Seasonal growth and biochemical cycle. *J. Exp. Mar. Biol. Ecol.* **1**, 220–235.

Ansell, A. D., Loosmore, F. A., and Lander, K. F. (1964). Studies on the hard-shell clam, *Venus mercenaria*, in British waters. II. Seasonal cycle in condition and biochemical composition. *J. Appl. Ecol.* **1**, 83–95.

Antheunisse, L. J. (1963). Neurosecretory phenomena in the zebra mussel *Dreissena polymorpha* Pallas. *Arch. Neerl. Zool.* **15**, 237–314.

Antony Raja, B. T. (1963). Observations on the rate of growth, sexual maturity and breeding of four sedentary organisms from the Madras harbour. *J. Mar. Biol. Assoc. India* **5**, 113–132.

Ashikaga, C. (1948). Biochemical studies on the pearl oyster, *Pinctada martensii*. I. The seasonal variation of the chemical composition, especially of glycogen constituents. *Seiri Seitai* **2**, 160–167.

Awati, P. R., and Rai, H. S. (1931). *Ostrea cucullata* (the Bombay oyster). *Indian Zool. Mem.* **3**, 1–107.

Bacci, G. (1965). Sex determination. *Int. Ser. Monogr. Pure Appl. Biol.*, Div. Zool. **26**, 1–306.

Baier, C. R. (1935). Studien zur Hydro-backteriologie stehenden Binnengerwasser. *Arch. Hydrobiol.* **29**, 183–264.

Baird, R. H. (1958). On the swimming behavior of escallops (*Pecten maximus* L.). *Proc. Malacol. Soc. London* **33**, 67–71.

Baird, R. H. (1966). Notes on an escallop (*Pecten maximus*) population in Holyhead Harbour. *J. Mar. Biol. Assoc. U.K.* **46**, 33–47.

Baird, R. H., and Gibson, F. A. (1956). Underwater observations on an escallop (*Pecten maximus*, L.) bed. *J. Mar. Biol. Assoc. U.K.* **35**, 555–562.

Balasubramanyan, R. (1970). Studies on the pholadid marine wood-borer *Martesia striata* (Linn.) *Proc. Symp. Mollusca, Cochin, 1968, Mar. Biol. Assoc. India, Symp. Ser.* **3**, Part 3, 707–711.

Ballantine, D., and Morton, J. E. (1956). Filtering, feeding and digestion in the lamellibranch *Lasaea rubra*. *J. Mar. Biol. Assoc. U.K.* **35**, 241–274.

Bargeton-Couteaux, M. (1942). Les variations saisonnières du tissu conjonctif vesiculeux chez l'huître. *Bull. Biol. Fr. Belg.* **76**, 175–191.

Battle, H. (1932). Rhythmic sexual maturity and spawning of certain bivalve molluscs. *Contrib. Can. Biol. Fish.* **7**, 255–276.

Bayne, B. L. (1963). Response of *Mytilus edulis* larvae to increases of hydrostatic pressure. *Nature (London)* **198**, 406–407.

Bayne, B. L. (1964a). The response of larvae of *Mytilus edulis* L. to light and gravity. *Oikos* **15**, 162–174.

Bayne, B. L. (1964b). Primary and secondary settlement in *Mytilus edulis* L. *J. Anim. Ecol.* **33**, 513–523.

Bayne, B. L. (1965). Growth and the delay of metamorphosis of the larvae of *Mytilus edulis* (L.) *Ophelia* **2**, 1–47.

Bayne, B. L. (1972). Some effects of stress in the adult on the larval development of *Mytilus edulis*. *Nature (London)* **237**, 459.

Bayne, B. L. (1975). Reproduction of bivalve molluscs under environmental stress. *In* "Physiological Ecology of Estuarine Organisms" (F. J. Vernberg, ed.), pp. 259–277. Univ. of South Carolina Press, Columbia.

Bayne, B. L. (1976a). Aspects of reproduction in bivalve molluscs. *In* "Estuarine Processes, Vol. 1: Uses, Stresses and Adaptation to the Estuary" (M. Wiley, ed.), pp. 432–448. Academic Press, New York.

Bayne, B. L., ed. (1976b). "Marine Mussels: Their Ecology and Physiology," International Biological Program, 10. Cambridge Univ. Press, London and New York.

Bayne, B. L. (1976c). The biology of mussel larvae. *In* "Marine Mussels: Their Ecology and Physiology" (B. L. Bayne, ed.), pp. 81–120. International Biological Program, No. 10. Cambridge Univ. Press, London and New York.

Bayne, B. L., and Thompson, R. J. (1970). Some physiological consequences of keeping *Mytilus edulis* in the laboratory. *Helgol. Wiss. Meeresunters.* **20**, 526–552.

Bayne, B. L., Gabbott, P. A., and Widdows, J. (1975). Some effects of stress in the adult on the eggs and larvae of *Mytilus edulis* L. *J. Mar. Biol. Assoc. U.K.* **55**, 675–690.

Beedham, G. E. (1958). Observations on the mantle of the Lamellibranchia. *Q. J. Microsc. Sci.* **99**, 181–197.

Belding, D. L. (1910). A report upon the scallop fishery of Massachusetts including the habitats, life history of *Pecten irradians,* its rate of growth and other factors of economic value. *Spec. Rep., Comm. Fish Game, Mass.* pp. 1–150.

Belding, D. L. (1930). The soft-shelled clam fishery of Massachusetts. *Mar. Fish. Ser., Div. Fish Game Mass.* No. 1.

Berg, W. E. (1949). Some effects of sperm extracts on the eggs of *Mytilus. Am. Nat.* **83**, 221–226.

Berg, W. E. (1950). Lytic effects of sperm extracts on the eggs of *Mytilus edulis. Biol. Bull. (Woods Hole, Mass.)* **98**, 128–138.

Berg, W. E., and Kutsky, P. B. (1951). Physiological studies of differentiation in *Mytilus edulis.* I. The oxygen uptake of isolated blastomeres and polar lobes. *Biol. Bull. (Woods Hole, Mass.)* **101**, 47–61.

Berner, L. (1935). La reproduction des moules comestibles (*Mytilus edulis* L. et *Mytilus galloprovincialis* Lmk) et leur répartition géographique. *Bull. Inst. Oceanogr.* Monaco No. 680, 1–7.

Berner, L. (1952). Biologie de *Pinnotheres pisum* Penn. (Decapode, Brachyvoure). *Bull. Soc. Zool. Fr.* 77, 344–349.

Blacknell, W. M., and Ansell, A. D. (1974). The direct development of the bivalve *Thyasira gouldi* (Philippi). *Thalassia Jugosl.* 10, 23–43.

Blake, N. J. (1972). Environmental regulation of neurosecretion and reproductive activity in the bay scallop, *Aequipecten irradians* (Lamarck). Ph.D. Thesis, Univ. of Rhode Island, Kingston.

Blaschko, H., and Milton, S. (1960). Oxidation of 5HT and related compounds by *Mytilus edulis* gill plates. *Br. J. Pharmacol. Chemother.* 15, 42–46.

Blegvad, H. (1914). Undersog 1 ser over Naering og Ernaerings for hold hos Havbundens invertebrate Dyresamfundi danske Fraevende. *Ber. Dan. Biol. Stn.* 22, 41–78.

Bloomer, H. H. (1939). A note on the sex of *Pseudanodonta* Bourguignat and *Anodonta* Lamarck. *Proc. Malacol. Soc. London* 23, 285–297.

Bourcart, C., and Lubet, P. (1965). Cycle sexuel et évolution des réserves chez *Mytilus galloprovincialis* Lmk (Mollusque Bivalve). *Rapp. P.-V. Reun., Comm. Int. Sci. Mer Mediterr.* 18, 155–158.

Bourcart, C., Lavallard, R., and Lubet, P. (1965). Ultrastructure du spermatozoïde de la moule (*Mytilus perna* von Jhering). *C. R. Acad. Sci.* 260, 5096–5099.

Bourne, N., and Quayle, D. B. (1970). Breeding and growth of razor clams in British Columbia. *Fish. Res. Board Can. Tech. Rep.* No. 232, 1–42.

Boyden, C. R. (1971). A comparative study of the reproductive cycles of the cockles *Cerastoderma edule* and *C. glaucum*. *J. Mar. Biol. Assoc. U.K.* 51, 605–622.

Boyle, P. J., and Turner, R. D. (1976). The larval development of the wood boring piddock *Martesia striata* (L.) (Mollusca: Bivalvia: Pholadidae). *J. Exp. Mar. Biol. Ecol.* 22, 55–68.

Breese, W. P., Millemann, R. E., and Dimick, R. E. (1963). Stimulation of spawning in the mussels, *Mytilus edulis* Linnaeus and *Mytilus californianus* Conrad, by Kraft mill effluent. *Biol. Bull. (Woods Hole, Mass.)* 125, 197–205.

Brenko, M. (1971). The reproductive cycle of *Mytilus galloprovincialis* Lmk in the northern Adriatic Sea and *Mytilus edulis* L. in Long Island Sound. *Thalassia Jugosl.* 7, 533–542.

Brenko, M. (1973). Gonad development, spawning and rearing of *Mytilus edulis* larvae. *Stud. Rev. Gen. Fish. Counc. Mediterr.* 53, 53–65.

Brenko, M. (1974). Temperature and salinity requirements for embryonic development of *Mytilus galloprovincialis* Lmk. *Thalassia Jugosl.* 10, 131–138.

Brenko, M., and Calabrese, A. (1969). The combined effects of salinities and temperature on larvae of the mussel, *Mytilus edulis*. *Mar. Biol.* 4, 224–226.

Brenko, M., Claus, C., and Bubic, S. (1977). Synergistic effects of lead, salinity and temperature on embryonic development of the mussel *Mytilus galloprovincialis*. *Mar. Biol.* 44, 109–115.

Brousseau, D. J. (1978). Spawning cycle, fecundity, and recruitment in a population of soft-shell clams, *Mya arenaria*, from Cape Ann, Massachusetts. *U.S. Fish Wildl. Serv. Fish. Bull.* 76, 115–166.

Brukema, J. J., and De Brunn, W. (1977). Seasonal changes in dry weight and chemical composition of the soft parts of the tellinid bivalve *Macoma balthica* in the Dutch Wadden Sea. *Neth. J. Sea Res.* 11, 42–55.

Buley, H. M. (1936). Consumption of diatoms and dinoflagellates by the mussel. *Bull. Scripps. Inst. Oceanogr., Tech. Ser.* 4, 19–27.

Bullis, R. H., and Ingle, R. M. (1958). A new fishery of scallops of western Florida. *Gulf Caribb. Fish. Inst., Univ. Miami, Proc. 11th Annu. Sess.* pp. 75–79.

Bullock, T. H. (1955). Compensation for temperature in the metabolism and activity of poikilotherms. *Biol. Rev. Cambridge Philos. Soc.* **30**, 311–342.

Butler, P. A. (1949). Gametogenesis in the oyster under conditions of depressed salinity. *Biol. Bull. (Woods Hole, Mass.)* **96**, 263–269.

Caddy, J. F. (1967). Maturation of gametes and spawning in *Macoma balthica* (L). *Can. J. Zool.* **45**, 955–965.

Cahour, A., and Lucas, A. (1968). Sex ratio et hermaphrodisme accidentel chez *Modiolus barbatus* (L.) Mollusque Bivalve. *C. R. Acad. Sci.* **162**, 221.

Calabrese, A. (1969a). Reproductive cycle of the coot clam, *Mulinia lateralis* (Say), in Long Island Sound. *Veliger* **12**, 265–269.

Calabrese, A. (1969b). Individual and combined effects of salinity and temperature on embryos and larvae of the coot clam, *Mulinia lateralis* (Say). *Biol. Bull. (Woods Hole, Mass.)* **137**, 417–428.

Calabrese, A. (1970). The pH tolerance of embryos and larvae of the coot clam, *Mulinia lateralis* (Say). *Veliger* **13**, 122–126.

Calabrese, A., and Davis, H. C. (1966). The pH tolerance of embryos and larvae of *Mercenaria mercenaria* and *Crassostrea virginica*. *Biol. Bull. (Woods Hole, Mass.)* **131**, 427–436.

Calabrese, A., and Davis, H. C. (1967). Effects of "soft" detergents on embryos and larvae of the American oyster (*Crassostrea virginica*). *Proc. Natl. Shellfish. Assoc.* **57**, 11–16.

Calabrese, A., and Davis, H. C. (1970). Tolerance requirements of embryos and larvae of bivalve molluscs. *Helgol. Wiss. Meersunters.* **20**, 553–564.

Calabrese, A., and Nelson, D. A. (1973). Inhibition of embryonic development of the hard clam, *Mercenaria mercenaria*, by heavy metals. *Bull. Environ. Contam. Toxicol.* **11**, 92–97.

Calabrese, A., and Rhodes, E. W. (1974). Culture of *Mulinia lateralis* and *Crepidula fornicata* embryos and larvae for studies of pollution effects. *Thalassia Jugosl.* **10**, 89–102.

Campbell, S. A. (1969). Seasonal cycles in the carotenoid content in *Mytilus edulis*. *Mar. Biol.* **4**, 227–232.

Carriker, M. R. (1961). Interrelation of functional morphology, behavior and autecology in early stages of the bivalve *Mercenaria mercenaria*. *J. Elisha Mitchell Sci. Soc.* **77**, 168–242.

Castagna, M., and Duggan, W. (1971). Rearing of the bay scallop, *Aequipecten irradians*. *Proc. Natl. Shellfish. Assoc.* **61**, 86–92.

Chanley, P. E. (1955). Possible causes of growth variations in clam larvae. *Proc. Natl. Shellfish. Assoc.* **45**, 84–94.

Chanley, P. E. (1965). Larval development of a boring clam, *Barnea truncata*. *Chesapeake Sci.* **6**, 162–166.

Chanley, P. E. (1966). Larval development of the large brood clam, *Noetia ponderosa* (Say). *Proc. Natl. Shellfish. Assoc.* **56**, 53–58.

Chanley, P. E. (1969). Larval development of the coquina clam, *Donax variabilis* Say, with a discussion of the structure of the larval hinge in the Tellinacea. *Bull. Mar. Sci.* **19**, 214–224.

Chanley, P. E., and Chanley, M. H. (1970). Larval development of the commensal clam, *Montacuta percompressa* Dall. *Proc. Malacol. Soc. London* **39**, 59–67.

Cheng, T. C. (1967). Marine molluscs as hosts for symbioses with a review of known parasites of commercially important species. *Adv. Mar. Biol.* **5**, 1–424.

Chipperfield, P. (1953). Observations on the breeding and settlement of *Mytilus edulis* (L.) in British waters. *J. Mar. Biol. Assoc. U.K.* **32**, 449–476.

Clark, G. R. (1976). Shell growth in the marine environment: Approaches to the problem of marginal calcification. *Am. Zool.* **16**, 617–626.

Coe, W. R. (1932). Development of the gonads and the sequence of the sexual phases in the California oyster (*Ostrea lurida*). *Bull. Scripps Inst. Oceanogr., Tech. Ser.* **3**, 119–144.

Coe, W. R. (1936). Sequence of functional sexual phases in *Teredo*. *Biol. Bull. (Woods Hole, Mass.)* **71**, 122–132.

Coe, W. R. (1943a). Sexual differentiation in mollusks. I. Pelecypods. *Q. Rev. Biol.* **18**, 154–164.

Coe, W. R. (1943b). Development of the primary gonads and differentiation of sexuality in *Teredo navalis* and other pelecypod mollusks. *Biol. Bull. (Woods Hole, Mass.)* **84**, 178–186.

Coe, W. R. (1945). Development of the reproductive system and variations in sexuality in *Pecten* and other pelecypod molluscs. *Trans. Conn. Acad. Sci.* **36**, 673–700.

Coe, W. R. (1948). Nutrition, environmental conditions and growth of marine bivalve molluscs. *J. Mar. Res.* **7**, 586–601.

Coe, W. R. (1955). Ecology of the bean clam, *Donax gouldii*, on the coast of California. *Ecology* **36**, 512–514.

Coe, W. R., and Fox, D. L. (1944). Biology of the California sea mussel (*Mytilus californianus*). III. Environmental conditions and rate of growth. *Biol. Bull. (Woods Hole, Mass.)* **87**, 59–72.

Coe, W. R., and Turner, H. J. (1938). Development of the gonads and gametes in the soft-shell clam (*Mya arenaria*). *J. Morphol.* **62**, 91–111.

Cole, H. A. (1937). Metamorphosis of the larva of *Ostrea edulis*. *Nature (London)* **139**, 413.

Cole, H. A. (1938). The fate of the larval organs in the metamorphosis of *Ostrea edulis*. *J. Mar. Biol. Assoc. U.K.* **22**, 469–484.

Collier, J. R. (1965). Morphogenetic significance of biochemical patterns in mosaic embryos. *In* "Biochemistry of Animal Development, Vol. 1: Descriptive Biochemistry of Animal Development" (R. Weber, ed.), pp. 203–244. Academic Press, New York.

Comely, C. A. (1974). Seasonal variation in the flesh weights and biochemical content of the scallop *Pecten maximus* (L.) in the Clyde Sea area. *J. Cons., Cons Perm. Int. Explor. Mer.* **35**, 281–295.

Conklin, E. G. (1897). The embryology of *Crepidula*. A contribution to the cell-lineage and early development of some gastropods. *J. Morphol.* **13**, 1–226.

Costello, D. P., and Henley, C. (1976). Spiralian development: A perspective. *Am. Zool.* **16**, 277–291.

Costello, D. P., Davidson, M. E., Eggers, A., Fox, H. M., and Henley, C. (1957). "Methods for Obtaining Marine Eggs and Embryos," Marine Biol. Lab., Woods Hole, Mass. Lancaster Press, Lancaster, Pennsylvania.

Cowden, R. R. (1976). Cytochemistry of oogenesis and early embryonic development. *Am. Zool.* **16**, 363–374.

Creek, G. A. (1960). The development of the lamellibranch *Cardium edule* L. *Proc. Zool. Soc. London* **135**, 243–260.

Culliney, J. L. (1974). Larval development of the giant scallop *Placopecten magellanicus* (Gmelin). *Biol. Bull. (Woods Hole, Mass.)* **147**, 321–332.

Culliney, J. L., and Turner, R. D. (1976). Larval development of the deep-water wood boring bivalve, *Xylophaga atlantica* Richards (Mollusca, Bivalvia, Pholadidae). *Ophelia* **15**, 149–161.

Dakin, W. J. (1909). *Pecten. Mem. Liverpool Mar. Biol. Comm.* **23**, 233–468.

Dakin, W. J. (1928). The eyes of *Pecten, Spondylus, Amussium* and allied lamelli-branchs, with a short discussion on their evolution. *Proc. R. Soc. London, Ser. B* **103**, 355–365.

Dalmon, J. (1935). Notes sur la biologie du petoncle *Chlamys varia* (L.) *Rev. Trav. Off. Sci. Tech. Peches Marit., Paris* **8**, 268–281.

Dan, J. C. (1956). The acrosome reaction. *Int. Rev. Cytol.* **5**, 365–393.

Dan, J. C. (1962). The vitelline coat of the *Mytilus* egg. I. Normal structure and the effect of acrosomal lysin. *Biol. Bull. (Woods Hole, Mass.)* **123**, 531–542.

Dan, J. C. (1967). Acrosome reaction and lysins, *In* "Fertilization" (C. B. Metz and A. Monroy, eds.), Vol. 1, pp. 237–293. Academic Press, New York.

Dan, J. C., and Wada, S. K. (1955). Studies on the acrosome. IV. The acrosome reaction in some bivalve spermatozoa. *Biol. Bull. (Woods Hole, Mass.)* **109**, 40–55.

Dan, J. C., Hashimoto, S., Kubo, M., and Yonehara, K. (1975). The fine structure of the acrosomal trigger. *In* "The Functional Anatomy of the Spermatozoon" (B. A. Afzelius, ed.), pp. 39–45. Pergamon, Oxford.

Daniell, R. J. (1920). Seasonal changes in the chemical composition of the mussel (*Mytilus edulis*). *Rep. Lancashire, Sea-Fish. Lab.* pp. 74–84.

Daniell, R. J. (1921). Seasonal changes in the chemical composition of the mussel (*Mytilus edulis*) continued. *Rep. Lancashire Sea-Fish. Lab.* pp. 205–221.

Daniell, R. J. (1922). Seasonal changes in the chemical composition of the mussel (*Mytilus edulis*) concluded. *Rep. Lancashire Sea-Fish. Lab.* pp. 27–50.

Dare, P. J., and Edwards, D. B. (1975). Seasonal changes in flesh weight and bio-chemical composition of mussels (*Mytilus edulis* L.) in Conway estuary, North Wales. *J. Exp. Mar. Biol. Ecol.* **18**, 89–97.

Davids, C. (1964). The influence of suspensions of micro-organisms of different concentrations on the pumping and retention of food by the mussel (*Mytilus edulis* L.). *Neth. J. Sea Res.* **2**, 233–249.

Davis, H. C. (1953). On food and feeding of larvae of the American oyster, *C. virginica. Biol. Bull. (Woods Hole, Mass.)* **104**, 334–350.

Davis, H. C. (1958). Survival and growth of clam and oyster larvae at different salinities. *Biol. Bull. (Woods Hole, Mass.)* **114**, 296–307.

Davis, H. C. (1960). Effect of turbidity producing materials in sea water on eggs and larvae of the clam [*Venus (Mercenaria) mercenaria*]. *Biol. Bull. (Woods Hole, Mass.)* **118**, 48–54.

Davis, H. C. (1961). Effects of some pesticides on eggs and larvae of oysters (*Crassostrea virginica*) and clams (*Venus mercenaria*). *Commer. Fish. Rev.* **23**, 8–23.

Davis, H. C., and Calabrese, A. (1964). Combined effect of temperature and salinity on development of eggs and growth of larvae of *M. mercenaria* and *C. virginica. U.S. Fish. Wildl. Serv., Fish. Bull.* **63**, 643–655.

Davis, H. C., and Chanley, P. E. (1956). Effects of some dissolved substances on bivalve larvae. *Proc. Natl. Shellfish. Assoc.* **46**, 59–74.

Davis, H. C., and Guillard, R. R. (1958). Relative value of ten genera of micro-organisms as foods for oyster and clam larvae. *U.S. Fish. Wildl. Serv., Fish. Bull.* **136**, 293–304.

Davis, H. C., and Hidu, H. (1969a). Effect of pesticides on embryonic development of clams and oysters and on survival and growth of larvae. *U.S. Fish Wildl. Serv., Fish. Bull.* **67**, 393–404.

Davis, H. C., and Hidu, H. C. (1969b). Effect of turbidity producing substances in sea water on eggs and larvae of three genera of bivalve molluscs. *Veliger* **11**, 316–323.

Davis, R. L., and Marshall, N. (1961). The feeding of the bay scallop, *Aequipecten irradians. Proc. Natl. Shellfish. Assoc.* **52**, 25–29.

de Block, J. W., and Geelen, H. J. (1958). The substratum required for the settling of mussels (*Mytilus edulis*). *Arch. Neerl. Zool.* **13**, 446–460.

de Longcamp, D., Lubet, P., and Drosdowsky, M. (1974). The *in vitro* biosynthesis of steroids by the gonad of the mussel (*Mytilus edulis*) *Gen. Comp. Endocrinol.* **22**, 116–127.

Deroux, G. (1960). Formation régulière de mâles, mûrs, de taille et d'organisation larvaire chez un Eulamellibranche commensal (*Montacuta phascolionis* Dautz.). *C. R. Acad. Sci.* **250**, 2264–2266.

Deroux, G. (1961). Rapports taxonomiques d'un Leptonace non decrit "*Lepton subtrigonum*" Jeffreys (nomen nudum–1873). *Cah. Biol. Mar.* **2**, 99–153.

De Schweinitz, E., and Lutz, R. A. (1976). Larval development of the horse mussel, *Modiolus modiolus* (L.) including a comparison with the larvae of *Mytilus edulis* L. as an aid in plankton identification. *Biol. Bull. (Woods Hole, Mass.)* **150**, 348–360.

De Wilde, P. A. W. (1975). Influence of temperature on behavior, energy metabolism and growth of *Macoma balthica. Proc. Eur. Symp. Mar. Biol., 9th*, Oban, Scotland, pp. 239–256.

De Zwann, A., and Zandee, D. I. (1972). Body distribution and seasonal changes in glycogen content of the common sea mussel *Mytilus edulis. Comp. Biochem. Physiol. A* **43**, 53–58.

Dickie, L. M. (1955). Fluctuations in abundance of the giant scallop, *Placopecten magellanicus* (Gmelin), in the Digby area of the Bay of Fundy. *J. Fish. Res. Board Can.* **12**, 797–857.

Dollfus, P. (1922). Resumé de nos principales connaissances pratiques sur les maladies et les ennemis de l'huître. *Notes Mem., Off. Sci. Tech. Peches Marit. Paris* No. 7, 1–58.

Drew, G. A. (1897). Notes on the embryology, anatomy and habits of *Yoldia limatula* Say. *Johns Hopkins Univ., Circ.* **17**, 11–14.

Drew, G. A. (1899). Some observations on the habits, anatomy and embryology of members of the protobranchia. *Anat. Anz.* **15**, 493–519.

Drew, G. A. (1900). *Yoldia limatula. Mem. Biol. Lab. Johns Hopkins Univ.* **4**, 1–37.

Drew, G. A. (1901). The life history of *Nucula delphinodonta* (Mighels). *Q. J. Microsc. Sci.* **44**, 313–391.

Drew, G. A. (1906). The habits, anatomy and embryology of the giant scallop (*Pecten tenuicostatus* Mighels). *Univ. Maine Stud.* No. 6.

Dupouy, J., and Martinez, J.-C. (1973). Action de *Protoeces maculatus* (Looss 1901) (Trematoda, Fellodistomatidae) sur le développement des gonades chez *Mytilus galloprovincialis* Lmk. *C. R. Seances Soc. Biol. Ses Fil.* **277**, 1889–1892.

Durve, V. S. (1964). Preliminary observations on the seasonal gonadal changes and spawning in the clam *Meretrix casta* (Chemnitz) from the marine fish farm. *J. Mar. Biol. Assoc. India* **6**, 241–248.

Duval, D. M. (1962). Observations on the annual cycle of *Barnea candida* (cl. Lamellibranchiata, fam. Pholadidae). *Proc. Malacol. Soc. London* **35**, 101–102.

Ehinger, R. E. (1978). Seasonal energy balance of sea scallop *Placopecten magellanicus* from Narragansett Bay. M. S. Thesis, Univ. of Rhode Island, Kingston.

Estabrooks, S. L. (1973). Seasonal variation in the glycogen and lipid content of the bay scallop, *Aequipecten irradians* Lamarck. M. S. Thesis, North Eastern Univ., Boston, Massachusetts.

Fahrmann, W. (1961). Licht und Electronenmikroskopische Untersuchungen des Nervenstystems von *Umio tumidus* (Phillipisson) unter besonderer Berucksichtigung der Neurosekretion. Z. *Zellforsch. Mikrosk. Anat.* **54**, 689–716.

Field, I. A. (1922). Biology and economic value of the sea mussel *Mytilus edulis. Bull. U.S. Bur. Fish.* **38**, 127–259.

Filatova, Z. A. (1957). General review of the bivalve mollusks of the northern seas of the U.S.S.R. (in Russian). *Trans. Inst. Oceanogr.* **20**, 1–44.

Firtel, R. A., and Monroy, A. (1970). Polysome and RNA synthesis during early development of the surf clam, *Spisula solidissima. Dev. Biol.* **21**, 87–104.

Fish, C. J. (1925). Seasonal distribution of the plankton of the Woods Hole region. *Bull. U.S. Bur. Fish.* **41**, 91–179.

Fox, D. L., ed. (1936). The habitat and food of the California sea-mussel. *Bull. Scripps Inst. Oceanogr., Tech. Ser.* **4**, 1–64.

Fox, D. L. (1957). Particulate organic detritus. *Geol. Soc. Am., Mem.,* **67**, 383–390.

Fox, D. L., and Coe, W. R. (1943). Biology of the California sea-mussel (*Mytilus californianus*). II. Nutrition, metabolism, growth and calcium deposition. *J. Exp. Zool.* **93**, 205–249.

Fraga, F. (1956a). Average seasonal variation of chemical constituents of the mussel (*M. edulis*). *Rapp. P.-V. Reun., Cons. Int. Explor. Mer* **140**, 35.

Fraga, F. (1956b). Variacion estacional de la composicion quimca del myillon (*Mytilus edulis*). *Invest. Pesq.* **4**, 109–125.

Franc, A. (1960). Classe des bivalves. *In* "Traité de Zoologie" (P.-P. Grassé, ed.), Vol. 2, pp. 1845–2133. Masson, Paris.

Franklin, L. E. (1970). Fertilization and the role of the acrosomal region in nonmammals. *Biol. Reprod., Suppl.* **2**, 159–176.

Franzén, Å. (1955). Comparative morphological investigations into spermatogenesis among mollusca. *Zool. Bidr. Uppsala* **30**, 399–456.

Fretter, V., and Graham, A. (1964). Reproduction. *In* "Physiology of Mollusca" (K. M. Wilbur and C. M. Yonge, eds.), Vol. 1, pp. 127–164. Academic Press, New York.

Fuji, A., and Hashizume, M. (1974). Energy budget for a Japanese common scallop *Patinopecten yessoensis* (Jay) in Mutusu Bay. *Bull. Fac. Fish. Hokkaido Univ.* **25**, 7–19.

Fujita, T. (1929). On the early development of the common Japanese oyster. *Jpn. Zool.* **2**, 353–358.

Fullarton, J. H. (1890). On the development of the common scallop (*Pecten opercularis* L.). *Rep. Fish. Board Scot., No. 8* **6**, 290–299.

Gabbott, P. A. (1975). Storage cycles in marine bivalve molluscs: A hypothesis concerning the relationship between glycogen metabolism and gametogenesis. *Proc. Eur. Mar. Biol. Symp., 9th,* Oban, Scotland, pp. 191–211.

Gabbott, P. A. (1976). Energy metabolism. *In* "Marine Mussels" (B. L. Bayne, ed.), pp. 293–355. Cambridge Univ. Press, London and New York.

Gabbott, P. A., and Bayne, B. L. (1973). Biochemical effects of temperature and nutritive stress on *Mytilus edulis* L. *J. Mar. Biol. Assoc. U.K.* **53**, 269–286.

Gabe, M. (1955). Particularités histologiques de cellules neurosécrétrices chez quelques lamellibranches. *C. R. Acad. Sci.* **240**, 1810–1812.

Gabe, M. (1965). La neurosécrétion chez les mollusques et ses rapports avec la reproduction. *Arch. Anat. Microsc. Morphol. Exp.* **54**, 371–385.

274 A. N. SASTRY

Gabe, M. (1966). "Neurosecretion." Pergamon, Oxford.

Gabe, M., and Rancurel, P. (1958). Caractères histologiques des cellules neurosécrétrices chez quelques Teredo. Bull. Inst. Fr. Afr. Noire 20, 73–78.

Gabrielli, F., and Baglioni, C. (1975). Maternal messenger RNA and histone synthesis in embryos of the surf clam Spisula solidissima. Dev. Biol. 43, 254.

Gage, J. (1966a). Observations on the bivalves Montacuta substriata and M. ferruginosa, "commensals" with spatangoids. J. Mar. Biol. Assoc. U.K. 46, 49–70.

Gage, J. (1966b). The life history of the bivalves Montacuta substriata and M. ferruginosa, "commensals" with spatangoids. J. Mar. Biol. Assoc. U.K. 46, 499–511.

Galtsoff, P. S. (1961). Physiology of reproduction in molluscs. Am. Zool. 1, 273–289.

Galtsoff, P. S. (1964). The American oyster Crassostrea virginica Gmelin. U.S. Fish. Wildl. Serv., Fish. Bull. 64, 1–480.

Ghiselin, M. T. (1969). The evolution of hermaphroditism among animals. Q. Rev. Biol. 44, 189–208.

Gibson, F. A. (1956). Escallops Pecten maximus L. in Irish waters. Sci. Proc. R. Dublin Soc. 27, 253–270.

Giese, A. C. (1959a). Comparative physiology: Annual reproductive cycles of marine invertebrates. Annu. Rev. Physiol. 21, 547–576.

Giese, A. C. (1959b). Reproductive cycles of some West Coast invertebrates. In "Photoperiodism and Related Phenomena in Plants and Animals" (R. Withrow, ed.), Publ. No. 55, pp. 625–638. Am. Assoc. Adv. Sci., Washington, D.C.

Giese, A. C. (1966). Lipids in the economy of marine invertebrates. Physiol. Rev. 46, 244–298.

Giese, A. C. (1967). Some methods for study of the biochemical constitution of marine invertebrates. Oceanogr. Mar. Biol. 5, 253–288.

Giese, A. C. (1969). A new approach to the biochemical composition of the molluscan body. Oceanogr. Mar. Biol. 7, 175–229.

Giese, A. C. (1976). Reproductive cycles of marine invertebrates. An. Acad. Bras. Cienc. 47, 49–67.

Giese, A. C., and Pearse, J. S. (1974). Introduction: General principles. In "Reproduction of Marine Invertebrates" (A. C. Giese and J. S. Pearse, eds.), Vol. 1, pp. 1–49. Academic Press, New York.

Giese, A. C., Hart, M. A., Smith, A. M., and Chung, M. A. (1967). Seasonal changes in body component indices and chemical composition in the Pismo clam, Tivela stultorum. Comp. Biochem. Physiol. 22, 549–561.

Gilbert, M. A. (1973). Growth rate, longevity and maximum size of Macoma balthica (L.). Biol. Bull. (Woods Hole, Mass.) 145, 119–126.

Gimazane, J. P. (1971). Introduction à l'étude expérimentale du cycle sexuel d'un mollusque bivalve Cardium edule L. Analyse des populations, évolution de la gonade et action de quelques facteurs: Nutrition, température, photopériode. Doctoral Thesis, Univ. de Caen (U.E.R. des Sciences de la Vie et due Comportement des Etres Vivants).

Gimazane, J. P. (1972). Etude expérimentale de l'action de quelques facteurs externes sur la reprise de l'activité génitale de la coque, Cerastoderma edule L., Mollusque Bivalve. C. R. Soc. Biol. 166, 587–589.

Golding, D. W. (1974). A survey of neuroendocrine phenomena in nonarthropod invertebrates. Biol. Rev. Cambridge Philos. Soc. 49, 161–224.

Golikov, A. N., and Scarlato, O. A. (1970). Abundance, dynamics and production properties of populations of edible bivalves Mizuhopecten yessoensis and Spisula *sachalinensis* related to the problem of organization of controllable submarine farms at the western shores of the Sea of Japan. *Helgol. Wiss. Meeresunters.* **20,** 498–513.

Graham, H. W., and Gay, H. (1945). Season of attachment and growth of sedentary marine organisms at Oakland, California. *Ecology* **26,** 375–386.

Granmo, A. (1972). Development and growth of eggs and larvae of *Mytilus edulis* exposed to a linear dodecylbenzenesulphonate. *Mar. Biol.* **15,** 356–358.

Grave, B. H. (1927a). An analysis of the spawning habits and spawning stimuli of *Cumingia tellinoides. Biol. Bull. (Woods Hole, Mass.)* **52,** 418–435.

Grave, B. H. (1927b). The natural history of *Cumingia tellinoides. Biol. Bull. (Woods Hole, Mass.)* **53,** 208–219.

Grave, B. H., and Smith, J. (1936). Sex inversion in *Teredo navalis* and its relation to sex ratios. *Biol. Bull. (Woods Hole, Mass.)* **70,** 332–343.

Grave, C. (1916). The process of feeding in the oyster. *Science* **44,** 178–181.

Greenfield, L. J. (1953). Observations on the nitrogen and glycogen content of *Teredo (Lyrodus) pedicellata* de Quatrefages at Miami, Florida. *Bull. Mar. Sci. Gulf Caribb.* **2,** 486–496.

Grice, G. D. (1953). A qualitative and quantitative seasonal study of the Copepoda and Cladocera of Alligator Harbor. M.S. Thesis, Florida State Univ., Tallahassee.

Grobben, C. (1892). Betrage zur Kentniss des Baues von *Cuspidaria (Naera) cuspidata.* Olivi nebst Betrachtungen uber das System der Lamellibranchiaten. *Arb. Zool. Inst. Wien* **10,** 101–146.

Gruffydd, L. D., and Beaumont, A. R. (1970). Determination of the optimum concentration of eggs and spermatozoa for the production of normal larvae in *Pecten maximus* (Mollusca, Lamellibranchia). *Helogl. Wiss. Meersunters.* **20,** 486–497.

Gunter, G. (1957). Temperature. *Geol. Soc. Am., Mem.* **67,** 159–184.

Gutsell, J. S. (1926). A hermaphroditic viviparous oyster of the Atlantic coast of North America. *Science* **64,** 450.

Gutsell, J. S. (1930). Natural history of the bay scallop (*Pecten irradians*). *Bull. U.S. Bur. Fish.* **46,** 569–632.

Haley, J. E., Stefano, G. B., and Catapane, E. J. (1976). Comparison of phospholipids in the pedal, cerebral and visceral ganglia of *Mytilus edulis. Am. Zool.* **16,** 184.

Hansen, B. (1953). Brood protection and sex ratio of *Transennella tantilla* (Gould), a Pacific bivalve. *Vidensk. Medd. Dan. Naturhist. Foren. Khobenhavn* **115,** 313–324.

Hatschek, B. (1881). Entwicklungsgeschichte von *Teredo. Arb. Zool. Inst. Wien* **3,** 1–44.

Hedgpeth, J. W., and Gonor, J. J. (1969). Aspects of the potential effect of thermal alteration on marine and estuarine benthos. *In* "Biological Aspects of Thermal Pollution" (P. A. Krenkel and F. L. Parker, eds.), pp. 80–118. Vanderbilt Univ. Press, Nashville, Tennessee.

Hennick, D. P. (1970). Reproductive cycle, size, maturity and sexual composition of commercially harvested weathervane scallops, (*Patinopecten caurinus*) in Alaska. *J. Fish. Res. Board Can.* **27,** 2112–2119.

Hidu, H. (1965). Effects of synthetic surfactants on the larvae of clams (*M. mercenaria*) and oysters (*C. virginica*). *J. Water Pollut. Control Fed.* **37,** 262–270.

Hill, C. L., and Kofoid, C. A. (1927). Marine borers and their relation to marine construction on the Pacific coast. *San Francisco Bay Mar. Piling Comm. Final Report.*

Hirschhorn, G. (1962). Growth and mortality rates of the razor clam (*Siliqua patula*) on Clatsop beaches, *Oreg. Fish. Comm. Oreg., Contrib.* No. 27.

Hornell, J. (1909). Report on the anatomy of *Placuna placenta* with notes on the distribution and economic uses. *Rep. Mar. Zool. Okahamandal Coast Kathiawar, India* Part I, 43–97.

Houtteville, P., and Lubet, P. (1974). Analyse expérimentale, en culture organo-typique, de l'action des ganglions cérébropleuraux et viscéraux sur l'manteau de la moule mâle *Mytilus edulis* (L.) (Mollusque Pelecypode). *C. R. Seances Soc. Biol. Ses Fil.* 278, 2469–2472.

Houteville, P., and Lubet, P. (1975). The sexuality of pelecypod molluscs. *In* "Intersexuality in the Animal Kingdom" (R. Reinboth, ed.), pp. 179–187. Springer-Verlag, Berlin and New York.

Howard, A. D. (1953). Some viviparous pelecypod mollusks. *Wasmann J. Biol.* 11, 233–260.

Hubner, E., and Anderson, E. (1976). Comparative spirilian oogenesis—Structural aspects: An overview. *Am. Zool.* 16, 315–343.

Hughes, R. N. (1969). A study of feeding in *Scrobicularia plana*. *J. Mar. Biol. Assoc. U.K.* 49, 805–823.

Hughes, R. N. (1970). An energy budget for a tidal-flat population of the bivalve *Scribicularia plana* (Da Costa). *J. Anim. Ecol.* 39, 357–381.

Hughes, R. N. (1971). Reproduction of *Scrobicularia plana* Da Costa (Pelecypoda: Semilidae) in North Wales. *Veliger* 14, 77–81.

Humphreys, W. J. (1962). Electron microscope studies on eggs of *Mytilus edulis*. *J. Ultrastruct. Res.* 7, 467–487.

Humphreys, W. J. (1964). Electron microscope studies of the fertilized egg and the two-cell stage of *Mytilus edulis*. *J. Ultrastruct. Res.* 10, 244–262.

Humphreys, W. J. (1967). The fine structure of cortical granules in eggs and gastrulae of *Mytilus edulis*. *J. Ultrastruct. Res.* 17, 314–326.

Humphreys, W. J. (1969). Initiation of shell formation in the bivalve, *Mytilus edulis*. *Proc., Electron Microsc. Soc. Am.* 27, 272–273.

Idler, R. D., Tamura, T., and Wainai, T. (1964). Seasonal variation in sterol, fat and unsaponifiable components of scallop muscle. *J. Fish. Res. Board Can.* 21, 1035–1042.

Isham, L. B., and Tierney, J. Q. (1953). Some aspects of the larval development and metamorphosis of *Teredo* (*Lyrodus*) *pedicellata* de Quatrefages. *Bull. Mar. Sci. Gulf Caribb.* 2, 574–589.

Ito, S., and Leuchtenberger, C. (1955). The possible role of the DNA content of spermatozoa for the activation process of the egg of the clam, *Spisula solidissima*. *Chromosoma* 7, 328.

Iwata, K. S. (1950). Spawning of *Mytilus edulis*. II. Discharge by electrical stimulation. *Nippon Suisan Gakkaishi* 15, 443–446.

Iwata, K. S. (1951a). Spawning of *Mytilus edulis*. IV. Discharge by KCl injection. *Nippon Suisan Gakkaishi* 16, 393–394.

Iwata, K. S. (1951b). Spawning of *Mytilus edulis*. VIII. Comparison of abilities of salts of alkali metals and of alkali earth metals to induce spawning. *Nippon Suisan Gakkaishi* 17, 94–95.

Iwata, K. S. (1951c). Autoactivation of eggs of *Mactra veneriformis* in sea water. *Annot. Zool. Jpn.* 24, 187–193.

Jacobson, M. K. (1955). Observations on *Donax fossor* Say at Rockaway Beach, N. Y. *Nautilus* **68**, 73–77.

Jenner, C., and McCrary, A. (1968). Sexual dimorphism in erycinacean bivalves. *Am. Malacol. Union Inc., Annu. Rep. Bull.* **35**, 43.

Jensen, A., and Sakshaug, E. (1970a). Producer–consumer relationships in the sea. I. Preliminary studies on phytoplankton density and *Mytilus* pigmentation. *J. Exp. Mar. Biol. Ecol.* **5**, 180–186.

Jensen, A., and Sakshaug, E. (1970b). Producer–consumer relationships in the sea. II. Correlation between *Mytilus* pigmentation and the density and composition of phytoplanktonic populations in inshore waters. *J. Exp. Mar. Biol. Ecol.* **5**, 246–253.

Jørgensen, C. B. (1960). Efficiency of particle retention and rate of water transport in undisturbed lamellibranchs. *J. Cons., Cons. Perm. Int. Explor. Mer* **26**, 94–116.

Jørgensen, C. B. (1966). "Biology of Suspension Feeding." Pergamon, New York.

Jørgensen, C. B. (1976). August Putter, August Krogh, and modern ideas on the use of dissolved organic matter in aquatic environments. *Biol. Rev. Cambridge Philos. Soc.* **51**, 291–328.

Jones, G. F. (1963). Brood protection in three southern California species of the pelecypod *Cardita*. *Wasmann J. Biol.* **21**, 141–148.

Kennedy, A. U., and Battle, H. I. (1964). Cyclic changes in the gonad of the American oyster, *Crassostrea virginica* (Gmelin). *Can. J. Zool.* **42**, 305–321.

Kennedy, V. S., Rosenburg, W. H., Castagna, M., and Mihursky, J. (1974a). *Mercenaria mercenaria* (Mollusca: Bivalvia): Temperature-time relationships for survival of embryos and larvae. *U.S. Fish Wildl. Serv., Fish. Bull.* **72**, 1160–1166.

Kennedy, V. S., Rosenburg, W. H., Zion, H. H., and Castagna, M. (1974b). Temperature-time relationships for survival of embryos and larvae of *Mulinia lateralis* (Mollusca: Bivalvia). *Mar. Biol.* **24**, 137–145.

Kidder, G. (1972a). Gene transcription in mosaic embryos. I. The pattern of RNA synthesis in early development of the coot clam, *Mulinia lateralis*. *J. Exp. Zool.* **180**, 55–74.

Kidder, G. (1972b). Gene transcription in mosaic embryos. II. Polyribosomes and messenger RNA in early development of the coot clam, *Mulinia lateralis*. *J. Exp. Zool.* **180**, 75–84.

Kidder, G. M. (1976). RNA synthesis and the ribosomal cistrons in early molluscan development. *Am. Zool.* **16**, 501–520.

Kinne, O. (1963). The effects of temperature and salinity on marine and brackish water animals. *Oceanogr. Mar. Biol.* **1**, 301–340.

Kinne, O. (1964). The effects of temperature and salinity on marine and brackish water animals. *Oceanogr. Mar. Biol.* **2**, 281–339.

Kinne, O. (1970). Temperature, animals, invertebrates. *In* "Marine Ecology, A Comprehensive Treatise on Life in Oceans and Coastal Waters" (O. Kinne, ed.), Vol. 1, Part 1, pp. 407–514. Wiley (Interscience), New York.

Kirby-Smith, W. W. (1970). Growth of the scallops, *Argopecten irradians concentricus* (Say) and *Argopecten gibbus* (Linne) as influenced by food and temperature. Ph.D. Thesis, Duke Univ., Durham, North Carolina.

Kirby-Smith, W. W. (1976). The detritus problem and the feeding and digestion of an estuarine organism. *In* "Estuarine Processes" (M. L. Wiley, ed.), Vol. 1, pp. 469–479. Academic Press, New York.

Knudsen, J. (1961). The bathyal and abyssal *Xylophaga* (Pholadidae, Bivalvia). *Galathea Rep.* **5**, 163–209.

Korringa, P. (1947). Relations between moon and periodicity in the breeding of marine animals. *Ecol. Monogr.* **17**, 347–381.

Korringa, P. (1952). Recent advances in oyster biology. *Q. Rev. Biol.* **27**, 266–308, 339–365.

Korringa, P. (1957). Lunar periodicity. *Geol. Soc. Am., Mem.* **67**, 917–934.

Kuenzler, E. J. (1961). Phosphorous budget of a mussel population. *Limnol. Oceanogr.* **6**, 400–415.

Lammens, J. J. (1967). Growth and reproduction in a tidal flat population of *Macoma balthica* (L.). *Neth. J. Sea Res.* **3**, 315–382.

Lane, C. E., Posner, G. S., and Greenfield, L. J. (1952). The distribution of glycogen in the shipworm, *Teredo* (*Lyrodus*) *pedicellata* Quatrefages. *Bull. Mar. Sci. Gulf Caribb.* **2**, 385–392.

Lankester, E. R. (1900). Mollusca. In "Treatise on Zoology" (P. Pelseneer, ed.), Part 5, pp. 205–284. Black, London.

Lawrence, J. M. (1976). Patterns of lipid storage in post-metamorphic marine invertebrates. *Am. Zool.* **16**, 747–762.

Lebour, M. V. (1938a). Notes on the breeding of some lamellibranchs from Plymouth and their larvae. *J. Mar. Biol. Assoc. U.K.* **23**, 119–144.

Lebour, M. V. (1938b). The life history of *Kellia suborbicularis*. *J. Mar. Biol. Assoc. U.K.* **22**, 447–451.

Lebour, M. V. (1947). Notes on the inshore plankton of Plymouth. *J. Mar. Biol. Assoc. U.K.* **26**, 527–547.

Lent, C. M. (1969). Adaptations of the ribbed mussel, *Modiolus demissus* (Dillwyn), to the intertidal habitat. *Am. Zool.* **9**, 283–292.

Leonard, U. K., Jr. (1969). Seasonal gonadal changes in two bivalve mollusks in Tomales Bay, California. *Veliger* **11**, 382–390.

Lillie, F. R. (1895). The embryology of the Unionidae. A study in cell-lineage. *J. Morphol.* **10**, 1–100.

Longo, F. J. (1976). Ultrastructural aspects of fertilization in spiralian eggs. *Am. Zool.* **16**, 375–394.

Longo, F., and Dornfeld, E. (1967). The fine structure of spermatid differentiation in the mussel, *Mytilus edulis*. *J. Ultrastruct. Res.* **20**, 462–480.

Loosanoff, V. L. (1937a). Development of the primary gonad and sexual phases in *Venus mercenaria* Linnaeus. *Biol. Bull.* (*Woods Hole, Mass.*) **72**, 389–405.

Loosanoff, V. L. (1937b). Seasonal gonadal changes in adult clams, *Venus mercenaria* (L). *Biol. Bull.* (*Woods Hole, Mass.*) **72**, 406–416.

Loosanoff, V. L. (1937c). Spermatogenesis in the hard-shell clam (*Venus mercenaria* Linneaus). *Yale J. Biol. Med.* **9**, 437–442.

Loosanoff, V. L. (1939). Effect of temperature upon shell movement of clams, *Venus mercenaria* (L). *Biol. Bull.* (*Woods Hole, Mass.*) **76**, 171–182.

Loosanoff, V. L. (1942). Shell movement of the edible mussels, *Mytilus edulis* (L.) in relation to temperature. *Ecology* **23**, 231–234.

Loosanoff, V. L. (1948). Gonad development and spawning of oysters (*O. virginica*) in low salinities. *Anat. Rec.* **101**, 55.

Loosanoff, V. L. (1952). Behavior of oysters in water of low salinities. *Proc. Natl. Shellfish. Assoc.* **43**, 135–151.

Loosanoff, V. L. (1953). Reproductive cycle in *Cyprina islandica*. *Biol. Bull.* (*Woods Hole, Mass.*) **104**, 146–155.

Loosanoff, V. L. (1954). New advances in the study of bivalve larvae. *Am. Sci.* **42**, 607–624.

Loosanoff, V. L. (1959). The size and shape of metamorphosing larvae of *Venus*

(*Mercenaria*) *mercenaria* grown at different temperatures. *Biol. Bull. (Woods Hole, Mass.)* 117, 308–318.

Loosanoff, V. L. (1962). Gametogenesis and spawning of the European oyster, *Ostrea edulis*, in the waters of Maine. *Biol. Bull. (Woods Hole, Mass.)* 122, 86–94.

Loosanoff, V. L. (1965). Gonad development and discharge of spawn in oysters of Long Island Sound. *Biol. Bull. (Woods Hole, Mass.)* 129, 546–561.

Loosanoff, V. L. (1971). Development of shellfish culture techniques. *Proc. Conf. Artif. Propag. Commer. Valuable Shellfish-Oysters, Coll. Mar. Stud., Univ. Delaware* pp. 9–40.

Loosanoff, V. L., and Davis, H. C. (1950). Conditioning *V. mercenaria* for spawning in winter and breeding its larvae in the laboratory. *Biol. Bull. (Woods Hole, Mass.)* 98, 60–65.

Loosanoff, V. L., and Davis, H. C. (1951). Delaying spawning of lamellibranchs by low temperature. *J. Mar. Res.* 10, 197–202.

Loosanoff, V. L., and Davis, H. C. (1952). Temperature requirements for maturation of gonads of northern oysters. *Biol. Bull. (Woods Hole, Mass.)* 103, 80–96.

Loosanoff, V. L., and Davis, H. C. (1963). Rearing of bivalve molluscs. *Adv. Mar. Biol.* 1, 1–136.

Loosanoff, V. L., Miller, W. S., and Smith, P. B. (1951). Growth and setting of larvae of *Venus mercenaria* in relation to temperature. *J. Mar. Res.* 10, 59–81.

Loosanoff, V. L., Davis, H. C., and Chanley, P. E. (1953). Behavior of clam larvae in different concentrations of food organisms. *Anat. Rec.* 117, 586–587.

Loosanoff, V. L., Davis, H. C., and Chanley, P. E. (1955). Food requirements of some bivalve larvae. *Proc. Natl. Shellfish. Assoc.* 45, 66–83.

Loosanoff, V. L., Davis, H. C., and Chanley, P. E. (1966). Dimensions and shapes of larvae of some marine bivalve mollusks. *Malacologia* 4, 351–435.

Lopez-Benito, M. (1955). Composicion de al vieira (*Pecten jacobeus*). *Invest. Pesq.* 1, 137–151.

Lough, R. G., and Gonor, J. J. (1971). Early embryonic stages of *Adula californiensis* (Pelecypoda: Mytilidae) and the effect of temperature and salinity on developmental rate. *Mar. Biol.* 8, 118–125.

Lough, R. G., and Gonor, J. J. (1973). A response-surface approach to the combined effects of temperature and salinity on the larval development of *Adula californiensis* (Pelecypoda: Mytilidae). I. Survival and growth of three- and fifteen-day old larvae. *Mar. Biol.* 22, 241–250.

Lubet, P. (1951). Sur l'émission des gamètes chez *Chlamys varia* L. (Moll. Lamellibr.). *C. R. Acad. Sci.* 235, 1680–1681.

Lubet, P. (1955a). Cycle neurosécrétoire chez *Chlamys varia* et *Mytilus edulis* L. (Mollusques Lamellibranches). *C. R. Acad. Sci.* 241, 119–121.

Lubet, P. (1955b). Effets de l'ablation des centres nerveux sur l'émission des-gamètes chez *Mytilus edulis* L. et *Chlamys varia* L. (Moll. Lamellibranches). *Ann. Sci. Nat., Zool. Biol. Anim.* 2, 175–183.

Lubet, P. (1957). Cycle sexuel de *Mytilus edulis* et de *M. galloprovincialis* dans le bassin d'Arcachon. *Annee Biol.* 33, 19–26.

Lubet, P. (1959). Recherches sur le cycle sexuel et l'émission des gamètes chez les Mytilides et les Pectinides. *Rev. Trav. Inst. Peches Marit.* 23, 389–548.

Lubet, P. (1965). Incidences de l'ablation bilaterale des ganglions cérébroides sur la gamétogenèse et le développement du tissu conjunctif chez la moule *Mytilus galloprovincialis* Lam. (Moll. Lamellibranches). *C. R. Soc. Biol.* 159, 397–399.

Lubet, P. (1966). Essai d'analyse expérimentale des peturbations produites par les ablations des ganglions nerveux chez *Mytilus edulis* L. et *Mytilus galloprovincialis* Lmk (Mollusques Lamellibranches). *Ann. Endocrinol.* **27**, 353–365.

Lubet, P. (1973). "Exposé Synoptique des Données Biologique sur la Moule *Mytilus galloprovincialis* (Lamarck 1819)," Synopsis FAO sur les Pêches No. 88 (SAST—Moule, 3, 16(10), 028, 08 pag. var.). FAO, Rome.

Lubet, P., and Bourcart, C. (1963). Nouvelles observations sur la physiologie sexuelle de *Mytilus galloprovincialis* Lamarck. *C. R. Soc. Biol.* **157**, 1996.

Lubet, P., and de Longcamp, D. (1969). Étude des variations annuelles des constituants lipidiques chez *Mytilus edulis* L. de la Baie de Seine (Calvados). *C. R. Seances Soc. Biol. Ses. Fil.* **163**, 1110–12.

Lubet, P., and Pujol, J. P. (1965). Incidence de la neurosécrétion sur l'euryhalinité de *Mytilus galloprovincialis* Lmk variation de la teneur en eau. *Rapp. P.-V. Reun. Comm. Int. Explor. Sci. Mer Mediterr.* **18**, 148–154.

Lucas, A. (1965). Recherche sur la sexualité des mollusques bivalves. Ph.D. Thesis, Fac. Sci. Univ. de Renne.

Lucas, A. (1966). Manifestation précoce de la sexualité chez quelques mollusques bivalves. *Estratto Lav. Soc. Malacol. Ital.* **3**, 153–158.

Lunetta, J. E. (1969). Fisiologia da reprodução dos mexilhões (*Mytilus perna* L. Molusca Lamellibranchia). *Zool. Biol. Mar. (Sao Paulo) Nova Ser.* **26**, 33–111.

MacBride, E. W. (1914). Invertebrata. *In* "Textbook of Embryology" (W. Heape, ed.), Vol. 1, pp. 291–371. Macmillan, London.

McLean, K. (1976). Some aspects of RNA synthesis in oyster development. *Am. Zool.* **16**, 521–528.

McMullen, J. C. (1967). Some aspects of the life history of razor clams *Siliqua patula* (Dixon) in Cook Inlet, Alaska. *Alaska, Dep. Fish Game, Inf. Leafl.* No. 110, 1–18.

McWhinnie, M. A. (1967). The heat responses of invertebrates (exclusive of insects). *In* "Thermobiology" (A. H. Rose, ed.), pp. 353–373. Academic Press, New York.

Mann, H. (1956). The influence of *Mytilicola intestinalis* (Copepoda Parasitica) on the development of the gonads of *Mytilus edulis*. *Rapp. P.-V. Reun. Cons. Perm. Inst. Explor. Mer* **140**, 57–58.

Marshall, N. (1956). Chlorophyll *a* in the phytoplankton in the coastal waters of the eastern Gulf of Mexico. *J. Mar. Res.* **15**, 14–32.

Marshall, N. (1960). Studies on the Niantic River, Connecticut, with special reference to the bay scallop, *Aequipecten irradians*. *Limnol. Oceanogr.* **5**, 86–105.

Marshall, N., and Wheeler, B. M. (1965). Role of coastal and upper estuarine waters contributing phytoplankton to the shores of the Niantic estuary. *Ecology* **46**, 665–673.

Martoja, M. (1972). Endocrinology of Mollusca. *In* "Chemical Zoology" (M. Florkin and B. T. Scheer, eds.), pp. 349–392. Academic Press, New York.

Mason, J. (1957). The age and growth of the scallop, *Pecten maximus* L., in Manx waters. *J. Mar. Biol. Assoc. U.K.* **36**, 473–492.

Mason, J. (1958). The breeding of the scallop, *Pecten maximus* L., in Manx waters. *J. Mar. Biol. Assoc. U.K.* **37**, 653–671.

Masson, M. (1977). Observations sur la nutrition des larves de *Mytilus galloprovincialis* avec des aliments inertes. *Mar. Biol.* **40**, 151–164.

Matveeva, T. A. (1953). On modes of reproduction in marine bivalves. *C. R. (Dokl.) Acad. Sci. URSS* **93**, 923–924.

Meisenheimer, J. (1901). Entwicklungsgeschichte von *Dreissensia polymorpha*. *Wiss. Zool.* **69**, 1–137.

Menzel, R. W. (1968). Chromosome numbers in nine families of pelecypod molluscs. *Nautilus* 82, 45–50, 53–58.

Menzel, R. W., and Menzel, M. Y. (1965). Chromosomes of two species of quahog clams and their hybrids. *Biol. Bull. (Woods Hole, Mass.)* 129, 181–188.

Merrill, A. S., and Burch, J. B. (1960). Hermaphroditism in the sea scallop, *Placopecten magellanicus* Gmelin. *Biol. Bull. (Woods Hole, Mass.)* 119, 197–201.

Metz, C. B., and Donovan, J. E. (1949). Fertilizin from the eggs of the clam, *Mactra solidissima. Biol. Bull. (Woods Hole, Mass.)* 97, 257.

Metz, C. B., and Monroy, A., eds. (1967). "Fertilization; Comparative Morphology, Biochemistry, and Immunology," Vol. 1. Academic Press, New York.

Mikheyev, V. P., and Sorokin Y. J. (1966). Quantitative studies of *Dreissensia* feeding habits by the radiocarbon method. *Zh. Obshch. Biol.* 27, 463–472.

Mileikovsky, S. A. (1970). Seasonal and daily dynamics in pelagic larvae of marine shelf bottom invertebrates in nearshore waters of Kandalaksha Bay (White Sea). *Mar. Biol.* 5, 180–194.

Milton, A. S., and Gosselin, R. E. (1960). Metabolism and cilio-accelerator action of 5-hydroxytryptamine (5-HTP) in gill plates of *Mytilus* and *Modiolus. Fed. Proc., Fed. Am. Soc. Exp. Biol.* 19, 126.

Minaur, J. (1969). Experiments on the artificial rearing of the larvae of *Pinctada maxima* (Jameson) (Lamellibranchia). *Aust. J. Mar. Freshwater Res.* 20, 175–187.

Miyazaki, I. (1938). On the incubatory habits and shelled larvae of bivalves. II *Bot. Zool., Tokyo* 6/7, 1213–1218.

Monroy, A. (1965a). Biochemical aspects of fertilization. *In* "The Biochemistry of Animal Development. Vol. I. Descriptive Biochemistry of Animal Development" (R. Weber, ed.), pp. 73–135. Academic Press, New York.

Monroy, A. (1965b). "Chemistry and Physiology of Fertilization." Holt, New York.

Montalenti, G. (1960). Perspectives in Marine Biology" (A. A. Buzzati-Traverso, ed.), pp. 589–602. Univ. of California Press, Berkeley.

Montalenti, G., and Bacci, G. (1951). Osservazioni e ipotesi sulla determinazione del sesso negli ermafrodite. *Sci. Genet.* 4, 5–12.

Moore, D. R., and Reish, D. J. (1969). Studies on the *Mytilus edulis* community in Alamitos Bay, California. IV. Seasonal variation in gametes from different regions of the bay. *Veliger* 11, 250–255.

Moore, J. K., and Marshall, N. (1967). An analysis of the movements of the bay scallop, *Aequipecten irradians,* in a shallow estuary. *Proc. Natl. Shellfish. Assoc.* 57, 77–82.

Moore, K. E., and Gosselin, R. E. (1962). Effects of 5-hydroxytryptamine on the anaerobic metabolism and phosphorylase activity of lamellibranch gill. *J. Pharmacol. Exp. Ther.,* 138, 145–53.

Moore, K. E., Milton, A. S., and Gosselin, R. E. (1961). Effect of 5-hydroxytryptamine on the respiration of excised lamellibranch gill. *Br. J. Pharmacol.* 17, 278–85.

Mori, S. (1950). Characteristic tidal rhythmic migration of a mussel, *Donax semignosus* DKR and the experimental analysis of its behavior. *Dobutsugaku Zasshi* 59, 88–89.

Morton, J. E. (1956). The tidal rhythm and action of the digestive system of the lamellibranch *Lasea rubra. J. Mar. Biol. Assoc. U.K.* 35, 563–586.

Morton, J. E. (1960). "Molluscs: An Introduction to Their Form and Functions." Harper, New York.

Morton, J. E., Boney, A. D., and Corner, E. D. S. (1957). The adaptation of *Lasea rubra* (Montagu), a small intertidal lamellibranch. *J. Mar. Biol. Assoc. U.K.* **36**, 383–404.

Nagabhushanam, R. (1961). Biochemical studies on the marine wood-boring mollusc, *Martesia striata. J. Sci. Ind. Res.* **20**, 171–173.

Nagabhushanam, R. (1962). Physiological experiments on neurosecretion in the oyster *Crassostrea virginica. Am. Zool.* **2**, 543–544.

Nagabhushanam, R. (1963). Neurosecretory cycle and reproduction in the bivalve *Crassostrea virginica. Indian J. Exp. Biol.* **1**, 161–162.

Nagabhushanam, R. (1964). Neurosecretory changes in the nervous system of the oyster, *Crassostrea virginica,* induced by various experimental conditions. *Indian J. Exp. Biol.* **2**, 1–4.

Nagabhushanam, R. (1969). Studies on neurosecretion in mollusca. *Proc. Symp. Mollusca, Cochin, 1968, Mar Biol. Assoc. India, Symp. Ser.* 3, Part 2, 572–579.

Nagabhushanam, R. (1970). Some aspects of biology of the shipworm, *Teredo furcillatus. Proc. Symp. Mollusca, Cochin, 1968, Mar. Biol. Assoc. India, Symp. Ser.* No. 3, Part 3, 755–757.

Nagabhushanam, R., and Dhamne, K. P. (1977). Seasonal gonadal changes in the clam, *Paphia laterisulca. Aquaculture* **10**, 141–152.

Nagabhushanam, R., and Mane, U. H. (1973). Neurosecretion in the clam, *Katelysia opima. Marthwada Univ. J. Sci.* **12**, 193–203.

Nagabhushanam, R., and Mane, U. H. (1975a). Reproduction and breeding of the clam, *Katelysia opima,* in Kalbdevi estuary at Ratnagiri, west coast of India. *Indian J. Mar. Sci.* **4**, 86–92.

Nagabhushanam, R., and Mane, U. H. (1975b). Reproduction in the mussel, *Mytilus viridis,* at Ratnagiri. *Bull. Dep. Mar. Sci., Univ. Cochin* **7**, 377–378.

Nagabhushanam, R., and Mane, U. H. (1975c). Seasonal variation in the bichemical composition of the clam, *Katelysia opima. Riv. Biol.* **67**, 279–301.

Naidu, K. S. (1970). Reproduction and breeding cycle of the giant scallop, *Placopecten magellanicus* (Gmelin) in Port au Port Bay, Newfoundland. *Can. J. Zool.* **48**, 1003–1012.

Naidu, K. S. (1971). Infection of the giant scallop *Placopecten magellanicus* from Newfoundland with an endozoic alga. *J. Invertebr. Pathol.* **17**, 145–157.

Nair, N. B., and Saraswathy, M. (1970). Some recent studies on the shipworms of India. *Proc. Symp. Mollusca, Cochin, 1968, Mar. Biol. Assoc. India, Symp. Ser.* 3. Part 3, 718–729.

Narasimham, K. A. (1969). Studies on some aspects of biology and fishery of the cockle, *Anadara granosa* (Linnaeus) from Kakinada Bay. *Proc. Symp. Mollusca, Cochin, 1968, Mar. Biol. Assoc. India, Symp. Ser.* 3, Part 2, 407–417.

Narchi, W. (1969). On *Pseudopythina rugifera* (Carpenter, 1864) (Bivalvia). *Veliger* **12**, 43–52.

Nayar, K. N. (1955). Studies on the growth of the wedge clam *Donax* (*latona*) *cuneatus* Linnaeus. *Indian J. Fish.* **2**, 325–348.

Nelson, T. C. (1928a). Relation of spawning of the oyster to temperature. *Ecology* **9**, 145–154.

Nelson, T. C. (1928b). On the distribution of critical temperature for spawning and ciliary activity in bivalve molluscs. *Science* **67**, 220–221.

Nelson, T. C. (1936). Water filtration by the oyster and a new hormone effect upon the rate of flow. *Proc. Soc. Exp. Biol. Med.* **34**, 189–190.

Newell, R. C. (1965). The role of detritus in the nutrition of two marine deposit feeders, the prosobranch *Hydrobia ulvae* and the bivalve *Macoma balthica*. *Proc. Zool. Soc. London* **144**, 25–45.

Newell, R. C. (1970). "Biology of Intertidal Animals." Am. Elsevier, New York.

Ockelmann, K. W. (1958). The zoology of East Greenland marine Lamellibranchiata. *Medd. Groenl.* **122**, 1–256.

Ockelmann, K. W. (1962). Developmental types in marine bivalves and their distribution along the Atlantic coast of Europe. *Proc. Eur. Malacol. Congr. 1st*, London, pp. 25–35.

Ockelmann, K. W. (1964). *Turtonia minuta* (Fabricius), a neotenous veneracean bivalve. *Ophelia* **1**, 121–146.

Ockelmann, K. W. (1965). Redescription, distribution, biology and dimorphous sperm of *Montacuta tenella* Loven (Mollusca, Leptonacea). *Ophelia* **2**, 211–222.

Odlaug, T. O. (1946). The effect of the copepod, *Mytilicola orientalis*, upon the Olympia oyster, *Ostrea lurida*. *Trans. Am. Microsc. Soc.* **65**, 311–317.

Oldfield, E. (1955). Observations on the anatomy and mode of life of *Lasaea rubra* (Montagu) and *Turtonia minuta* (Fabricus). *Proc. Malacol. Soc. London* **31**, 226–249.

Oldfield, E. (1961). The functional morphology of *Kellia suborbicularis* (Montagu), *Montacuta ferruginosa* (Montagu) and *M. substriata* (Montagu), (Mollusca, Lamellibranchiata). *Proc. Malacol. Soc. London* **34**, 255–295.

Oldfield, E. (1964). The reproduction and development of some members of the Erycinidae and Montacutidae (Mollusca, Eulamellibranchiata). *Proc. Malacol. Soc. London* **36**, 79–120.

Olsen, A. M. (1955). Underwater studies on the Tasmanian commercial scallop, *Notovola meridionalis* (Tate) (Lamellibranchiata Pectinidae). *Aust. J. Mar. Freshwater Res.* **6**, 392–409.

Orton, J. H. (1920). Sea temperature, breeding and distribution of marine animals. *J. Mar. Biol. Assoc. U.K.* **12**, 339–366.

Orton, J. H. (1927a). Observations and experiments on sex change in the European oysters (*O. edulis*). Part I. The change from female to male. *J. Mar. Biol. Assoc. U.K.* **14**, 967–1045.

Orton, J. H. (1927b). A note on the physiology of sex and sex determination. *J. Mar. Biol. Assoc. U.K.* **14**, 1047–1055.

Owen, G. (1966a). Feeding. *In* "Physiology of the Mollusca" (K. M. Wilbur and C. M. Yonge, eds.), Vol. 2, pp. 1–52. Academic Press, New York.

Owen, G. (1966b). Digestion. *In* "Physiology of the Mollusca" (K. M. Wilbur and C. M. Yonge, eds.), Vol. 2, pp. 53–96. Academic Press, New York.

Owen, G. (1974). Feeding and digestion in the Bivalvia. *Adv. Comp. Physiol. Biochem.* **5**, 1–35. Academic Press, New York.

Palombi, A. (1934). Gli stadi larvali dei trematodi del Golfo di Napoli. 1. Contributo allo studio della morfologia, biologia e sistematica delle cercarie marine. *Pubbl. Staz. Zool. Napoli* **14**, 51–94.

Panikkar, N. K., and Aiyyar, R. J. (1939). Observations on breeding in brackish-water animals of Madras. *Proc. Indian Acad. Sci., Sect. B* **9**, 343–364.

Paparo, A. (1972). Innervation of the lateral cilia in the mussel, *Mytilus edulis* L. *Biol. Bull. (Woods Hole, Mass.)* **143**, 592–604.

Paparo, A., and Aiello, E. L. (1970). Cilio-inhibitory effects of branchial nerve stimulation in the mussel, *Mytilus edulis. Comp. Gen. Pharmacol.* 1, 241–250.

Pasteels, J. J. (1931). Recherches sur le déterminisme du mode de segmentation des
mollusques lamellibranches (actions des rayons ultra-violets sur l'oeuf de
Barnea candida). *Arch. Biol.* 42, 389.

Pasteels, J. J., and Harven, E. de. (1962). Étude au microscope électronique du
spermatozoïde d'un mollusque bivalve, *Barnea candida. Arch. Biol.* 73, 463–465.

Pathansali, D. (1964). Notes on the biology of the cockle, *Anadara granosa* L. *Indo-
Pac. Fish. Counc. Proc.* 8, 26–31.

Paul, M. D. (1942). Studies on the growth and breeding of certain sedentary organisms in the Madras harbor. *Proc. Indian Acad. Sci. Sect. B* 15, 1–42.

Pelseneer, P. (1912). L'hermaphrodisme chez les lamellibranches. *Verh. Int. Zool.
Kongr., 8th, Graz.* pp. 444–446.

Pelseneer, P. (1935). Essai d'ethologie zoologique d'après l'étude des mollusques.
Acad. R. Belg., Bl. Sci., Publ. Fond. Agathon de Potter 1, 416–596.

Pfitzenmeyer, H. T. (1962). Periods of spawning and setting of the soft-shelled clam,
Mya arenaria, at Solomons, Maryland. *Chesapeake Sci.* 3, 114–120.

Pfitzenmeyer, H. T. (1965). Annual cycle of gametogenesis in the soft-shelled clam,
Mya arenaria, at Solomons, Maryland. *Chesapeake Sci.* 6, 52–59.

Popham, M. L. (1940). The mantle cavity of some of the Erycinidae, Montacutidae,
and Galeommatidae, with special reference to the ciliary mechanisms. *J. Mar.
Biol. Assoc. U.K.* 24, 549–587.

Porter, H. (1964). Seasonal gonadal changes of adult clams, *Mercenaria mercenaria*
(L.), in North Carolina. *Proc. Natl. Shellfish. Assoc.* 55, 35–52.

Posgay, J. A. (1950). Investigations of the sea scallop, *Pecten grandis. Mass. Dep.
Nat. Resour., Div. Mar. Fish., Invest. Methods Improv. Shellfish Resour. Mass.,
Rep.* No. 3, 24–30.

Posgay, J. A. (1953). Sea scallop investigations. *Mass. Dep. Nat. Resour. Div. Mar.
Fish., Invest. Shellfish. Mass., Rep.* No. 6, 9–24.

Posgay, J. A., and Norman, K. D. (1958). An observation on the spawning of the
sea scallop, *Placopecten magellanicus* (Gmelin) on Georges Bank. *Limnol.
Oceanogr.* 3, 478.

Potts, F. A. (1923). The structure and function of the liver of *Teredo*, the shipworm.
Proc. Cambridge Philos. Soc. Biol. Sci. 1, 1–17.

Pratt, D. M. (1959). The phytoplankton of Narragansett Bay. *Limnol. Oceanogr.* 4,
425–440.

Prosser, C. L. (1955). Physiological variation in animals. *Biol. Rev. Cambridge
Philos. Soc.* 30, 229–262.

Puppi, A. (1963). Electrophysiological and pharmacological analysis of serotonin of
the inhibitory mechanism of the posterior adductor muscle of Lamellibranchiata.
Acta Physiol. Acad. Sci. Hung. 23, 259–268.

Purchon, R. D. (1941). On the biology and relationships of the lamellibranch
Xylophaga dorsalis (Turton). *J. Mar. Biol. Assoc. U.K.* 25, 1–39.

Purchon, R. D. (1968). "The Biology of the Mollusca." Pergamon, Oxford.

Quayle, D. B. (1943). Sex, gonad development and seasonal gonad changes in
Paphia staminea Conrad. *J. Fish. Res. Board Can.* 6, 140–151.

Quayle, D. B. (1952). Structure and biology of the larva and spat of *Venerupis
pullastra* (Montagu). *Trans. R. Soc. Edinburgh* 62, 255–297.

Quayle, D. B. (1959). The early development of *Bankia setacea* Tryon. In "Marine

Boring and Fouling Organisms" (D. L. Ray, ed.), pp. 157–174. Univ. of Washington Press, Seattle.

Rae, J. G., III. (1978). Reproduction in two sympatric species of *Macoma* (Bivalvia). *Biol. Bull. (Woods Hole, Mass.)* **155**, 207–219.

Rahaman, A. A. (1965a). Nitrogen content of the lamellibranch *Donax cuneatus* Linnaeus. *Curr. Sci.* **34**(7), 217–218.

Rahaman, A. A. (1965b). The chemical composition of the lamellibranch *Donax cuneatus* Linn. *Proc. Indian Acad. Sci., Sect. B* **62**, 188–194.

Rai, H. S. (1932). The shell fisheries of the Bombay Presidency. *J. Bombay Nat. Hist. Soc.* **32**, 826–847.

Rao, K. P. (1953). Rate of water propulsion in *Mytilus californianus* as a function of latitude. *Biol. Bull. (Woods Hole, Mass.)* **104**, 171–181.

Rao, K. P. (1954). Tidal rhythmicity of the rate of water propulsion in *Mytilus* and its modifiability by transplantation. *Biol. Bull. (Woods Hole, Mass.)* **106**, 353–359.

Rao, K. S. (1967). Annual reproductive cycle of the wedge clam, *Donax cuneatus* Linnaeus. *J. Mar. Biol. Assoc. India* **9**, 141–146.

Rao, K. V. (1951). Observations on the probable effects of salinity on the spawning, development, and setting of the Indian backwater oyster, *Ostrea madrasensis* Preston. *Proc. Indian Acad. Sci. Sect. B* **33**, 231–256.

Rao, K. V. (1952). Studies on the growth of *Katelysia opima* (Gmelin). *Indo-Pac. Fish. Counc., Proc., Sect. II*, pp. 94–102.

Rao, K. V., Narasimham, K. A., and Alagarswami, K. (1962). A preliminary account of the biology and fishery of the razor shell, *Solen kempi* Preston, from Ratnagiri in Maharashtra State. *Indian J. Fish.* **9**, 542–579.

Rattenbury, J. C., and Berg, W. E. (1954). Embryonic segregation during early development of *Mytilus edulis*. *J. Morphol.* **95**, 393–414.

Raven, C. P. (1958). Morphogenesis: The Analysis of Molluscan Development." Pergamon, New York.

Raven, C. P. (1961). "Oogenesis: The Storage of Development Information." Pergamon, New York.

Raven, C. P. (1964). Development. *In* "Physiology of Mollusca" (K. M. Wilbur and C. M. Yonge, eds.), Vol. 1, pp. 165–195. Academic Press, New York.

Raven, C. P. (1972). Chemical embryology of Mollusca. *In* "Chemical Zoology" (M. Florkin and B. T. Scheer, eds.), pp. 155–185. Academic Press, New York.

Rebhun, L. I. (1962). Dispersal of the vitelline membrane of the eggs of *Spisula solidissima* by alkaline, isotonic NaCl. *J. Ultrastruct. Res.* **6**, 123–134.

Reddiah, K. (1962). The sexuality and spawning of manx pectinids. *J. Mar. Biol. Assoc. U.K.* **42**, 683–703.

Rees, C. B. (1950). The identification and classification of lamellibranch larvae. *Hull Bull. Mar. Ecol.* **3**, 73–104.

Rees, C. B. (1951). Continuous plankton records: First report on the distribution of lamellibranch larvae in the North Sea. *Hull Bull. Mar. Ecol.* **3**, 105–334.

Rees, C. B. (1954). The distribution of lamellibranch larvae in the North Sea, 1950–51. *Hull Bull. Mar. Ecol.* **4**, 21–46.

Reid, R. G. B. (1969). Seasonal observations on diet and stored glycogen and lipids in the horse clam, *Tresus capax* (Gould, 1850). *Veliger* **11**, 378–381.

Reinboth, R., ed. (1975). "Intersexuality in the Animal Kingdom." Springer-Verlag, Berlin and New York.

Renzoni, A., and Giusti, F. (1972). Further observations on the reproductive cycle of *Mytilus galloprovincialis* Lam. In "Proc. Eur. Mar. Biol. Symp. 5th, Venice." (B. Battaglia, ed.), pp. 295–300. Piccin Editore, Padua, Italy.

Retzius, G. (1904). Zur Kenntnis der Spermien der Evertebraten I. *Biol. Untersuch.* 11, 1–32.

Reverberi, G. (1966). Electron microscopy of some cytoplasmic structures of the oocytes of *Mytilus. Exp. Cell Res.* 42, 392–394.

Reverberi, G. (1967). Some observations on the ultrastructure of the ovarian *Mytilus* egg. *Acta Embryol. Morphol. Exp.* 10, 1–14.

Reverberi, G. (1971). *Mytilus.* In "Experimental Embryology of Marine and Fresh Water Invertebrates" (G. Reverberi, ed.), pp. 175–187. North-Holland Publ., Amsterdam.

Reverberi, G., and Mancuso, V. (1961). The constituents of the egg of *Mytilus* as seen in the electron microscope. *Acta Embryol. Morphol. Exp.* 4, 102–121.

Rokop, F. J. (1974). Reproductive patterns in the deep-sea benthos. *Science* 186, 743–745.

Ropes, J. W. (1968a). Hermaphroditism in the surf clam, *Spisula solidissima. Proc. Natl. Shellfish. Assoc.* 58, 63–65.

Ropes, J. W. (1968b). Reproductive cycle in the surf clam, *Spisula solidissima* in offshore New Jersey. *Biol. Bull. (Woods Hole, Mass.)* 135, 349–365.

Ropes, J. W., and Stickney, A. P. (1965). Reproductive cycle of *Mya arenaria* in New England. *Biol. Bull. (Woods Hole, Mass.)* 128, 315–327.

Runnström, S. (1927). Über die Thermopathie der Fortpflanzung und Entwicklung mariner Tiere in Beziehung zu ihrer geographischen Verbreitung. *Bergens Mus. Arbok, Naturvitensk. Rekke* No. 2, 1–67.

Runnström, S. (1936). Die Anpassung der Fortpflanzung und Entwicklung mariner Tiere an die Temperaturverhaltnisse verscheidener Verbreitungsgebiete. *Bergens Mus. Arbok Naturvitensk. Rekke* No. 3, 1–36.

Runnström, S., Hagstrom, B. E., and Perlmann, P. (1959). Fertilization, In "The Cell" (J. Brachet and A. E. Mirsky, eds.), Vol. 1, pp. 327–397. Academic Press, New York.

Ryther, J., and Yentsch, C. S. (1958). Primary production of continental shelf waters off New York. *Limnol. Oceanogr.* 3, 327–335.

Sagara, J. (1958). Artificial discharge of the reproductive elements of certain bivalves caused by treatment of sea water and by injection with NH_4OH. *Nippon Suisan Gakkaishi* 23, 505–510.

Saleuddin, A. S. M. (1964). The gonads and reproductive cycle of *Astarte sulcata* (Da Costa) and sexuality in *A. elliptica* (Brown). *Proc. Malacol. Soc. London* 36, 141–148.

Saleuddin, A. S. M. (1965). The mode of life and functional anatomy of *Astarte* spp. (Eulamellibranchia). *Proc. Malacol. Soc. London* 36, 229–257.

Sanders, H. L., and Hessler, R. R. (1969). Ecology of the deep-sea benthos. *Science* 163, 1419–1424.

Sastry, A. N. (1955). Studies on the biology of some Mollusca. M.S. Thesis, Andhra Univ., Waltair, India.

Sastry, A. N. (1961). Studies on the bay scallop, *Aequipecten irradians concentricus* Say, in Alligator Harbor, Florida. Ph.D. Thesis, Florida State Univ., Tallahassee.

Sastry, A. N. (1962). Some morphological and ecological differences in two closely related species of scallops, *Aequipecten irradians* Lamarck and *Aequipecten gibbus* Dall, from the Gulf of Mexico. *Q. J. Fla. Acad. Sci.* 25, 89–95.

Sastry, A. N. (1963). Reproduction of the bay scallop, *Aequipecten irradians* Lamarck. Influence of temperature on maturation and spawning. *Biol. Bull. (Woods Hole, Mass.)* **125**, 146–153.

Sastry, A. N. (1965). The development and external morphology of pelagic larval and post-larval stages of the bay scallop, *Aequipecten irradians concentricus* Say, reared in the laboratory. *Bull. Mar. Sci.* **15**, 417–435.

Sastry, A. N. (1966a). Temperature effects in reproduction of the bay scallop, *Aequipecten irradians* Lamarck. *Biol. Bull. (Woods Hole, Mass.)* **130**, 118–134.

Sastry, A. N. (1966b). Variation in reproduction of latitudinally separated populations of two marine invertebrates. *Am. Zool.* **6**, 374–375.

Sastry, A. N. (1968). Relationships among food, temperature and gonad development of the bay scallop, *Aequipecten irradians* Lamarck. *Physiol. Zool.* **41**, 44–53.

Sastry, A. N. (1970a). Reproductive physiological variation in latitudinally separated populations of the bay scallop, *Aequipecten irradians* Lamarck. *Biol. Bull. (Woods Hole, Mass.)* **138**, 56–65.

Sastry, A. N. (1970b). Environmental regulation of oocyte growth in the bay scallop, *Aequipecten irradians* Lamarck. *Experientia* **26**, 1371–1372.

Sastry, A. N. (1975). Physiology and ecology of reproduction in marine invertebrates. *In* "Physiological Ecology of Estuarine Organisms" (F. J. Vernberg, ed.), pp. 279–299. Univ. of South Carolina Press, Columbia.

Sastry, A. N., and Blake, N. J. (1971). Regulation of gonad development in the bay scallop, *Aequipecten irradians* Lamarck. *Biol. Bull. (Woods Hole, Mass.)* **140**, 274–283.

Savage, N., and Goldberg, R. (1976). Investigation of a practical means of distinguishing *Mya arenaria* and *Hiatella* sp. larvae in plankton samples. *Proc. Natl. Shellfish. Assoc.* **66**, 42–53.

Sawada, N. (1952). Experimental studies on the maturation division of eggs in *Mactra veneriformis*. I. Inhibitory effects of the body fluid. *Mem. Ehime Univ. Sect. II* **1**, 231.

Sawada, N. (1954a). Experimental studies on the maturation division of eggs in *Mactra veneriformis*. IV. On the effect of certain polysaccharides. *Mem. Ehime Univ. Sect. II* **B2**, 89.

Sawada, N. (1954b). Experimental studies on the maturation division of eggs in *Mactra veneriformis*. V. On the activation by periodate. *Mem. Ehime Univ. Sect. II* **B2**, 93.

Sawada, N., and Murakami, T. H. (1959). Experimental studies on the maturation division of eggs in *Mactra veneriformis*. VI. Histochemical study on eggs. *Mem. Ehime Univ. Sect. II* **3**, 235–241.

Schechter, V. (1956). The effect of water upon gametes, upon maturation and upon fertilization and cleavage. *Exp. Cell Res.* **10**, 619–630.

Segal, E. (1970). Light. *In* "Marine Ecology, A Comprehensive Treatise on Life in Oceans and Coastal Waters" (O. Kinne, ed.), Vol. 1, Part 1, pp. 159–211. Wiley (Interscience), New York.

Segal, E., Rao, K. P., and James, T. W. (1953). Rate of activity as a function of intertidal height within populations of some littoral molluscs. *Nature (London)* **172**, 1108–1109.

Sellmer, G. (1967). Functional morphology and ecological life history of the gem clam, *Gemma gemma* (Eulamellibranchia; Veneridae). *Malacologia* **5**, 137–223.

Shaw, W. N. (1965). Seasonal gonadal cycle of the male soft-shell clam, *Mya arenaria*, in Maryland. *U.S. Fish Wildl. Serv., Spec. Sci. Rep.—Fish.* No. 508, 1–5.

Sindermann, C. J. (1970). "Principal Diseases of Marine Fish and Shellfish." Academic Press, New York.

Sleigh, M. A., ed. (1974). "Cilia and Flagella." Academic Press, New York.

Smayda, T. (1957). Phytoplankton studies in Lower Narragansett Bay. *Limnol. Oceanogr.* **2**, 342–359.

Sniesziko, E., ed. (1970). A symposium on diseases of fishes and shellfish. *Am. Fish. Soc. Spec. Publ.* No. 5.

Soot-Ryen, T. (1959). Reports of the Lund University Chile Expedition 1948–49. 35. Pelecypoda. *Acta Univ. Lund, Sect. 2* **55**.

Sprague, L. (1965). Observations on *Chytridiopsis mytilovum* (Field), formerly *Haplosporidium mytilovum* Field, (Microsporida). *J. Protozool.* **12**, 385–389.

Srinivasan, V. V. (1963). The distribution of glycogen in *Martesia fragilis*, a common woodborer of Madras. *Curr. Sci.* **32**, 211–213.

Srinivasan, V. V., and Krishnaswamy, S. (1964). The chemical composition of *Martesia fragilis*, a wood-boring pholad of Madras. *Zool. Jahrb., Abt. Allg. Zool. Physiol. Tiere* **70**, 539–546.

Stefano, G. B., and Aiello, E. (1975). Histofluorescent of serotonin and dopamine in the nervous system and gill of *Mytilus edulis* (Bivalvia). *Biol. Bull. (Woods Hole, Mass.)* **148**, 141–156.

Stephens, G. C. (1968). Dissolved organic matter as potential source of nutrition for marine organisms. *Am. Zool.* **8**, 95–106.

Stephens, G. C., and Schinske, R. A. (1961). Uptake of amino acids by marine invertebrates. *Limnol. Oceanogr.* **6**, 175–181.

Stephenson, A. (1934). The breeding of reef animals. Pt. 2. Invertebrates other than corals. *Rep. Great Barrier Reef Comm.* **3**, 247–272.

Stevenson, R. N., and South, G. R. (1975). Observations on phagocytosis of *Cocomyxa parasitica* (Cocomyxacea, Chlorococcales) in *Placopecten magellanicus*. *J. Invertebr. Pathol.* **25**, 307–312.

Stickney, A. P. (1963). Histology of the reproductive system of the soft-shell clam (*Mya arenaria*). *Biol. Bull. (Woods Hole, Mass.)* **125**, 344–351.

Stickney, A. P. (1964). Salinity, temperature and food requirements of soft-shell clam larvae in laboratory culture. *Ecology* **45**, 283–291.

Stoll, E. (1938). Sur la mode de locomotion de quelques mollusques marins. *Trav. Sta. Biol. Roscoff* **16**, 3–33.

Sugiura, Y. (1959). Seasonal changes in sexual maturity and sexuality of *Mytilus edulis*. *Nippon Suisan Gaikkaishi* **25**, 1–6.

Sugiura, Y. (1962). Electrical induction of spawning in two marine invertebrates (*Urechis unicinctus* and hermaphroditic *Mytilus edulis*). *Biol. Bull. (Woods Hole, Mass.)* **123**, 203–206.

Sullivan, C. M. (1948). Bivalve larvae of Malpeque Bay, Prince Edward Island. *Bull., Fish. Res. Board Can.* **77**, 1–36.

Sweeney, D. (1963). Dopamine: Its occurrence in molluscan ganglia. *Science* **139**, 1051.

Swift, H., Rebhun, L. I., Rasch, E., and Woodward, J. (1956). The cytology of nuclear RNA. *In* "Cellular Mechanisms in Differentiation and Growth" (D. Rudnik, ed.), pp. 45–59. Princeton Univ. Press, Princeton, New Jersey.

Takahashi, K., and Mori, K. (1971). Seasonal variation in metabolism of lipid and glycogen in scallops, *Patinopecten yessoensis* (Jay). I. Biochemical studies. *Tohoku J. Agric. Res.* **22**, 114–133.

Takatsuki, S. I. (1934). On the nature and functions of the amoebocytes of *Ostrea edulis*. *Q. J. Microsc. Sci.* **76**, 379–431.

Tanaka, S., and Hatano, H. (1952). Studies on the seasonal changes in the chemical constituents of the pearl oyster. *Publ. Seto Mar. Biol. Lab.* **2**, 341–355.

Tang, S. I. (1941). The breeding of the escallop (*Pecten maximum* L.) with a note on the growth rate. *Proc. Liverpool Biol. Soc.* **54**, 9–28.

Theede, H. (1963). Experimentelle Untersuchunger über die Filtration sleistung der Miermuschel, *Mytilus edulis* L. *Kiel. Meeresforsch.* **19**, 20–41.

Thompson, R. J. (1977). Blood chemistry, biochemical composition of tissues and the annual reproductive cycle in the giant scallop, *Placopecten magellanicus* (Gmelin) from southeast Newfoundland. *J. Fish. Res. Board Can.* **34**, 2104–2116.

Thorson, G. (1935). Biologische Studien über die Lamellibranchier *Modiolaria discors* L. and *Modiolaria nigra* Gray in Ostgronland. *Zool. Anz.* **111**(11/12), 297–304.

Thorson, G. (1936). The larval development, growth and metabolism of arctic marine bottom invertebrates, etc. *Medd. Groenl.* **100**, 1–155.

Thorson, G. (1946). Reproduction and larval development of Danish marine bottom invertebrates, with special reference to planktonic larvae in the sound (Oresund). *Medd. Dan. Fisk.- Havunders. (Ser. Plankton)* **4**, 1–523.

Thorson, G. (1950). Reproduction and larval ecology of marine bottom invertebrates. *Biol. Rev. Cambridge Philos. Soc.* **25**, 1–45.

Thorson, G. (1964). Light as an ecological factor in the dispersal and settlement of marine bottom invertebrates. *Ophelia* **1**, 167–208.

Tiffany, W. J. (1971). The tidal migration of *Donax variabilis* Say (Mollusca: Bivalvia). *Veliger* **14**, 82–85.

Tilney, L. G., and Mooseker, M. S. (1976). Actin filament-membrane attachment: Are membrane particles involved? *J. Cell Biol.* **71**, 402–416.

Tranter, D. J. (1958a). Reproduction in Australian pearl oysters (Lamellibranchia). III. *Pinctada albina* (Lamarck): Breeding season and sexuality. *Aust. J. Mar. Freshwater Res.* **9**, 191–216.

Tranter, D. J. (1958b). Reproduction in Australian pearl oysters (Lamellibranchia). I. *Pinctada albina* (Lamarck) Primary gonad development. *Aust. J. Mar. Freshwater Res.* **9**, 135–143.

Tranter, D. J. (1958c). Reproduction in Australian pearl oysters (Lamellibranchia). IV. *Pinctada margaritifera* (Linneaus). *Aust. J. Mar. Freshwater Res.* **9**, 509–525.

Tranter, D. J. (1958d). Reproduction in Australian pearl oysters (Lamellibranchia). II. *Pinctada albina* (Lamarck): Gametogenesis. *Aust. J. Mar. Freshwater Res.* **9**, 144–158.

Tranter, D. J. (1959). Reproduction in Australian pearl oysters (Lamellibranchia). V. *Pinctada fucata* (Gould). *Aust. J. Mar. Freshwater Res.* **10**, 45–66.

Trevallion, A. (1971). Studies on *Tellina tenuis* Da Costa. III. Aspects of general biology and energy flow. *J. Exp. Mar. Biol. Ecol.* **7**, 95–122.

Turner, H. J., and George, C. J. (1955). Some aspects of the behavior of the quahoug, *Venus mercenaria*, during the early stages. *Mass. Dep. Nat. Resour., Div. Mar. Fish., Invest. Shellfish. Mass., Rep.* No. 8, 5–14.

Turner, H. J., and Hanks, J. E. (1960). Experimental stimulation of gametogenesis in *Hydroides dianthus* and *Pecten irradians* during the winter. *Biol. Bull. (Woods Hole, Mass.)* **119**, 145–152.

Turner, R. D. (1954). The family Pholadidae in the western Atlantic and eastern Pacific. Part I:–Pholadidae. *Johnsonia* **3**, 33–162.

Turner, R. D., and Johnson, A. C. (1971). Biology of wood-boring molluscs. *In* "Marine Borers, Fungi and Fouling Organisms of Wood" (E. B. G. Jones and S. K. Eltingham, eds.), pp. 259–301. Org. Econ. Co-op. Dev., Paris.

Tyler, A. (1949). Properties of fertilizin and related substances of eggs and sperm of marine animals. *Am. Nat.* **83**, 195–219.

Ukeles, R. (1971). Nutritional requirements in shellfish culture. *Proc. Conf. Artif. Propag. Commer. Valuable Shellfish-Oysters, Coll. Mar. Stud., Univ. Delaware* pp. 43–64.

Ukeles, R. (1975). Views in bivalve larvae nutrition. *Proc. Int. Conf. Aquacult. Nutr., 1st, Sea Grant Program Univ. Delaware Co-op. U.S.-Jpn. Aquacult. Panel* pp. 127–162.

Umiji, S. (1969). Neurosecreção em *Mytilus perna* (Molusco–Lamelibrânquio). *Zool. Biol. Mar. (Sao Paulo) Nova Ser.* **26**, 181–254.

Uzmann, J. R. (1952). *Cercaria myae* nov. sp., a forktailed larva from the marine bivalve, *Mya arenaria*. *J. Parasitol.* **38**, 161–164.

Uzmann, J. R. (1953). *Cercaria milfordensis* nov. sp., a microcercous trematode larva from the marine bivalve, *Mytilus edulis* L., with special reference to its effect on the host. *J. Parasitol.* **39**, 445–451.

van Dam, L. (1954). On the respiration in scallops (Lamellibranchia). *Biol. Bull. (Woods Hole, Mass.)* **107**, 192–202.

van Weel, P. B. (1961). The comparative physiology of digestion in molluscs. *Am. Zool.* **1**, 245–252.

Vassallo, M. T. (1973). Lipid storage and transfer in the scallop *Chlamys hericia*. *Comp. Biochem. Physiol. A* **44**, 1169–1175.

Venkataraman, R., and Chari, T. (1951). Studies on oysters and clams: Biochemical variation. *Indian J. Med. Res.* **39**, 533–541.

Vernberg, F. J. (1962). Comparative physiology: Latitudinal effects on physiological properties of animals. *Annu. Rev. Physiol.* **25**, 517–546.

Verwey, J. (1952). On the ecology of distribution of cockle and mussel in the Dutch Wadden Sea. Their role in sedimentation and the source of their food supply. With a short review of the feeding behavior of bivalve molluscs. *Arch. Neerl. Zool.* **10**, 171–239.

Verwey, J. (1954). De mossel in zijn eisen. *Faraday 24e*, 13 pp. No. 2. (With English summary.)

Verwey, J. (1957). Discussion. Colloque International de Biologie Mer, St. Roscoff. *Annee Biol.* **33**, 238.

Von Oertzen, J. A. (1972). Cycles and rates of reproduction of six Baltic Sea bivalves of different zoogeographic origin. *Mar. Biol.* **14**, 143–149.

Wada, K. (1978). Chromosome karyotypes of three bivalves, *Isognomon alatrus* and *Pinctada imbricata*; and the bay scallop, *Argopecten irradians*. *Biol. Bull. (Woods Hole, Mass.)* **155**, 235–245.

Wada, S. K. (1947). Fertilization in *Mytilus edulis*, with special reference to the sperm acrosome reaction. *Mem. Coll. Fish., Kagoshima Univ.* **4**, 105–112. (With English summary.)

Wada, S. K. (1954). Spawning in the tridacnid clams. *Jap. J. Zool.* **11**, 273–285.

Wada, S. K. (1968). Mollusca. In "Invertebrate Embryology" (M. Kumé and K. Dan, eds.; J. C. Dan, transl.), pp. 485–525. Publ. for U.S. Natl. Libr. Med., Washington, D.C. by Nolit Publ. House, Belgrade.

Wada, S. K., and Wada, R. (1953). On a new pearl oyster from the Pacific coast of Japan, with special reference to the cross-fertilization with another pearl oyster, Pinctada martensii (Dunker). J. S. Oceanogr. Soc. Jpn. 8, 127–138.

Wada, S. K., Collier, J. R., and Dan, J. C. (1956). Studies on the acrosome. V. An egg-membrane lysin from the acrosome of Mytilus edulis spermatozoa. Exp. Cell Res. 10, 168–180.

Walne, P. R. (1956). Experimental rearing of larvae of Ostrea edulis L. in the laboratory. Fish. Invest. Minist. Agric. Fish. Food (G.B.) Ser. II Salmon Freshwater Fish. 20, 1–23.

Walne, P. R. (1963). Breeding of the Chilean oyster (Ostrea chilensis Philippi) in the laboratory. Nature (London) 197, 676.

Walne, P. R. (1964). The culture of marine bivalve larvae. In "Physiology of Mollusca" (K. M. Wilbur and C. M. Yonge, eds.), Vol. 1, pp. 197–210. Academic Press, New York.

Walne, P. R. (1965). Observations on the influence of food supply and temperature on the feeding and growth of the larvae of Ostrea edulis L. Fish Invest. Minist. Agric. Fish. Food (G.B.) Ser. II Salmon Freshwater Fish. 24, 1–45.

Wells, H. W., and Gray, J. E. (1960). The seasonal occurrence of Mytilus edulis on the Carolina coast as a result of transport around Cape Hatteras. Biol. Bull. (Woods Hole, Mass.) 119, 550–559.

Welsh, J. H. (1961). Neurohormones of mollusca. Am. Zool. 1, 267–272.

Werner, B. (1939). Über die Entwicklung und Artsunterscheidung von Muschellarven des Nordseeplanktons, unter besonderer Berucksichtigung der Schalenentwicklung. Zool. Jahrb., Abt. Anat. 66, 237–270.

White, K. M. (1937). Mytilus. Mem. Liverpool Mar. Biol. Comm. 37, 1–117.

Wiborg, K. F. (1946). Undersokelser over obskjellet (Modiola modiolus). Fiskeridir. Skr. Ser. 8, 1–85.

Widdows, J. (1973). Effect of temperature and food on the heart beat, ventilation rate and oxygen uptake of Mytilus edulis. Mar. Biol. 20, 269–276.

Widdows, J. (1978a). Combined effects of body size, food concentration and season on the physiology of Mytilus edulis. J. Mar. Biol. Assoc. U.K. 58, 109–124.

Widdows, J. (1978b). Physiological indices of stress in Mytilus edulis. J. Mar. Biol. Assoc. U.K. 58, 125–142.

Wilbur, K. M. (1964). Shell formation and regeneration. In "Physiology of the Mollusca" (K. M. Wilbur and C. M. Yonge, eds.), Vol. I, pp. 243–282. Academic Press, New York.

Williams, C. S. (1969). The effect of Mytilicola intestinalis on the biochemical composition of mussels. J. Mar. Biol. Assoc. U.K. 49, 161–173.

Williams, R. B., and Murdoch, M. B. (1966). Phytoplankton production in the Beaufort Channel, North Carolina. Limnol. Oceanogr. 11, 73–82.

Wilson, B. R. (1968). Survival and reproduction of the mussel Xenostrobus securis (Lamarck) (Mollusca, Bivalvia, Mytilidae) in Swan estuary, Western Australia, Part I. Salinity tolerance. J. Nat. Hist. 2, 307–328.

Wilson, B. R. (1969). Survival and reproduction of the mussel Xenostrobus securis (Lamarck) (Mollusca, Bivalvia, Mytilidae) in Western Australia. Part II. Reproduction, growth and longevity. J. Nat. Hist. 3, 93–102.

Wilson, B. R., and Hodgkin, E. P. (1967). A comparative account of the reproductive cycles of five species of marine mussels (Bivalvia: Mytilidae) in the vicinity of Fremantle, Western Australia. *Aust. J. Mar. Freshwater Res.* **18**, 175–203.

Wilson, D. P. (1960). Some problems in larval ecology related to the localized distribution of bottom animals. In "Perspectives in Marine Biology" (A. A. Buzzati-Traverso, ed.), pp. 87–103. Univ. of California Press, Berkeley.

Winter, J. (1969). Uber den Einfluss der Nahrungs-konzentration und anderer Faltoren auf Feltrierleistung und Nahrungoausnutzung der Muscheln *Arctica islandica* und *Modiolus modiolus*. *Mar. Biol.* **4**, 87–135.

Winter, J. (1970). Filter feeding and food utilization in *Arctica islandica* L. and *Modiolus modiolus* L. at different food concentrations. In "Marine Food Chains" (J. H. Steele, ed.), pp. 196–206. Univ. of California Press, Los Angeles.

Woods, F. H. (1931). History of the germ cells in *Sphaerium striatinum* (Lam.). *J. Morphol.* **51**, 545–595.

Woods, F. H. (1932). Keimbahu determinants and continuity of the germ cells in *Sphaerium striatinum* (Lam). *J. Morphol.* **53**, 345–365.

Wright, P. B., and Moore, H. B. (1970). A contribution to the ecology of *Cyclinella tenuis* (Mollusca:Bilvalvia). *Bull. Mar. Sci.* **20**, 793–801.

Yamamoto, G. (1951a). On acceleration of maturation and ovulation of the ovarian eggs *in vitro* in the scallop, *Pecten yessoensis* Jay. *Sci. Rep. Tohoku Univ., Ser.* 4 **19**, 161–166.

Yamamoto, G. (1951b). Ecological study on the scallop, *Pecten (Patinopecten) yessoensis* in Mutsu Bay. *Nippon Suisan Gakkaishi* **17**, 53–56.

Yentsch, C. S. (1963). Primary production. *Oceanogr. Mar. Biol.* **1**, 157–175.

Yonge, C. M. (1928). Structure and function of the organs of feeding and digestion in the septibranchs *Cuspidaria* and *Poromya*. *Philos. Trans. R. Soc. London, Ser.* B **216**, 221–263.

Yonge, C. M. (1930). "A Year on the Great Barrier Reef." Putnam, London.

Yoshida, H. (1953). Studies on larvae and young shells of industrial bivalves in Japan. *J. Shimonoseki College Fish.* **3**, 1–106.

Young, R. T. (1941). The distribution of the mussel (*Mytilus californianus*) in relation to the salinity of its environment. *Ecology* **22**, 379–386.

Young, R. T. (1942). Spawning season of the California mussel, *Mytilus californianus*. *Ecology* **23**, 490–492.

Young, R. T. (1946). Stimulation of spawning in the mussel, *Mytilus californianus*. *Ecology* **26**, 58–69.

Zobell, C. E., and Feltham, C. B. (1938). Bacteria as food for certain invertebrates. *J. Mar. Res.* **1**, 312–327.

Zobell, C. E., and Landon, W. A. (1937). The bacterial nutrition of the California mussel. *Proc. Soc. Exp. Biol. Med.* **36**, 607–609.

Zs Nagy, I., Rosa, K., Foldes, I., Perenyi, L., Salanki, J., and Demeter, D. (1965). Subcellular localization of 5-hydroxytryptamine in the central nervous system of lamellibranchiates. *J. Neurochem.* **12**, 245–251.

Chapter 6

PELECYPODA: OSTREIDAE

Jay D. Andrews

6.1 Introduction 293
6.2 Sexual Reproduction 298
    6.2.1 Anatomy and Sexuality 298
    6.2.2 Fecundity 301
    6.2.3 Reproductive Behavior 303
    6.2.4 Spawning 305
6.3 Development 307
    6.3.1 Embryonic Development 307
    6.3.2 Larval Life 308
6.4 Setting (Spatfall or Settlement) 322
    6.4.1 Distribution of Spatfall 324
    6.4.2 Gregariousness and Delay of Metamorphosis 331
    6.4.3 Survival of Spat 334
    References 335

6.1 Introduction

Oysters of the two common genera exhibit marked adaptations in reproductive methods to climate, turbidity, and salinity of their habitats. *Ostrea* is essentially marine (polyhaline) and occurs most abundantly in rich clear-water bays of the upwelling west coasts of continents in oceanic climates. These coasts are strongly moderated by oceanic currents and provide cool summers for breeding and relatively warm winters for nearly continuous growth and glycogen storage. *Crassostrea* is estuarine (euryhaline, turbid waters) in habitat, hence tolerates silty, low-salinity waters where hot summers and cold winters prevail. Continental climates, where this genus predominates, provide warmer waters for breeding but also lower winter temperatures that usually induce a period of dormancy among oysters. Estuaries, although usually rich, fluctuate widely in nutrient supply and in temperature and salinity regimes, hence in kinds and quantities of food available. Numerous exceptions to these generalizations may be cited, but the evolution of the two genera reflects these environmental patterns (Andrews, 1971). Importation and culture of exotic species have tended to obscure nature's long-term patterns of adaptation and evolution.

The taxonomy and pertinent morphology for separating oyster genera are summarized by Gunter (1950); general distributions are given for some species. Four of the 10 contrasting characters given for the two genera are related to breeding activities: (1) The average size is smaller in *Ostrea*; (2) eggs are large in *Ostrea* and small in *Crassostrea*, with corresponding variations in number; (3) eggs are brooded in *Ostrea* and released freely in the water by *Crassostrea*; and (4) gill openings (ostia) are large and small, respectively. These characteristics may reflect adaptations to different environments, which ensure successful reproduction.

Oysters exhibit striking variations in reproductive traits as specialized bivalves with a great lability of sex organs. Care of young ranges widely to parallel tropical and cold-water marine animals in dependence on holoplanktonic larvae and retention of larvae until metamorphosis, respectively. Species in the family Ostreidae may produce enormous numbers of small eggs with completely planktotrophic larvae in *Crassostrea*, or relatively fewer large eggs and full brooding of larvae in the mantle cavity of some *Ostrea*. The oviparous species depend upon warm waters and tidal transport mechanisms to maintain populations, as do most tropical sedentary animals. Brooding species in colder waters and open circulation release advanced larvae, hence reduce periods of free-swimming larvae and avoid scarcity of food for early larval stages. Consequently, in *Ostrea*, eggs are large and fewer in number in proportion to the completeness of brooding and the prospect of failure of free-living larvae. The key factors are temperature and openness of marine water that affects dispersion. Tolerance of low salinities, so common in fish larvae, has not become a "solution to dispersion" in the evolution of *Ostrea* larvae. The opportunistic *Crassostrea* species are much more likely to form reefs and to accumulate huge populations as fugitives in the harsh physical environments of estuaries where interspecific competition is suppressed. Neither group spurns off-bottom attachment, but *Ostrea* species are more likely to be adapted to utilize other organisms and natural surfaces away from bottom silt. Gunter (1950) lists common species by genus and discusses the habitats of the two types of oysters first compared by Orton (1928).

Much taxonomic confusion has been caused by the use of shell (conchological) characters in naming genera and species of oysters. Many ecotypes and ecomorphic forms are produced by varying environmental conditions that include temperature, salinity, food, bottom type, intertidal exposure, and density. The plasticity of shell growth in response to microhabitats and various symbionts and epibionts increases variability. Ranson (1939, 1948, 1967) used prodissoconchs to define and distinguish extant genera. Thomson (1954) reviewed the conchological, malacological, and physiological characteristics relating to the taxonomy of oysters. The first two categories are only remotely related to reproduction, except for the size of ostia. Thomson notes the lack of comparative physiology but describes adaptations of incubatory (*Ostrea*) and nonincubatory (*Crassostrea* and *Pycnodonte*) genera to turbidity and salinity. *Pycnodonte* is best known as a fossil genus but prefers clear marine waters, as does *Ostrea*. He points out that only *Crassostrea* is adapted for intertidal life, although his statement that certain species seldom occur elsewhere may be a reflection of survival rather than preference. Thomson gives generic descriptions for living oysters and descriptions of 10 species native to Australia. This variety of Australian species, including 5 *Crassostrea*, 4 *Ostrea*, and 1 *Pycnodonte*, seems to exceed that found for other continents [North America —1, 3, and 1, respectively (Gunter, 1951)], although more may be expected from tropical regions. Also, the observation of smaller size in *Ostrea* species based on North American fauna is not clear in Australian oysters—size and age of breeding oysters are important in determining relative fecundity. Yonge (1960) reports 15 species of oysters in eastern Asia but only 1 species of *Ostrea*. Revisions of oyster taxonomy have been proposed by Stenzel (1971).

All ostreids are adherent species with shells cemented to firm substrates by the left valve. Although derived from lamellibranch stocks of ancient lineage, they are highly specialized for their sedentary habit. Their complex morphological and physiological adaptations indicate an advanced rather than a primitive systematic status. Unlike other monomyarians, they have evolved with a loss of foot and byssal apparatus, except briefly in larval life, and have relied upon increased reproductive effort to perpetuate the species. Most monomyarians have become motile (scallops of the genera *Pecten* and *Chlamys*) or, if sedentary (pearl oysters, *Pinctada* with a byssus, and thorny oysters, *Spondylus*, cemented), are limited to tropical waters. Once committed to a sedentary life, ostreids specialized in food collection and reproductive adaptations. The efficiency of filtration (up to 400 liters/day in *Crassostrea*) permits rapid growth, early sexual maturity, and probably release of more sexual products than by any other invertebrates. Nevertheless, while playing the "numbers game," some ostreids evolved the habit of incubation in the branchial chamber. Incubation of eggs and larvae is not common in marine bivalves (see Section 5.4.4.3), although usually necessary in freshwater groups for lack of food and tidal currents. Yonge (1960) considers oviparity in *Crassostrea* to be primitive, but to me it appears to be a further specialization from probable incubatory ancestors. If oviparity in oysters were really primitive, would not egg expulsion be through the exhalant chamber, the common method in most bivalves, rather than by the difficult and elaborate process of forcing eggs back through the gill openings and out the inhalant chamber? In tidal currents, the presumed advantages of ejection by clapping the valves over expulsion in a vigorous exhalant current are not evident to me as necessary for dispersion. A determination of the method used by the tropical, oviparous *Pycnodonte* may be enlightening (Yonge, 1960). All three ostreid genera are old and appear unchanged in conchological characters as far back as the Miocene. All three were living together in warm, polyhaline Miocene seas on the Virginia coast, as evidenced by fossils in the Yorktown cliffs.

The literature on oysters is voluminous. It is often stated that oysters are the best known of all invertebrate animals. Three extensive bibliographies (Baughman, 1948; Ranson, 1951; Joyce, 1972) with thousands of entries lend this impression but, as oysters are commercial species, these bibliographies include many localized, field-type surveys of limited general value. Critical subjects such as food preferences and utilization, dispersion of larvae and aggregation of spat, and genetics are poorly known or have only recently become fields of inquiry. For example, a flurry of papers on sex change in the late 1920s and 1930s followed belatedly the discovery of consecutive sexuality in *Crepidula* by Orton (1909). Yet the physiological causes and the ecological purposes of protandry and hermaphroditism still are not clear.

Several recent works provide excellent accounts of oyster biology. "Oysters" (Yonge, 1960) is a superb semipopular short book by an author with wide research experience with molluscs and a personal acquaintance with most commercial oyster species. The book offers a selected bibliography of key papers by region. A comprehensive 480-page monograph by Galtsoff (1964) on *Crassostrea virginica* gives the biology in detail with 80 pages on reproduction. It is a scholarly, technical work, profusely illustrated, and includes major original contributions by the author representing a lifetime of work on oysters. An important critical review of the status of oyster biology by Korringa (1952) is still a prime source consulted by ostreologists. Two older semipopular books provide excellent introductions to oyster biology, particularly for the incubatory *Ostrea edulis* of western Europe. Ranson's (1951) "Les Huitres, Biologie-Culture, Bibliographie," in French, includes a large bibliography. Orton's (1937) book entitled "Oyster Biology and Oyster Culture" is a short easy-reading account but is somewhat dated and no longer readily available.

The "Nelson school" of marine biologists centered at Rutgers University in New Jersey has had a major influence on American shellfish research and culture activities throughout the twentieth century. Nelson (1957a) provides a combination of history, review, and his current views near the end of a long life devoted primarily to oysters. A glimpse of a lifetime of personal experiences and an introduction to the literature may be obtained from this popular article, which is the closest he came to a long-promised book on the field ecology of oysters.

Recent publications on oyster culture, often concerned with the manipulation of oyster populations to ensure reproductive success and a supply of seed oysters, include an excellent bulletin, "Pacific Oyster Culture in British Columbia," by Quayle (1969). This is a thorough account of the culture of *Crassostrea gigas*, a continental climate species introduced into the oceanic climate of western North American coasts. Matthiessen (1970) provides a review and compilation of oyster culture and the oyster industry in North America by region. Bardach *et al.* (1972) reviewed oyster culture in major producing countries, including information on cultching and hatchery techniques, in a book on aquaculture. Other regional accounts of culture operations emphasizing reproduction have also appeared (Curtin, 1971).

The operations of shellfish hatcheries have been reviewed recently (Loosanoff and Davis, 1963a,b; Landers, 1968; Davis, 1969; Hidu *et al.*, 1969; Walne, 1974; Hidu and Richmond, 1974). Extensive studies on salinity and temperature tolerances and food requirements of larvae have been made at the Milford Laboratory by Loosanoff and Davis and associates (Davis and Guillard, 1958; Ukeles, 1971). Detailed studies on the culture requirements of *O. edulis* larvae have been made at Conway, Wales, by Walne (1964b, 1966, 1974). Scottish investigations at Millport of the breeding cycles of *O. edulis* and the effects of water quality on the growth of larvae were partly field-oriented (Millar, 1964; Millar and Scott, 1968). Imai *et al.* (1950) cultured larvae in Japan at low densities in large tanks using filtered water and added boiled starch. The colorless, naked flagellate *Monas*, which feeds on bacteria, was used to inoculate outdoor tanks and provide food for larvae.

Genetic studies of oysters and experiments to crossbreed species and genera are few and as yet mostly inconclusive. Imai and Sakai (1961) did some of the earliest work with races of *C. gigas*. Breeding temperatures, size, and growth rates varied from northern to southern races in

Japan, much as was found in *C. virginica* on the North American Atlantic coast. Japanese biologists have cultured successfully most of the important oyster species of both genera, but successful crossbreeding has been limited to *C. gigas* and *Crassostrea angulata*. Menzel (1968a), working with five species of *Crassostrea*, obtained hybrids of four species including a cross of *C. virginica* and *C. gigas* that earlier investigators had failed to obtain. This last cross has been repeated frequently by Dupuy *et al.* (1977) under hatchery conditions with successful crosses using males and females of each species. The diploid number for all species of oysters examined has been 20 chromosomes with no evident morphological variation (Menzel, 1968b; Longwell and Stiles, 1970). The prospects for interspecific hybrids are good as methods of culture improve and laboratories become equipped to handle foreign imports safely in quarantine. However, the inbreeding of siblings appears to reduce the fertility of some pairs and stocks—presumably through lethal genes (Longwell and Stiles, 1970, 1973). Dupuy and Andrews (unpublished data) have bred siblings through five generations in several lines with marked stunting of growth and sterility of some progeny.

Some subjects and many references, reviewed adequately by Korringa (1941, 1952), Yonge (1960), Galtsoff (1964), and others just cited, are not repeated in this chapter. Moreover, reproductive events found in bivalves in general are covered in detail by Sastry in Chapter 5 of this volume. Rather, summary accounts of phases of reproduction in areas of uncertainty and detailed presentations of new information are given. Some statements throughout the text are based upon observations and research of the author during 32 years of oyster studies. An introduction to the comparative reproductive physiology of invertebrates is given by Giese (1959) and Giese and Pearse (1974). A review of reproductive and larval ecology of marine bottom invertebrates is presented by Thorson (1950).

6.2 Sexual Reproduction

6.2.1 Anatomy and Sexuality

The reproductive organs of ostreids are exceedingly simple in structure and function, and sexuality is highly labile. In the resting stage of the annual cycle of reproduction, primordia are undifferentiated and inconspicuous. The space to be occupied by gonads is filled with Leydig cells containing glycogen. No discrete sex organs persist. Sex is difficult to determine unless relict sex products persist or developing gonia are present. In the warm season, branching gonadal tubules fill the visceral mass around the digestive system and out to the mantle epithelia. Follicular cells are poorly developed, hence nutrient transfer is apparently direct from the blood to germinal cells. The nourishment of gametocytes probably occurs directly from Leydig cell storage tissues permeated by blood sinuses that bathe all the tissues with blood. The digestive tubules and gonadal system penetrate storage or connective tissues without discrete organ boundaries or separating membranes. The gonadal tubules in the periphery of the visceral mass become gonoducts, often visible through the mantle epithelia, and these are lined with cilia along the outer halves of their lumina for transport of sex products. These tubules end in a pair of genital pores that open into the exhalant chamber ventral to the adductor muscle. Accessory sex organs are absent. When temperatures become too low for gametogenesis, blood cells quickly invade and digest disorganized follicles and remaining sex products, and storage of glycogen begins for the next cycle. If temperature cycles are slow to change or ambivalent, oysters may retain sex products for long periods after failure to spawn completely. The repair of injured or fragmented tissues is extremely rapid in oysters of all ages—gill and gonadal tissues are two notable examples of this facile capacity. Entry to recent literature on the reproductive cycles of various oyster species may be obtained from Dinamani (1974).

The anatomy of sex organs, cytology of gametogenesis, and microstructure of eggs and sperm are presented in detail by Galtsoff (1964) (see also Sections 5.3.3 and 5.3.5) with comparisons among ostreid species and with other bivalves. Much of this well-illustrated study represents original work by Galtsoff.

Most bivalves are unisexual; hence, the occurrence of protandry and consecutive sexual change in ostreids and a few other genera has stimulated much interest (see also Section 5.3.2). True functional hermaphroditism probably does not occur in oysters, for the genetic handicap of self-fertilization is too great for a group committed to outcrossing and to a high reproductive potential for survival. Protandry is firmly established in both genera and usually occurs in a very high proportion of oysters. Most investigators have reported sex ratios of populations by age, but this has little comparative meaning in species with variable growth rates and ranging from tropical to cold-temperate zones. Where temperature regimes are favorable for reproduction, size is probably more important than age. Early maturity is notable, for spat less than

1 in. in length and only 6 weeks of age have been observed with ripe gonads in Chesapeake Bay, and earlier sexual development occurs in warm Gulf of Mexico waters. Mostly males but some females were found among first-summer spat. Sex ratios in the James River change from 90% males in yearlings (35 mm) to about the reverse ratio in old oysters (Table I). It has been suggested that favorable nutritive

TABLE I

Sex Ratios of Oysters (*Crassostrea virginica*) by Sizes in Virginia as Determined by Fresh Smears [a]

Size groups		Oysters examined				Percentages	
Average length (mm)	Average weight (gm)	Total	♂	♀	? [b]	♂	♀
		James River					
		Wreck Shoal, 6 July 1951					
36	9	156	129	21	6	83	17
45	21	127	73	52	2	57	43
58	30	114	43	70	1	38	62
69	46	37	5	32	0	14	86
		Brown Shoal, 15 June 1951					
36	10	27	15	2	10	88	12
42	17	45	23	6	16	79	21
51	30	74	44	18	12	71	29
54	35	12	6	4	2	60	40
61	42	10	5	5	0	50	50
		Nansemond Ridge, 15 June 1951					
48	21	117	95	5	17	95	5
60	40	151	93	36	22	72	28
70	60	43	19	19	5	50	50
80	—	90	14	47	29	23	77
		Rappahannock River					
		Hoghouse Bar, 2 August 1951					
35	7	113	101	12	0	89	11
42	13	95	79	16	0	83	17
55	33	41	29	12	0	71	29
64	50	123	71	52	0	58	42
79	81	65	27	38	0	42	58
88	116	58	22	36	0	38	62
103	113	34	7	27	0	21	79

[a] Andrews, unpublished data.
[b] Undetermined for lack of sex products.

conditions may induce the female phase initially in some oysters, and it has been shown in other molluscs (e.g., *Crepidula*) that proximity to females inhibits change in males (Coe, 1953; see also, Volume 4, Section 1.3.2). A normal pattern of sex ratios with increase in size was found in oysters in the nutritively poor James River despite severe stunting.

In *Crassostrea*, individuals usually function as males or females for a reproductive season and sex changes occur following winter resting periods. Fewer than 1% of *C. virginica* were found to possess both male and female sex products simultaneously (Andrews, unpublished data). These specimens were obviously changing sexes, and some gametes remained in follicles from a previous spawning while developing new gonia were attached to epithelia. As oysters mature in size and growth slows, only about 10% of oviparous oysters exhibit sex changes annually, as demonstrated by sampling and observation of spawning in marked individuals (Needler, 1942; Galtsoff, 1964). Populations of large, old oysters usually exhibit a reversal of sex ratios, with many more females than males, although changes in sex may occur in either direction.

Oysters of the genus *Ostrea* exhibit regular alternating sexuality, with mature individuals usually functioning consecutively as females and males in the same reproductive season if temperatures and time permit. The change from female to male gametogenesis is very rapid and, if the gonads are not emptied with each spawning, both sex products may be found. These repeated alternations of sex appear to continue throughout life. This situation is commonly called hermaphroditism, whereas it appears to be merely the overlapping of two consecutive phases and has no functional significance in nature. As genetic work with oysters accelerates, this genus may offer unique opportunities for self-fertilization, which can only be accomplished in *Crassostrea* by freezing sperm and hoping for a change in sex. One would expect *Ostrea* to be "programmed" for complete release of gonadal products by the female where brooding prevents multiple spawning so common in *Crassostrea*.

6.2.2 Fecundity

Fecundity of oysters is exceptionally great, but its importance in oyster culture is nebulous and deceptive. Success of reproduction depends more upon physical and biological attributes of habitats than the number of gametes released. Reproductive activity of oysters is wasteful and tends to produce too few or too many spat in most areas and regions. In Europe some estimates of the number of millions of brood oysters involved in a particular fishery area have been made. In the United States, few attempts have been made to inventory brood stocks, and intensive sets often occur when oyster stocks are lowest in abundance. Empirical observations over a period of years provide the best evidence for the reproductive potential of an area.

Among factors affecting fecundity are size and age, nutrition, diseases and parasites, and duration of the breeding period. *Crassostrea* is much more prolific than *Ostrea*. A healthy oyster usually converts all stored glycogen into sex products during seasons when temperatures are high enough for maturation. In native *C. virginica* along the western Atlantic coast, spawning may be intermittent throughout the year in the warmest climates, or it may occur only once in northern climes and be suppressed in some years. The most favorable conditions for culture would be expected to be somewhere between the extremes, such as Chesapeake Bay where there are 4 months when the waters are too cold for significant feeding (the marketing season), 3 months that are too warm for production of the best kinds and quantities of food, and only 5 months of rapidly changing temperatures for oysters to grow and store energy for reproduction. Oceanic climates with moderated cold and warm seasons are more favorable for growth and reproduction of adapted species. If regularity of successful spatfalls is the criterion for reproductive success, the numerous small oysters with many spawnings south of Chesapeake Bay are more effective than the big, plump oysters to the north. Yet diseases and predators are far more numerous and active in the warmer regions. Studies on fecundity are more of biological interest than practical use.

Counts of eggs or larvae for oysters of various ages and sizes have yielded widely varying totals. In general, specimens of *Crassostrea* spawn or produce about 100 times as many eggs as those of the genus *Ostrea*. Seasonal egg counts of *Ostrea* are more reliable than those of *Crassostrea*, for presumably new spawning does not occur while eggs are being incubated in the branchial chamber. It is not known to what extent losses or partial abortions occur. Embryos or larvae may be removed from the branchial chamber, stained, and counted by sampling. The counts range from about 100,000 in 1-year-old *O. edulis* to nearly 2 million in 4-year-olds, but less than a million is a more common total (Cole, 1941). The eggs are large enough (150 μm) to provide for about 8–10 days of incubation. *Ostrea lurida* yielded about 250,000 larvae from 2-in. oysters with an egg size of about 100 μm (Hopkins, 1937).

Spawning is rarely completed in one release in *Crassostrea*, except perhaps in cool waters where minimal temperatures delay spawning and permit only one opportunity. Although spawnings of 10–20 million eggs are routine in *C. virginica*, Galtsoff (1964) reported about 100 million on occasion. It is impossible to estimate how many eggs (only 50 μm in diameter) are released by a large southern oyster with multiple spawnings and successive gonadal buildups over periods of 6 months or more.

On the western coast of North America, *C. gigas* grows much larger than the eastern oyster, but temperatures in most areas barely reach the minimum for spawning, hence massive gonads 10–12 mm thick develop in the rich waters. However, egg counts have yielded relatively low numbers, perhaps because of partial spawnings.

6.2.3 Reproductive Behavior

Spawning and larval behavior are very complex in comparison with the simplicity of reproductive organs in oysters. A large literature exists on the commercial species. Many apparent contradictions of timing and behavior probably reflect the wide variations in environmental parameters from one locality or region to another. Hence it is difficult to generalize. It is important to note that most commercial species in the Northern Hemisphere have wide geographic ranges. *Crassostrea virginica* occurs from the Gulf of St. Lawrence to Mexico and Latin America. This wide range is permitted and enhanced by special hydrographic features such as shallow, enclosed waters, e.g., Prince Edward Island surrounded by the cold waters of the Gulf of St. Lawrence. *Crassostrea gigas* occurs widely along the eastern coast of Asia, and it has been introduced and produced commercially in milder climates such as Australasia (imported 1947 to 1952), the western coast of North America (imported late 1800s), and more recently western Europe (imported 1966) (Menzel, 1974; Marteil, 1976). *Ostrea edulis* is found from Norwegian pols (oyster ponds) south and through the Mediterranean Sea, whereas *O. lurida* has a wide distribution from Alaska to lower California.

Generalizing about reproductive methods is complicated by the occurrence of races or ecotypes within species. For example, Chesapeake Bay races of *C. virginica* usually do not spawn in Long Island Sound because temperatures are too low (Loosanoff and Nomejko, 1951a). They do fatten and may develop large gonads and will spawn if moved inshore to warmer waters of more shallow, enclosed estuaries.

There are also exceptions to the general pattern of incubatory species prevailing on oceanic-type coasts and nonincubatory ones in continental climates. *Crassostrea angulata* occurs naturally on the coast of Spain and Portugal but became established in France about 1868 and was grown in Great Britain as relaid stock. It was ravaged in France by a gill disease and replaced by importations of *C. gigas* in the early 1970s. Yonge (1960) remarks that the distribution of this species is puzzling, but he concludes that it was in southern Europe too early to be introduced from Asia where the closely related *C. gigas* is indigenous (Menzel, 1974). There is no indigenous *Crassostrea* on the western coast of North America; *C. virginica* was imported regularly for many decades from the east coast but did not reproduce successfully except in a few localities. It is no longer of commercial importance on the west coast. However, *C. gigas*, with continued imports of seed from Japan, has become the commercial oyster. *Ostrea* occurs in the warm, high-salinity southeastern regions of North America (*O. equestris* and *O. frons*) and of Asia (*O. denselamellosa*).

It is not accurate to infer that *Ostrea* are cold water oysters and that *Crassostrea* prefer warm waters simply because some species of *Ostrea* occur in high latitudes. The incubatory genus breeds at cooler temperatures (usually 15°–20°C) but cannot tolerate freezing or long exposure intertidally summer or winter, and it is not adapted for long periods of winter hibernation, particularly if these are attended by the lowered salinities typical of estuaries.

The famous diked oyster parks of France, described by Dean (1890) and apparently adapted for Olympia oysters (*O. lurida*) by oystermen on the western coast of North America, are a concession to these requirements. The shallow, ponded flats are usually diked near low-tide level to hold water when tides are out. This prevents excessive warming in summer and freezing in winter. In contrast, *Crassostrea* requires warm waters (usually 20°–25°C) for breeding and tolerates some freezing, intertidal exposure, and long periods of winter dormancy (up to 7 months). Menzel (1955) presents a long list of contrasting characters of the two groups in one of the few studies comparing the genera growing together in the same habitat. To his list could probably be added that *Crassostrea* grows faster and breeds earlier (and at a smaller size) than *Ostrea* in the same environment. The smallness and slow growth of *Ostrea* are illustrated by beds of Olympia oysters about 2 in. in length that are 4 or 5 years old. This long period is for marketing and not to ensure successful reproduction in *O. lurida*.

6.2.4 Spawning

The spawning of oysters has been described frequently from laboratory observations (Galtsoff, 1964; Yonge, 1960), although few have observed it in nature in open waters. However, photographs and observations were made in Pendrell Sound, British Columbia, by Quayle (1969). Females of *Ostrea* spawn unnoticed, for no eggs are released into the water, although careful observations reveal the rhythmic contractions required to transfer eggs from the suprabranchial chamber to the gill chamber. Males release clumps of sperm or "sperm balls" in a steady flow into the exhalant current without adductor muscle contractions. The eggs released in the suprabranchial chamber are forced through the water tubes and hormone-relaxed ostia of the gills into the branchial chamber by a series of contractions. Relaxation of the adductor muscle permits the valves of the shell to open by ligament expansion, and the mantle edges extend as curtains to close the exhalant chamber. Once in the gill chamber, eggs are expelled through a narrow opening in the mantle closure of the inhalant chamber in *Crassostrea* but are retained on the gills for incubation in *Ostrea*. Eggs of *Crassostrea* are expelled in batches by contractions of the adductor muscle and are dispersed by currents.

Rapidity of gonadal maturation is quite variable, as reported in the literature (see Section 5.3.7). High temperatures increase the rate of ripening, but the 3–5 days observed by Loosanoff and Davis (1952) for Long Island oysters to mature from winter dormancy is far shorter than the usual 4–6 weeks required in Chesapeake Bay.

Males tend to mature sex products earlier than females and are more easily stimulated to spawn, usually by a rise in temperature or exposure to eggs or sperm. In nature females probably do not spawn until sperm enter the water transport system and provide stimuli. Prolonged manipulation and chemicals (Walne, 1974) may be used in the laboratory to induce spawning (see Section 5.3.9.1). In *Crassostrea*, oysters from colder regions of their range mature and spawn more readily in the laboratory than those from warmer areas where temperature appears less effective as a stimulus (Loosanoff, 1969). The occurrence of physiological races by regions has led to much variation in the rate of maturation, the success of artificial spawning, and the degree and duration of stimuli required to effect it (Loosanoff and Nomejko, 1951a). Galtsoff (1964) gives a detailed account of spawning for *C. virginica*, and information on other species. Males spawn much easier and more quickly

than females. Southern oysters (*C. virginica*) are more refractory to
stimuli than northern races and often require many hours of manipulation in the laboratory. Usually only a few of any group of ripe oysters
spawn upon stimulation, and culturists prefer to begin with at least
20–30 oysters. This ensures the spawning of both males and females,
for only a fraction of the oysters can be spawned unless they have
been held in a ripe gonad condition at reduced temperatures for a
period. The inability to spawn individual oysters at will is a serious
handicap in genetic studies. It is a common practice in laboratories
and hatcheries to hold ripe oysters at temperatures below the spawning
level for the local race to facilitate spawning. The short periods of
latency from stimulation to spawning, reported by Galtsoff to be as
short as 6–38 min, are not usually found in southern oysters. Storage at
moderate refrigerator temperatures usually permits renewal of spawning upon return to warmed water, particularly in males (Andrews,
unpublished data).

The reproductive period tends to be longer in *Ostrea* than in *Crassostrea* because spawning begins at lower temperatures. In regions that
may be considered about the middle of its range, *Ostrea* typically releases larvae for about 5 months beginning in May, and *Crassostrea*
spawns for 3 months (Chesapeake Bay) beginning in late June. Larvae
are found in plankton samples or in brooding larviparous oysters
throughout these long breeding seasons. This means that multiple spawnings of populations occur in both genera, at least weekly in *Crassostrea*,
even if larvae life exceeds the usual 10–20 days. From observations of
the condition of gonads, the extent and timing of spawning may be
estimated in oviparous oysters. To obtain these many broods of larvae,
repeated spawnings by the same individuals are supplemented by those
of individuals of various ages and gonadal condition, influenced by the
effects of depth and temperature of waters on timing of maturation.
All these factors contribute to multiple, localized, mass spawnings. In
Crassostrea, gametogenesis continues as long as nutritive state and
temperature conditions permit. Hence many oysters are found in the
fall with residual sex products which may not all be resorbed before
winter. These observations are based on sampling of larvae, monitoring
of spatfalls, and tens of thousands of stained sections of oysters for
disease studies (Andrews, unpublished data).

A tendency for lunar periodicity of spawning in *O. edulis* has been
reported by Orton (1926), Korringa (1941), and Walne (1974) with
swarming (release of larvae) occurring about 10 days after spring tides.
The patterns of larval abundance are not sharply defined, for some
larvae are found throughout the breeding season. Several investigators

have reported no periodicity in larval abundance in *C. virginica* (Loosanoff and Nomejko, 1951b) and my own unpublished data on setting patterns for larvae also show no periodicity. The occurrence of larvae or spat is an unsatisfactory method of studying reproductive periodicity, however, because so many external factors may alter the timing of larval growth and spatfall.

One advantage of the simple structure of the organs of reproduction in oysters is the feasibility of stripping or removing ripe sex products from the gonads without spawning. This has been practiced for many years at the Virginia Institute of Marine Science (VIMS), with larvae and juveniles maturing as readily as from spawned oysters. It is a particularly simple method of obtaining larvae on schedule in hatcheries, although the loss of parents is a serious handicap for genetic studies. The gonads are slashed with a razor blade, and sex products pressed or washed out of the gonadal tubules. In many other bivalve molluscs stripping has not produced viable embryos.

6.3 Development

6.3.1 Embryonic Development

Mass spawning, induced by mutual stimulation of the sexes through hormones associated with sex products, ensures rapid fertilization and embryonic development (Nelson and Allison, 1940). In oviparous species, it is essential that sperm be quickly available in tidal waters that soon carry eggs away from the spawners. It is probable that spawning occurs during slack-water periods where tides are not strong. In incubatory species, fertilization must occur within the branchial or suprabranchial chambers. Ciliary motion in the gills and in the water tubes is not reversed during the transfer of eggs from exhalant to inhalant chambers, hence a current bearing released sperm continues through the water transport system, interrupted only by valve clapping to transfer or expel eggs. Males usually spawn first, and an abundance of sperm ensures quick fertilization of eggs. It is probable that, in *Ostrea*, fertilization occurs almost as rapidly as eggs are transferred to the branchial chamber, and possibly sperm from passing water currents attach to the vitelline membrane when eggs enter the suprabranchial chamber. The apparent requirement that sperm be passed through the water transport organs to induce normal female spawning suggests closer observations of spawning *Crassostrea* to determine if sperm attachment occurs within the branchial chambers of females. In artificial spawning in hatcheries, it is necessary to limit the number of sperm added in closed containers, or polyspermy may occur and interfere with normal development. There is little evidence of failure of fertilization in natural spawnings.

Early divisions occur rapidly at normal spawning temperatures, and swimming trochophore larvae appear in 6–8 hr. Shelled veligers develop in 24–30 hr in *Crassostrea*, whereas the appearance of a shell requires 3–4 days in *Ostrea*. In the laboratory, temperature shock and toxins are most likely to cause abnormal larvae which may be misshapen or exhibit defective swimming.

A detailed description of the structure and physiology of eggs and sperm and of the early embryology of oysters is given by Galtsoff (1964) in his monograph on *C. virginica*, in which he includes notes on, and references to, other species. See Sastry in this volume (Section 5.4.1) for a detailed account of embryonic development in bivalves in general.

6.3.2 Larval Life

The reproductive system of oysters is highly efficient physiologically and anatomically in producing large numbers of fertilized eggs. In contrast, larval life constitutes a period of enormous losses, and the literature indicates great confusion about the causes. The problem of loss of young is not peculiar to oysters (Thorson, 1950), but the great disparity between potential and success of reproduction focuses effort and attention on ostreids. Theories and contradictions about larval behavior and settlement almost match the number of authors or studies. Galtsoff (1964) notes the inadequacy of sampling techniques for field studies and the lack of knowledge of larval response to environmental parameters. He concludes that only two general assumptions can be made about the behavior of larvae—that their own power of locomotion is very limited and that they are dispersed beyond the vicinity of spawning grounds.

The confusion about larval life is probably related to the great variety of habitats that oysters successfully occupy, hence there may be more reality to different published observations and conclusions than reviews imply. The range of habitats utilized by oysters is astonishing, and comparisons may be informative concerning larval ecology and behavior. Following Thorson's (1957) dictum, simpler marine ecosystems occur in cold climates in restricted habitats or on stable substrates (e.g., flat, uniform sea bottoms). The Bras d'Or lakes in Nova Scotia (Medcof, 1955), Pendrell Sound, (Quayle, 1969), and the fiord pols of Norway are good examples of oyster communities with ecosystems simplified by physical parameters that have been studied. All these cold habitats provide stratified, shallow, warm surface waters, a short breeding period, and limited tidal and wind turbulence where single broods (spawnings) of larvae may be studied effectively. Major oyster-producing centers are located in warmer climates, more brackish habitats, and areas with stronger wind or tidal mixing, which greatly complicate larval ecology in time and space. Such areas exhibit, respectively, a longer breeding period, greater salinity and turbidity changes, and increased flushing and mixing rates, resulting in greater losses and dispersions.

The incubatory *Ostrea* species of western Europe and western North America tend to inhabit estuaries dominated by oceanic waters rather than freshwater runoff. These regions provide the relatively clear, salty waters this genus prefers for growth. Reproduction usually occurs in more enclosed river estuaries with freshwater inflow and lower salinities. *Ostrea* often lives in regions of large tidal ranges and utilizes warm surface waters of deep estuaries or fiords where strong vertical stratifications occur. This generalized description of the endemic habitats of *Ostrea* does not preclude survival, with human help, in less favorable areas (e.g., the Gulf of Maine).

Crassostrea virginica of eastern North American coasts is typical of this genus and lives in shallow, vigorously mixed waters that fluctuate widely from hot to cold temperatures, low to high salinities, and clear to muddy waters. Most endemic areas are shallow, but estuaries, lagoons, open oceanic bays, and nearly closed salty lakes provide extremely varied habitats both physically and biologically, including intertidal zones in warm climes. Oysters have probably evolved many more local races than are suggested by the three minimal breeding temperatures of Stauber (1950) for *C. virginica*. Therefore, published tolerances for parameters such as salinity or temperature for larvae and adults cannot be relied upon throughout the ranges of species.

A penetrating review of the vicissitudes of pelagic life is given by Korringa (1952), who conducted many experiments in the Oosterschelde in the Netherlands (Korringa, 1941). High fecundity and pelagic life have evolved to attain wide dispersal, but payment is made in high losses of larvae. These losses are directly dependent upon duration of larval life, which is determined mainly by temperature but secondarily by availability of food and attachment sites (see Section 5.4.3). Allowing for racial variations, most *Ostrea* species need temperatures of 20°C or higher for rapid growth of larvae, whereas *Crassostrea* prefer temperatures of 25°C or higher. Hatcheries that routinely set *C. virginica* larvae in 10 days from spawning—to exceed 15 days is costly and unsatisfactory—usually hold water temperatures near 30°C. A few degrees of temperature change alter the larval period drastically. Food deficiency probably causes many prolonged cultures in hatchery operations, but this factor has not been evaluated in nature. It is perhaps symptomatic of culture problems that the naked flagellates (*Monochrysis* and *Isochrysis*) used to feed *Crassostrea* larvae in North American laboratories were isolated in Great Britain and must be grown at temperatures of about 20°C. Most laboratory cultures of larvae grown in Virginia on these algal species for food have required 15 or more days, whereas a nearby hatchery using centrifuged natural water reports that 10 days of culture to setting is routine. Food species recently isolated from endemic waters (Dupuy et al., 1977) have reduced larval culture to 8–10 days. The wide variability in size of larvae in a given culture may also reflect food problems, for it is much less apparent in plankton samples and in incubated larvae in brooding oysters.

Oysters of the genus *Ostrea*, which are known to inhabit more open, oceanic types of areas than the oviparous genus, do not seem to gain much in reduction of larval losses and effective dispersal due to lower temperatures and slower growth during planktonic larval life. After 8–10 days of incubation the 180 μm *O. edulis* larvae must spend another 10 days (temperature-dependent) of pelagic life to acquire a setting size of about 300 μm (Walne, 1964a). Yet, *O. equestris* sets abundantly on oil riggings in the open Gulf of Mexico (Gunter, 1951) and in dry seasons invades salty coastal lagoons (Menzel, 1955). This is a subtropical species in distribution, yet it breeds primarily in the Gulf at times when temperatures are typical (20°C) for partially incubating *Ostrea* species (Menzel, 1955).

In contrast, *Ostrea* species that breed in cold waters (10°–12°C) exhibit full incubation of very large eggs and larvae. Hollis (1963) gives fecundity of *Ostrea lutaria* as 1 million eggs from 200 to 270 μm in diameter, and about 20 days' incubation. Larvae of 440 μm settle promptly after a very short free-swimming phase. Walne (1963) found in *O. chilensis* many fewer large eggs (323 × 264 μm) and suggested an incubation period of 5–6 weeks at temperatures of 13°–15°C. These cold-water species apparently breed throughout much of the year—8 months according to Hollis. He states that *O. lutaria* becomes hermaphroditic (functional?) after the second and third breeding seasons. In these species of *Ostrea*, pelagic life has almost been eliminated in response to cold, open waters, as exhibited by many other invertebrates in similar environments (see, for example, Sections 2.3.9 and 5.4.4.3.3).

6.3.2.1 LARVAL BEHAVIOR

The behavior of oyster larvae is a controversial subject. Enormous wastage of marine planktonic larvae is accepted, yet many investigators feel obliged to invent for them patterns of behavior that fit transport systems such as Bousfield (1955) reports for barnacles. Certain basic premises may be derived from laboratory and field observations of larvae. Larvae swim continuously night and day in standing water containers in the laboratory and do not need currents to keep them in suspension. Larvae swim upward in laboratory containers, predominantly in loose spirals, at a rate of nearly 1 cm/sec. Swimming is interrupted by periods of rest and sinking at about the same rate if the velum is withdrawn. In warm surface waters of fiords, larvae must swim continuously until setting, for the underlying waters are cold and often sulfurous. Larvae have a definite capacity to change their vertical position in open waters, and this ability increases with size.

In strong light, larvae close their valves and sink, but will resume swimming if the light persists. In moderate or indirect light, larvae raft in shadows or under shade-producing objects and often align themselves in peculiar vertical patterns. These irregular patterns of distribution in containers appear to be negative responses to light, for uniformity is observed in darkness. Light clearly has a stimulatory effect on larvae in clear waters but, in natural waters with silt and plankton, the effect is often limited to a few meters at the surface.

A second factor known to stimulate and inhibit larval movement is salinity. Haskin (1964) has shown that increasing salinity causes larvae of *C. virginica* to swim in small laboratory containers. In laboratory cultures changing the salinity suddenly from 30‰ to 20‰ causes larvae to drop to the bottom, but the next day they are swimming again. Haliclines probably affect vertical movement of larvae, but in river estuaries gradients are seldom very sharp or strong during the breeding season.

Temperature affects rate of growth and duration of larval life through rate of feeding. In shallow habitats, temperature gradients are small during warm breeding seasons, but in deep fiords larvae are restricted to warm surface layers by strong thermoclines and must respond to thermal gradients to survive. Larvae in nature have a strong capability for remaining in masses or layers of water with favorable temperatures, as exemplified by Westley (1956) in Dabob Bay, Washington.

Many other factors believed by various investigators to influence vertical and horizontal distribution have been discussed by Korringa (1941). Among these are current velocity, tidal stage, wave action or turbulence, pH, wind action, and the three fctors (light, temperature, and salinity) already discussed as limiting larval distribution under special circumstances. Korringa concludes that *O. edulis* larvae are passively distributed by tides and mixing, at least until the setting stage is reached.

A theory that larvae rise from the bottom on flood tides and drop back to the bottom on ebb tides was proposed long ago by Julius Nelson (1917) of New Jersey. This has been elaborated and adjusted to the presumed two-layered transport system of these estuaries by his successors, as summarized by Carriker (1951) and Nelson (1957b). The behavior of larvae has long been a topic of disagreement, hence an analysis of new and old studies is presented here in considerable detail.

Most investigators assume that oyster larvae deliberately descend to the bottom for setting, hence have attempted to show that mature larvae are more abundant there. In my experience larvae tend to set more intensively off the bottom if suitable cultch is available. Where oysters occur on natural reefs, there probably is a premium on setting high on the reef, thereby avoiding siltation and smothering and ensuring the best supply of moving water for food collection. The occurrence of *Crassostrea* oysters on intertidal structures such as pilings is commonplace and very impressive in areas of high setting because of reduced predation. Quayle (1964) seems to support these views, since he states that *C. gigas* is selective as to the type of surface on which larvae set but exercises little choice as to where the surface is. *Crassostrea gigas* settled in quantity up to the level of the highest tide, although mortality ensued there. In the clear waters of Pendrell Sound, it would be interesting to relate the vertical distribution of larval stages to setting patterns in respect to day and night periods. The logical conclusion is that in most oyster habitats scarcity of off-bottom surfaces forces larvae to search the bottom for settlement, but they may swim up many meters and be transported many kilometers between searches. There are also many opinions about time of day and tidal stage at which oyster larvae prefer to set. Hidu (1967) believes that increased temperatures cause larvae to set on tidal flats during the daytime. In areas where mature larvae are extremely abundant (Cape May, New Jersey, and Pendrell Sound), setting occurs immediately after the exposure of cultch without special requirements of time, place, or conditions such as bacterial coatings (Quayle, 1964; Hidu, 1967).

6.3.2.2 LOSSES FROM PREDATION

Losses of larvae during pelagic life are obviously high, but assessment of causes is difficult. Predation and dispersion are probably major causes, although disease has not been evaluated. Predation on bivalve larvae by a great variety of filter and particulate feeders is accepted by most biologists, but there is little information on the importance of these losses. Cerruti (1941) gives a long list of plankton-feeding organisms observed to eat larvae or to contain them in their digestive tracts. Bivalve molluscs, tunicates, barnacles, annelid and crustacean larvae, young herrings, cyprinodonts, cnidarians, ctenophores, and even polychaetes catch and eat bivalve larvae or entangle them with mucus. In preliminary experiments, adult oysters in 2-liter beakers removed most of several thousand vitally stained oyster larvae in 2–4 hr (Andrews, unpublished data). Other filter feeders including bivalve molluscs, tunicates, and barnacles were less efficient but were effective. Most invertebrates with ciliary and mucus systems of feeding are enemies of bivalve larvae. Cyprinodont fishes snap up larvae (individually) in small aquaria. It appears that oyster larvae incur much risk in attempting to settle on bottoms with populations of oysters, barnacles, and sea squirts. Predation on larvae deserves more attention than it has received. Thorson (1950) discusses transport of larvae by currents, and enemies of invertebrate larvae, and agrees with Cerruti (1941) and Korringa (1941) that predation is probably the most important cause of wastage of planktonic larvae.

6.3.2.3 DISPERSION OF LARVAE

Dispersal of oyster larvae is poorly understood because following larval broods always involves many problems and difficulties. Usually hydrographic regimes have not been known or appreciated sufficiently to plan sampling of larvae. Attempts to fit all oyster areas into common patterns of larval distribution are futile. The physical parameters of the shallow, stratified lagoons of the New Jersey marshlands and the open, turbulent, deep waters of Long Island Sound almost preclude similarity of behavior and abundance of larvae. Larval studies of the shallow, clear, pondlike lakes of Bras d'Or yield results quite different from those for the muddy, tidal James River estuary. Tides, winds, freshwater flow, and morphometry of the basin determine the amount of stratification and mixing and rates of flushing that occur. The abundance of larvae is extremely variable in time and place, hence only favored areas retain larvae in swarms available for sampling in adequate numbers. Sampling usually has been too sporadic and generally inadequate to indicate distribution of larvae and occurrence of stages in time and space. Therefore relationships between larval behavior and transport systems are poorly understood. A comparison of studies on trap-type estuaries and lagoons with those on open, flushing-type river estuaries may clarify larval dispersion problems.

The most satisfactory studies of larval life have involved trap-type bodies of water. This designation encompasses a wide variety of habitats characterized by low flushing rates or strong stratifications that limit vertical mixing. Typical trap-type seed-producing areas in Chesapeake Bay include St. Mary's River, Maryland (Manning and Whaley, 1954), Great Wicomico River and Piankatank River in Virginia (Andrews, unpublished data), and others in Maryland that exhibit low runoff, small tidal amplitude, and restricted entrances (e.g., a sill). Shallow flats, marshlands, and coves also contribute to low flushing rates, hence larvae are trapped. It is characteristic of these seed oyster areas that larvae tend to accumulate toward the upper ends of the saltwater sector of the tributaries.

Other habitats in the low-flushing category that have been studied successfully are the shallow lagoons or ponds of New Jersey (Carriker, 1961), the Bras d'Or lakes, (Medcof, 1955), Pendrell Sound, (Quayle, 1964 and 1969), and the Oosterschelde (Korringa, 1941). These range from almost tideless lakes to deep, stratified fiords in physical characteristics. The essential feature is a low loss of larvae per tidal cycle, which Korringa estimated at 4% for the Oosterschelde. The Oosterschelde is thoroughly mixed, whereas Pendrell Sound is strongly stratified, although both have large tidal ranges.

The seaside of the eastern shore of Virginia and similar areas in the Carolinas and Georgia are high-salinity habitats that appear superficially to be distinctive because of moderately high tidal amplitudes for this coast. These areas exhibit intensive spatfalls where spat predators are numerous. Flushing directly into the ocean, these habitats might be expected to experience large larval losses because of tidal amplitude, but low runoff and numerous shallow lagoons or bays with wide fringes of emergent salt marshes apparently compensate for flushing.

Studies on the larval distribution and abundance of C. gigas in Willapa Bay, Washington, by Sayce and Larson (1965) demonstrate roughly the rate of losses in a bay or lagoonal system with considerable tidal amplitude. Working with essentially a single brood of larvae at minimal temperatures, they showed the progression of stages to setting with daily or less frequent samples at several fixed stations. Although sample counts fluctuated widely, progression of the brood was evident at each of five stations from straight-hinge to setting larvae in 10–14

days. The number of larvae declined from about 10,000/60-gal sample
to less than 1/1-gal of setting size larvae usually. Hourly samples at
seven depths from the surface to 40 ft over a tidal cycle did not appear
to show any firm patterns with time or depth except a scarcity of
larvae at the surface. It is unfortunate that the authors did not report
counts of larvae of *O. lurida* which set concurrently with the Pacific
oyster and often in far greater abundance.

River estuaries have been much more difficult to study and less
successful for obtaining regular spatfalls. The James River and Delaware
Bay are examples that exhibit large freshwater runoff, strong tidal
mixing, rather steep horizontal salinity gradients, and strong upriver
transport systems. It is widely recognized that certain invertebrates,
e.g., crabs and barnacles, are successful in moving far up estuaries
without the ability to swim there (Bousfield, 1955). Larvae and juveniles
of several species of fishes (sciaenids) ride bottom currents from the
ocean to fresh water with well-known circulatory patterns (Pritchard,
1952). Both the James River and Delaware Bay have oyster seed areas
in low-salinity and mostly predator-free sectors. These seed areas were
highly productive until overharvesting and increased salinity and disease
reduced recruitment levels. Delaware Bay has recovered but the
James River has not. Oyster larvae use transport systems in river
estuaries, but what reactions or factors entrain them. Studies by VIMS
on the James River are used here as a basis for the discussion of larval
dispersion in open estuaries.

The James River is a type-B estuary (Pritchard, 1955) with a salinity
gradient from surface to bottom. It is essentially a vertically stratified,
two-layered tidal system with lower salinities and a net flow seaward
in the upper layer which compensates for freshwater input, whereas
the lower layer is saltier and has a net flow upriver. During late summer
and fall dry seasons, both the lower James River and Delaware Bay
tend to become type-C estuaries (horizontally stratified) with saltier
net inflow on the left side (looking downriver) and less saline waters
moving downriver on the opposite side. The effects of these circulation
patterns on oysters are shown by heavier spatfalls on the left shores
over the short term (annually) and the development of major natural
oyster beds on this side over the long term (thousand of years) (Figs.
1 and 2). With this flow pattern (type C) the vertical distribution of
oyster larvae becomes less decisive in their transport. Sharp haloclines
occur in spring but do not persist long because of vigorous tidal and
wind mixing and, in late summer when oyster larvae are present, there
is usually little or no vertical temperature or salinity gradient in the
lower James River and very little upriver.

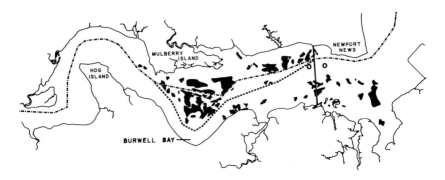

Fig. 1. Map of natural oyster beds (Baylor survey) of the James River. The present artificial ship channel between Mulberry Island and Burwell Bay was cut straight through oyster bars. The old natural channel follows the curve of Burwell Bay; hence, most natural bars above the bridge are on the Mulberry Island side of the old channel (left side looking downriver). The two sampling stations for larval data in Fig. 3 are shown as circles near the bridge.

The James River is a river estuary with a large freshwater flow, vigorous tidal mixing, and a strong tendency to flush larvae out of the system. Oyster setting was regular but moderate until 1960 when a disease destroyed much of the brood stock. In 1950, hourly pumped plankton samples of 100 and 500 liters yielded few larvae at any depth in the channel adjacent to the most productive seedbed (Wreck Shoal) in the river, and over the bed, during a period of intensive setting. In the 1960s setting failed in the seed area, presumably because of loss of brood oysters to disease. Also, proposals to dredge deeper channels raised questions of the effect on larval transport and setting patterns. A three-dimensional study of larval distribution with simultaneous hourly samples from five boat stations to assess transport systems produced 3000 samples of 300 liters each in 1965. Problems encountered were (1) scarcity of oyster larvae, especially advanced ones, (2) the decision to count all bivalve larvae caused difficulties in identification, (3) the task of quantitative counting of total samples was not solved satisfactorily, and (4) data were interpreted diversely by participants, hence were not conclusive. The data revealed regular rhythms of larval abundance with tidal stage (Fig. 3), and larvae were 5–10 times as abundant in peak periods near high slack water as at low times near low slack water. Oyster larvae predominated, but mostly early stages were found throughout the 12-day period from the earliest to the latest samples, although advanced larvae of other bivalve species were present. I concluded that larvae were being recruited continuously

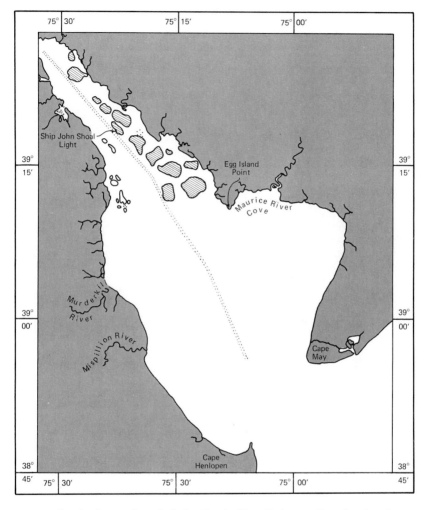

FIG. 2. Sketch of natural seeds beds (hatched) in Delaware Bay showing the pre-dominance of bars on the New Jersey (northeast) side of channel.

by newly spawned broods, since no increase in the size of larvae was observed. No vertical stratification of larvae occurred, except that fewer larvae were found in surface and bottom samples (J. D. Andrews, unpublished data).

These data and experiences suggest that the distribution of bivalve larvae in the James River is passive and dependent upon tidal currents and mixing. Waters in the lower river are homogeneous in late summer,

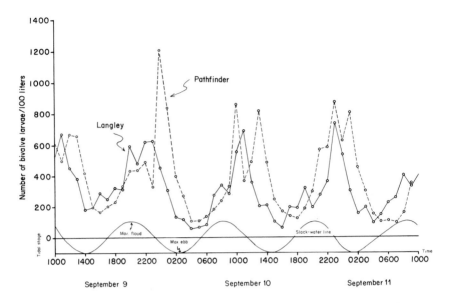

FIG. 3. Patterns of abundance of molluscan bivalve larvae in the James River at the peak period of spawning for 1965. Advanced larvae were scarce and very little setting occurred. Five stations were sampled simultaneously. Two channel stations 1 mile apart, above and below the bridge (see Fig. 1), are compared for changes in abundance of bivalve larvae with tidal stage and current. Four samples of about 300 liters each were taken simultaneously at depths of 1, 4, 7, and 10 m hourly at each station. All bivalve larvae were counted by species in the entire samples and the results adjusted to 100 liters for comparison with the literature. Patterns of abundance were similar for all species of bivalve larvae and for all depths of water. Surface and bottom samples had fewer larvae than those from mid-water depths. Oyster larvae in early stages predominated in the samples. Each point on the graph represents total bivalve larvae for four depths adjusted to 100 liters each (400 liters of water). Larvae were approximately five times as abundant between maximum flood and maximum ebb as during the other half of tidal cycles. Oyster larvae did not increase in size during 12 days of monitoring, therefore they were being recruited steadily (mostly straight hinge stage).

therefore this area serves as a huge mixing bowl from which larvae are randomly dispersed and carried both up- and downriver by tidal flows. The scarcity of advanced oyster larvae implies high losses, and this is critical whether caused by dispersion or predation. The James River apparently requires massive populations of brood oysters to sustain regular, adequate recruitment under intensive harvesting. Before 1960, spratfalls were dependable and adequate to provide over 2 million bushels of seed oysters each year. Only occasional light sets have occurred in the 1960s and 1970s yielding only one-fifth the seed supplies produced in earlier decades. The cessation of private oyster planting in Hampton Roads and off the mouth of the James River has greatly reduced the stocks of brood oysters available for spawning. Delaware Bay disease caused by the protozoan *Minchinia nelsoni* was the immediate cause of abandonment of these beds which have not been planted in 20 years since 1960.

James River studies suggest that swarms of oyster larvae are being constantly dispersed, with upper and inshore sectors of the swarm becoming less dense. Evidence from weekly setting data, wherein newly set spat represent the final distribution and abundance of eyed larvae, indicates that swarms become riverwide before setting begins. A grid of 19 stations, where test cultch collectors were exposed weekly, showed that abundance of spatfall increased or decreased synchronously at all stations throughout the river from week to week. This could occur only if the swarms were very large in areal extent and covered all stations during a week of tidal excursions. Spatfall nearly always decreased with distance upriver from the mouth, therefore the swarm resembled a cloud with variable densities of larvae, the lowest being upriver and inshore, or the periphery of the swarm.

The brief summary of larval studies in Delaware Bay by Haskin (1964)—in which he concludes that mature larvae drop to the bottom during ebb tides, as his predecessors Nelson and Perkins (1931) and Carriker (1951) did for coastal areas of New Jersey—is difficult to analyze without more background. This is another open system, apparently with hydrographic regimes that favor a concentration of larvae on the New Jersey shore (Hidu and Haskin, 1971). The hourly counts of eyed larvae during a tidal cycle, as illustrated by Fig. 1 in Haskin (1964), show a pattern very similar to that observed in the James River. The highest counts occurred at high slack water, and densities of larvae changed rapidly before and after peaks. He could have been sampling the upstream sector of a larval swarm carried by tidal excursions without the vertical migrations this school of investigators so firmly believes in. The lags in bottom counts can be explained by physical hydrography— longer flood tides in deeper waters and earlier ebb tides in surface layers. The details of these extensive larval studies on Delaware Bay would help clarify larval behavior in open flushing systems.

Wood and Hargis (1971), in a paper derived from the 1965 James River study, support the hypothesis of the Nelson school that oyster larvae rise from the bottom during flood tide but consider that stimulation to rise is limited to an increase in salinity rather than possibly currents too. This interpretation differs from my previous description of events and is based on a separate set of similar data from the same collection of plankton samples. Because of personal disagreements and involvement, critiques will be left to others. One could ask how the larvae received signals to drop to the bottom at the time of strongest ebb tides, and why some continued swimming through low slack water? Also, if larvae from one brood were being transported upstream, why were nearly all oyster larvae in the same early stages (1–4 days old) throughout the 12-day sampling period? Whatever the transport system, most species of bivalve larvae appeared to utilize the same one.

6.3.2.4 LARVAL ECOLOGY

To emphasize the complexity of larval ecology, some observations by Quayle (1964) are helpful. In Pendrell Sound, single larval broods of C. gigas have been followed quantitatively through pelagic and setting periods. After some spawnings early larvae exceeded 400/liter (nearly half a million per cubic meter) from which mature larvae showed "survival levels of up to 2%, which must be considered quite high." Yet there are virtually no planktonic predators except a few ctenophores. The main sessile predator is the adult oyster, which is the dominant intertidal animal. Since adult oysters are limited to the steep, rocky, intertidal shores of the sound by deep, cold waters, predation may occur only when larvae approach the shores, as they must under natural conditions to set. Commercial spatfalls are collected on shell strings suspended from floats. Dispersal by currents may be more important than predation in this greatly simplified ecosystem. Situations such as this suggest that larvae dropping to bottoms covered with sessile filter feeders risk predation; yet gregariousness and choice of populated oyster beds in preference to planted shells are easily observed in practice.

In summary, I concur with Galtsoff that present knowledge of larval behavior and dispersion is unsatisfactory. The problems are difficult, and many studies are inconclusive. Korringa (1941) concluded that larval dispersal was passive in the well-mixed Oosterschelde, and in 1952, after reviewing Carriker's long study, Korringa was still asking if passive dispersal could not be invoked in most situations. Nearly three decades later with several more studies to consider, this condition appears to be sustained. There has been a tendency to lump all oyster habitats in one broad category as estuaries, but these are often quite different and complex hydrographically. Wind and/or tide may control circulation. Many studies were made in complex habitats (e.g., Prytherch, 1929), whereas simpler ecosystems have been more enlightening. It is a conspicuous fact that oyster larvae set upstream from spawning sources, where they often concentrate in unbelievable numbers. In river estuaries, the salt wedge provides a transport system whereby fish larvae and probably other invertebrates move upstream in season. When freshwater flow causes stratification, deeper waters flood longer and faster than they ebb, and this transports some larvae upstream with random vertical distribution. Eddies, varying currents, and stratified grazing by animals cause patchiness of larvae.

Oyster larvae exhibit behavioral patterns that keep them in warm, shallow, and otherwise favorable masses of water when confronted with distinct changes in physical parameters. Some invertebrate larvae reach upriver sites by seasonal circulatory patterns (barnacles), by setting within one or two tidal cycles before dispersion occurs (tunicates), and by larval behavior (barnacles) more regularly than bivalve molluscs. This is indicated by successes of setting. One envies the facile fitting of the two-layed transport system for barnacles reported by Bousfield (1955) in the Miramichi estuary—based on weekly samples. Bivalves, particularly oysters, have great fecundity, and the history of oyster communities suggests that populations are sustained more effectively with many reproductive efforts per year in a warm region than by one or two in a cold habitat—especially when humans are ruthless and uncontrolled predators. This implies that oysters are heavily dependent upon hydrographic situations and less upon larval behavior for success in reproduction. This conclusion was reached by Banse (1956) for other invertebrate larvae (polychaete and echinoderm) and by Quayle (1964) for lamellibranchs.

6.3.2.5 ATTACHMENT AND METAMORPHOSIS

The attachment of mature larvae to substrates is a critical phase in the life cycle of oysters that is difficult to follow in nature. Large numbers of larvae fail to find suitable substrates in nature, and even in artificial tanks and hatcheries losses are high. Mature, eyed larvae are vigorous swimmers and have been observed setting at sizes from about 260 to 350 μm for both genera. The usual size is about 300 μm, but this may be increased by growth in cool waters or delay in settlement. Two cold-water species of incubatory oysters, *Ostrea chilensis* and *O. lutaria*, have very short pelagic phases and large eggs (about 300 μm) and eyed larvae (about 500 μm) (Millar and Hollis, 1963; Walne, 1963). See Section 5.4.3.1 for a comparison with other bivalves.

Mature larvae develop a pair of pigmented eyes and a slender elongate foot containing a large byssal gland. A pediveliger (Carriker, 1961) swims with the foot protruding to grip any firm surface on contact. A period of crawling and searching in a limited circular area may be followed by permanent attachment. A droplet of liquid cement is exuded through a pore at the heel of the foot, and the left valve of the larva is pressed into it. The homogeneous organic cement hardens in a few minutes, and metamorphosis proceeds rapidly. The large velum collapses quickly, suggesting that fragmentation and digestion or loss predominate over resorption, whereas the foot is reduced more slowly by resorption (Cole, 1938). The anatomy of larvae and early spat is diagrammed by Stafford (1913), Yonge (1960), and Galtsoff (1964).

The behavior of larvae and the physiology of attachment have been disclosed slowly by studies on *Ostrea* and *Crassostrea* over 100 years. Stafford (1913) discovered the byssal gland and deduced the method of attachment observed by Nelson (1924) and redescribed by Prytherch (1934). Cole and Knight-Jones (1939) confirmed the setting behavior in *O. edulis*. Prytherch described a fine byssal thread trailed by crawling larvae that is presumed to aid in retaining contact with the surface. He noted the preferences of larvae for small pits and irregularities on surfaces. The tendency of barnacles to orient and set in fine scratches and irregularities on surfaces has been called rugotropism by Crisp and Barnes (1954). Most investigators have noted the preferences of oyster larvae for microscopic rugosity, hence ground-glass plates and cement–sand-coated tiles and plates have been used more effectively than smooth glass and plastic as cultch.

Once attachment has occurred, larvae are not easily dislodged by currents. However, within 12–24 hr after setting, spat may be brushed off smooth surfaces without injury. Reattachment occurs only by mantle deposition of a dissoconch shell around the shell margin, and subsequent dislodgement without breakage depends upon the use of flexible cultch material with some rugosity such as fine-mesh screens and plastics. The tiny "free" spat are difficult to handle and grow until a length of a few millimeters is reached, but in hatchery and experimental operations the advantages of dispensing with cultch are numerous. A new set of problems is encountered in handling cultchless spat in nurseries and nature, for they are subject to smothering and severe predation. Commercial use of cultchless spat is in the experimental phase and requires adaptation to local ecosystems (Andrews, unpublished data).

6.4 Setting (Spatfall or Settlement)

A large literature exists on setting of oysters, and results and interpretations are often conflicting. The most extensive and useful experiments and summaries are those reported by Hopkins (1937), Cole and Knight-Jones (1939, 1949), and Korringa (1941). Earlier works, by the Nelsons and by Prytherch, among others, are reviewed in these papers. Korringa's detailed paper is a stimulating analysis of the factors involved in monitoring larvae and spatfall for prediction of setting. He also provides a running critique of other important papers. Except for a few factors such as gregariousness (Cole and Knight-Jones, 1949), advances in basic knowledge of setting beyond the excellent work of the period 1920–1940 seem meager. Technological advances in materials and methods—hatcheries, food cultures, antibiotics, new cultches, and plastics for containers and cultch—have provided ready sources of cultured larvae and spat for critical experiments, but the effects of light, silt, current velocity, predation, and ectocrines have not been effectively studied.

Predictions of setting may be based upon the monitoring of larvae or spatfall. Larval sampling is the preferred direct, concurrent method but provides only limited advance notice for large cultching operations, whereas the spatfall method is dependent upon historical records of setting in previous years, thus more subject to failure. The purpose is to ensure optimal exposure of cultch to coincide with time and intensity of setting. The success of predictions has varied inversely with length of spawning seasons and complexity of ecosystems. Where broods are few, larvae abundant, and areas limited in size and complexity, predictions based upon larval sampling are feasible (Pendrell Sound, Quayle, 1969; Dabob Bay, Westley, 1956; Willapa Bay, Lindsay et al., 1959). Trap-type estuaries such as the Oosterschelde and the St. Mary's River should be amenable to prediction by larval sampling once hydrographic regimes are understood and times of maximal spawning are known.

In open bays and warm waters, dispersal and prolonged breeding seasons have usually prevented adequate sampling of mature larvae. Local formulas for spawning, cultching, and spatting, based on accumulated experience and summing of heat in degree-days, have been proposed. For oviparous oysters, these calculations provide approximate times of first spawning, but predicting successful spatfalls usually depends upon empirical knowledge of previous setting patterns. Incubatory oysters growing in cool oceanic climates are more difficult to follow, because temperatures often are critically low for spawning or larval growth and breeding periods may be longer with irregular releases of larvae and sex changes. In only a few studies have adequate numbers of mature larvae been found to justify predictions of spatfall.

Methods of measuring intensity and timing of setting are extremely varied, and results are seldom comparable even if ecological conditions are similar. Bags and strings of shells are commonly used for measuring the relative abundance of spatfall by place or year. These have the

merit of being favorable substrates similar to those used in most commercial plantings (but not in western Europe for O. edulis). For
scientific studies, shells are too variable in age, shape, position (angle),
and induced turbulence to permit comparisons. Korringa comments that
using shells is "not exactly smart," yet most ostreologists must plead
guilty. Several investigators have used flat surfaces (Schaefer, 1937),
held at fixed angles in the water, such as glass plates (Hopkins, 1935,
1937; Korringa, 1941), slates (Cole and Knight-Jones, 1939), and more
recently cement board or asbestos board (Butler, 1954) and plastics.
Larvae do not attach readily to smooth glass and plastic surfaces, hence
these are often obtained in roughened form or may be dipped in a
slurry of lime, cement, and sand, as are commercial tiles in Europe.

6.4.1 Distribution of Spatfall

The distribution of spatfall is clearly dependent upon the vertical
and horizontal occurrence of mature larvae. Since larvae exhibit patchy
distributions because of the effects of numerous physical and biotic
factors, initial spatfalls are clumped in distribution also. Intensity and
distribution of initial spatfalls are affected also by quality and position
of cultch, gregariousness, and repeated setting. Often test panels are
exposed for periods of a week or longer, and this introduces the factors
of survival from predation and silting. Numbers of surviving spat may
bear little relation to the abundance of larvae and initial numbers of spat.
The literature offers few studies where simple ecosystems permit
analysis of variables that affect larval behavior and setting. Cole and
Knight-Jones (1939, 1949) reared larvae to setting in outdoor concrete
tanks at Conway, Wales, with standing water exposed to ambient sun-
light. The water in these 6-ft-deep tanks was not circulated or changed
for months, yet for several consecutive years cultures were reared
successfully. Tides, currents, turbulence, predators, and temperature
and salinity gradients were eliminated as factors affecting larval behavior and setting. The major uncontrolled factor was light, although
the authors considered the large tanks (85,000 gal) to represent semi-
natural conditions except for tides. Little information was obtained
about the distribution of larvae in the tanks. Sampling with bottles
indicated that vertical distribution was "not quite uniform," with more
larvae in the upper than in the lower half of the tanks. Trays of shells
9 in. below the surface obtained heavy spatfalls night and day, with
few spat at middepth and still fewer 9 in. from the bottom (7 hr exposures). Eyed larvae in abundance had the same vertical distribution

as smaller larvae. The authors make no mention of rafts or drifts of larvae oriented to utilize the shade of corners or trays of shells that would be expected in culture containers. Larvae probably aggregated under trays of shells in proportion to the distance above bottom and the shadows cast, and eyed larvae then attached to face-down surfaces (the only surface counted). Larvae swim vertically mostly, but they are not incapable of horizontal adjustments in standing-water tanks. The short time of exposure limits the effects of gregariousness—as discovered by these authors later. Experiments with dark-painted and clear glass plates could be subjected to the same analysis of larval behavior. Spat always settled in greater numbers in the daytime, hence shells and plates may have served as shade traps, thereby causing differences between night and day exposures. At night, relatively higher spatfalls on surface trays are not readily explained, unless during darkness larvae swam to the surface and a boundary or surface film effect was invoked in the placid tanks. In view of heavier setting in daytime than at night, the drastic differences in spatfall with the angle of the plates (undersurface = $0°$ and upper surface = $180°$) in favor of undersurfaces may be mostly explainable in terms of negative reactions of mature larvae to light. Larvae continue to swim in strong light despite an aversion to it, as demonstrated in Pendrell Sound and Norway pol where they may be restricted to near-surface layers as thin as 1 m.

The Bras d'Or lakes of Nova Scotia are natural oyster habitats that closely resemble the Conway tanks in physical and biotic factors. Medcof (1955), in a paper titled "Day and Night Characteristics of Spatfall and Behavior of Larvae," presents data on setting but no direct information on larvae. Except for the size of habitats, the same factors eliminated in Conway tanks were ineffective in the clear, shallow lakes with imperceptible tides. Strong light with the bottom visible at 4 m was a variable. Collectors of concrete-coated cardboard (9×5.7 cm) were suspended by wires to avoid shadows and turbulence. Spatfall increased directly with depth (six depths from 0 to about 2 m), and undersurfaces caught more spat than upper surfaces (the ratio increasing with depth, although seldom more than 2:1). Spatfall by day averaged three times as high as at night, although the night period included some evening and morning light. The patterns of spatfall were consistent in that deeper collectors caught more spat day and night. No information is available on the distribution of larvae. It is difficult to deduce from Medcof's data why he concludes that mature larvae are benthic and that they leave the bottom only when induced to do so.

Two simple habitats, one natural and the other artificial, exhibit apparent similarities in factors that may affect larval behavior and spatfall patterns. With similar depths and light conditions, they agree in that larvae tend to set on undersurfaces and more intensively in daytime than at night. The vertical distribution of spatfall was reversed in the two habitats, and the intensity gradient was much higher in the Conway tanks where light was probably more intense. It appears that strong light may stimulate activity in larvae to seek areas of reduced intensity. Larvae of the two genera probably exhibit similar reactions to environmental stimuli.

Korringa (1941) studied oyster setting in the Oosterschelde, where a greatly different set of factors was involved. This large tidal estuary has vigorous vertical mixing, strong currents, turbid waters, 3–4 m tidal ranges and maximal depths of 8 m. Although about two-thirds of the basin water leaves on each ebb tide, only about 3.7% new water is returned per tidal cycle, therefore the basin approximates a closed system for oyster larvae. Tidal mixing prevents vertical stratification of waters, hence temperature and salinity fluctuations are minimal. Strong currents, turbulence, heavy silting, and darkness are the predominant physical factors involved in the maturing and setting of larvae. The contrast of habitats between the Oosterschelde and the Conway tanks or Bras d'Or lakes provides an opportunity for comparison of larval behavior and setting.

Korringa recognized the importance of knowledge of larval distribution in studies of setting. He reported a fairly uniform vertical distribution of larvae day and night and correlations between abundance of mature larvae and intensity of spatfall. Most larvae settled during slack-water periods when currents were weak. More spat settled near the bottom than near the surface, which he attributed to lower current velocities and not vertical distribution of larvae. Spat settled on upper surfaces more intensively than on undersurfaces, with vertical plates being chosen least. Microscopical roughness enhanced setting. Light appeared to have no orientation effect on larvae in turbid Oosterschelde waters. Korringa found heavier setting during days than during nights, although light and temperature, the stimulatory factors usually invoked, were apparently not involved. Predation losses probably exceeded those from dispersion in the Oosterschelde.

These three studies illustrate the complexity of larval ecology and spatfall in systems where many physical factors appear to be uniform or ineffective. "Controlled" habitats often introduce unexpected or unrecognized variables, hence conflicting results permeate the literature. Many field studies on setting patterns of C. virginica have been made with little or no knowledge of larval distribution. Monitoring of setting often involves exposure of cultch for periods of a week or longer. This introduces additional variables that affect level and survival of initial spatfall such as silting, predation, and interference from other organisms.

Butler (1954) presents strong evidence that *C. virginica* larvae prefer upper surfaces at Pensacola, Florida. His studies lack larval distribution data and are based on weekly exposures, thereby presenting the problem of survival versus initial set. He found 64% of 70,000 spat on upper surfaces (180°) and 36% on lower surfaces of cement plates. However, in a vertical series, he reported 50% of the set on the bottom plate and 96% of these on the undersurface which was very near the bottom. This agrees with Hopkins' (1935) rsults with glass plates, which were also on the bottom, but conflicts with the results of Pomerat and Reiner (1942) with suspended cultch at Pensacola, too. Butler concludes that a fine layer of silt is a deterrent to attachment on the upper surface, which is normally preferred. It is probable that silt is involved in most field experiments, especially close to the bottom, for several millimeters may be deposited during one night in stormy weather. Butler also calls attention to the heavier setting of barnacles on undersides and their effect in causing oyster larvae to be repelled from lower surfaces. He obtained similar distributions of spatfall in Hopkins-type, vertical, eggcrate collectors.

Setting of oyster spat at Gloucester Point, Virginia, on 6 × 6 in cement plates (asbestos-board) suspended about 1 ft off the bottom in 2 m of turbid waters is shown in Table II. One upper (180°) and one lower (0°) surface of two plates held 3 in. apart were examined routinely. Distances between cultch units are critical, as Thomson (1950) observed with bundles of slats. Spat strongly preferred the upper surfaces of plates in weekly exposures. Plates were washed vigorously to remove silt and unattached organisms, yet many more flatworms (*Stylochus ellipticus*) were found on the undersides, hence selective predation may have occurred.

Data from Hidu (1967) provide conclusive proof to me that *C. virginica* in natural waters attaches more intensively on upper than on lower surfaces. He reports 91% of spatfall in 1965 on the upper surfaces of weekly cement boards and 86% in 1966. Also, he reports from 64 to 74% of spatfall on upper surfaces in 6 years of routine daily monitoring at Cape May with clam (*Mercenaria*) shells (Haskin, unpublished data). Those who have obtained heavier sets on the undersurfaces seem to have additional factors affecting their results, such as survival of spat because of predation or silting over a period of time, or light affecting the behavior of larvae. There is no evidence to suggest that species or generic differences are involved.

Quayle's (1952, 1953) studies in Pendrell Sound, where single broods of larvae may be followed, illustrate the complexity of larval behavior

JAY D. ANDREWS

TABLE II

SPATFALL ON ASBESTOS PLATES HELD IN A HORIZONTAL POSITION IN A WOODEN
FRAME NEAR THE BOTTOM AT THE VIMS PIER, GLOUCESTER POINT, VIRGINIA, 1970[a]

Dates	Frame 1		Frame 2	
	0°	180°	0°	180°
July				
14–21	0	0	0	0
21–28	0	1	—	—
28–4	0	0	0	0
August				
4–11	0	0	0	4
11–18	0	7	2	0
18–26	12	407	19	245
26–1	1	38	2	26
September				
1–8	1	41	0	1
8–15	7	82	1	12
15–22	19	422	10	203
22–29	3	33	11	34
29–6	18	15	0	13
October				
6–14	2	—	1	0
14–22	4	17	4	10
22–2	1	1	0	0
November				
2–17	0	1	0	0
Total	68 (6%)	1065 (94%)	50 (8%)	548 (92%)

[a] Surface angle 0° = down side and 180° = upper side.

and setting in habitats that may be visualized as simple ecosystems. He
found all stages of larvae concentrated in the warm upper 10 ft of water,
but in patches, for 12- to 14-ft tides often effected wide dispersal. In a
series of fixed-elevation, vertical shell strings exposed over several weeks
before counting, he found maximal setting in the lower 12 ft of the
intertidal zone. Shell strings from a float showed spatfall down to 38 ft,
with intensity decreasing from the surface downward. The deeper spat
were in cold waters when strings were removed from floats and had
resulted from larvae in warm water mixed downward by a storm. Rather
perplexing is his finding of twice as many surviving spat on lower shell
surfaces as on upper surfaces consistently at all depths in 1951 and just
the opposite ratio in 1952 and 1953. Even more perplexing are his spatfall studies on angles of surfaces in which the lower of pairs of cement-

coated laths gave results contrary to those from the upper laths, both fixed at the same angle but possibly too close together, consequently producing eddies and trapping larvae.

Two additional recent studies, one involving a trap-type estuary and the other involving an open system, suggest that hydrography must be carefully studied in each habitat before special attributes of larval behavior are invoked to explain distributions of spatfall. The first study, an attempt by Carter (1967) to predict locations and quantities of oyster brood stock required to produce commercial spatfalls in the Manokin River, reveals the hydrographic attributes of trap-type estuaries. Assuming that oyster larvae are advected and dispersed like conservative constituents of salt water, Carter made a point release of rhodamine dye and followed the distribution for 14 days. In July the Manokin River had negligible freshwater flow and vertical and horizontal homogeneity of salt content in the area of experimentation, hence southwesterly winds were the predominant cause of advection. This was an oceanographic study with biotic parameters assumed from the literature. The hydrography was very similar to that reported for the St. Mary's River, with the lower sector controlled by winds producing net upriver surface flows and then a gradual transition to weak salinity-gradient and river-flow control of circulation in the upper sectors. Lateral boundaries introduced turbulent diffusion. After 14 days the dye was most concentrated in the upper reaches of the river and on the left shore looking downriver. This is exactly what happens to larvae in the St. Mary's River (Manning and Whaley, 1954) and the Great Wicomico River, as evidenced by intensity of spatfalls—regardless of the orientation of rivers in respect to prevailing wind direction (See Section 6.3.2.3). The location of brood stock in these trap-type estuaries does not appear to be critical, for dye and larvae appear to become dispersed unevenly throughout the system in 14 days, with further physical hydrographic losses confined to the lower sectors. The dilution factor in this relatively low-flushing Manokin River was very great, with maximum concentration of dye after 14 days of about 0.10 ppb. Carter estimates from the literature that only 0.01% or 100 larvae per million eggs spawned will set, therefore it would require 50,000 bushels of brood oysters for this one small river to produce 1000 spat per bushel (defined as a commercial set). There are many intangible factors in larval survival including predation, repeated attempts to set, and multiple spawnings. The physical background has been established with far more clarity than the knowledge of larval biology needed for management decisions. It should be noted that, in the trap-type estuaries of Chespeake Bay, there is no simple two-layered transport system in midsummer when oyster larvae are being dispersed.

The simulation of larval distribution by dye suggests that physical parameters of circulation may be dominant in respect to larval dispersion in trap-type estuaries such as the St. Mary's River and the Manokin River. Biotic losses may be superimposed on dispersion by physical factors but affect primarily the level of setting and not its distribution. If most early invertebrate larvae move to surface waters for feeding, as Thorson (1964) concludes, they would tend to be transported upstream in the lower sectors of the two rivers studied. Once started upriver, larvae tend to become trapped in the upperriver sectors with weak recycling hydrographic regimes. Larvae are known to be relatively abundant in these seed areas, and very intensive setting occurs in the narrow upper reaches of the rivers (Andrews, unpublished data).

The Cape May shore of New Jersey is on the open system of Delaware Bay, where the Nelsons and their successors have studied C. virginica for over six decades. It is one of the favored areas where high concentrations of mature larvae and very intensive spatfalls occur regularly (see Fig. 2). The lower Delaware Bay in summer is a type-C estuary (Pritchard, 1955) with little vertical stratification because of tidal (5–6 ft range) and wind mixing of the wide, rather shallow lower sector. The salt balance is maintained by net tidal flow upriver on the eastern side along the New Jersey coast to compensate for net downriver flow along the western shore. This results in a counterclockwise, circular, flow that appears to be ideal for concentrating oyster larvae on the Cape May shore. Many larval distribution and setting studies in the area are briefly reviewed by Haskin (1964).

A recent study of larval behavior and spatfall across a short sector of the Cape May shore was conducted by Hidu and Haskin (1971). They established two 4000-ft transects across the wide intertidal and steep slope zones to the deeper channel zone with six piling stations in each replicate line. Horizontal asbestos-board panels were suspended at several depth intervals for weekly spatfall monitoring. Pumped plankton samples of 200 liters were taken at three depths—bottom, middle, and surface—eight times during each of two 12-hr tidal cycles. A 160-μm plankton net caught only advanced and mature larvae.

Spatfall was light at Cape May in 1965, but in 1966 good weekly sets occurred between 25 June and 25 July, and over 100,000 spat were counted on the panels. Mature oyster larvae ran as high as 24/liter, a density much in excess of any previous samples in Delaware Bay. Larval density was highest on the slope between the channel and inshore flats, and intensive spatfalls were also found on the slope. Inshore larval densities were comparatively low, with maximums of about 1/liter. Disregarding current velocities, which were high offshore (means of mostly

1–2 ft/sec) and low over inshore flats (0.25–0.75 ft/sec), Hidu and Haskin calculated that "inclination to set" was 10 times greater over the flats. They discussed temperature increases and gregarious setting as possible factors stimulating the heavy inshore setting by larvae.

The Hidu and Haskin study demonstrates that mature larvae and setting occur at all levels in the water. Most dramatic is the concentration of larvae by tidal transport and the Coriolis force on the edge of the channel from which they obviously spill over onto tidal flats on flood tides. Conditions were more inducive to settlement on the flats, but whether daylight effects (light), reduced currents, trapping, temperatures, or other factors are most responsible is not clarified yet. This is one of the most interesting field studies on a complex habitat in the literature.

Intensive spatfalls on the Cape May tidal flats may be related to increases in temperature and intensity of light. Lutz et al. (1970) found increased setting rates in laboratory experiments when temperatures were increased. Ritchie and Menzel (1969) found larvae seeking the dark undersurfaces of shells under laboratory conditions with 50 fc of ambient or manipulated light. These factors of temperature and light may induce larvae to set to escape unfavorable situations in shallow and intertidal habitats, but in most natural situations neither factor is effective. The orientation of cultch is an important factor (Thomson, 1950).

6.4.2 Gregariousness and Delay of Metamorphosis

Gregariousness in spatfall was discovered fortuitously by Cole and Knight-Jones (1939) in their tanks at Conway. The results of a fine series of experiments are reported in a paper on setting behavior of larvae of *O. edulis* (Cole and Knight-Jones, 1949). In crowded tanks with most variables eliminated, larvae set more readily on shells bearing spat than on controls. The relative attractiveness of spatted shells was the inverse of the intensity of setting although up to a density of 50–100 spat per shell, larvae set more readily on shells bearing large numbers than on those with few spat. Expressed simply, gregariousness is more important and more effective in light setting. The behavior of larvae on oyster beds was investigated by Knight-Jones (1952), who reported greater numbers of recently settled spat on densely stocked grounds than on barren or poorly stocked beds. Larvae seem to be genetically programmed to aggregate in setting to ensure breeding success later by mass spawning. There is no apparent limit to the number of spat that will set on a given surface area, for counts of 800 per smooth shellface in Virginia and thousands per shellface at Cape May in Delaware Bay have been recorded (T. C. Nelson, personal communication). Survival is relatively unimportant in heavy spatfalls, and territorial attributes are not invoked.

Confirmation and application of the findings on gregariousness have been slow to occur in other areas and species. Galtsoff *et al.* (1930) noted the patchiness of spatfall where light setting occurred. It is always striking to note on shell plantings in Virginia that those few shells with spat often have 2 or more when most shells have none. It requires an average of about 5 spat per shell to ensure that 90% of exposed shells in bags obtain spat (Fig. 4). The requirement on commercial planted shells may rise to 15 per shell. It is impressive to note that during years of light setting commercial beds of newly planted shells obtain far fewer spat than adjacent stocked natural seedbeds. It appears that American biologists have assumed gregariousness without attempting to exploit it. Harvesting requirements affect the use of seed, which may account in part for the reluctance of management personnel to sprinkle seed oysters on shell plantings or shells on seedbeds to enhance recruitment. Oystermen hold simple concerns about the needs for broodstock and often plant a few old oysters at the heads of small creeks, unrealistically expecting this to improve setting in their vicinity.

Crisp (1967) noted the remarkable lack of interest in gregariousness in other species after the discovery by Cole and Knight-Jones (1949). He was interested in the mechanisms of gregarious behavior, which have been explored in other invertebrates (barnacles and polychaete worms, particularly). He found that removing the periostracum and outer organic matrix of *C. virginica* shells with sodium hypochlorite made them unattractive to larvae for attachment. He also noted, as have several others, that soaking shells in oyster fluids and tissue extracts increased the level of settlement. He believes that oyster larvae recognize the insoluble organic matrix on shells as do barnacles (Crisp, 1965). He postulated that they may also recognize soluble materials from metabolic activity of living oysters. Unfortunately, in his short-term experiments at VIMS, he did not test any of the many artificial substrates that oyster larvae attach to readily, including cement boards in use at VIMS for monitoring natural spatfall. None of these have organic matrixes or soluble oyster organics to attract larvae.

Concurrently, Hidu (1969) explored gregarious setting in *C. virginica* at Cape May. With very high densities of eyed larvae (10,000/15-liter container), in a few hours he obtained two to five times as many spat on shells with 24-hr or 2 month-old preset spat as on controls. He believed that a water-borne pheromone was indicated when shells around

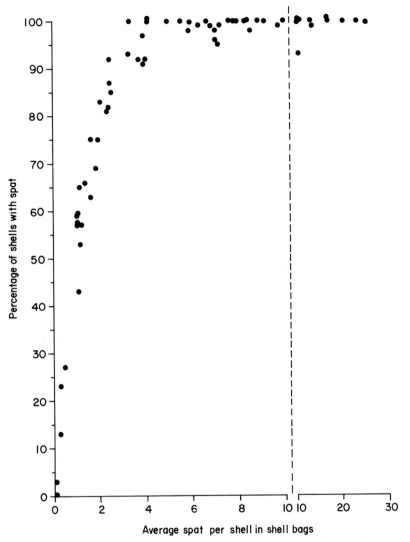

FIG. 4. Intensity of spatfall required to obtain satisfactory distribution of spat on clean shells exposed during a setting season in quarter-bushel chicken-wire bags.

2 month-old spat inside a fine plankton net bag obtained heavier spat-falls than controls. Bayne (1969) reports no apparent response of larvae to substrates with or without spat until contact is made with the foot. Also, Bayne (1965) has shown with *Mytilus edulis* that pediveligers tend to set more quickly, as delay in metamorphosis is prolonged, hence

timing of experimentation in a given brood of larvae may be important. In nature this delay in setting may be prolonged inversely with temperature (Bayne, 1965), and it becomes an important factor in the success of attachment and metamorphosis. Prolonged search for a suitable substrate tends to use up the food stored as oil globules for the rapid reorganization that must occur in changing from a pelagic to a sedentary mode of life. This period of metamorphosis is short (probably about 48 hr) and temperature-dependent but is characterized by a lack of feeding organs. Foot and velum are lost, and gills and digestive tract undergo rapid development and reorganization, respectively (see also Sections 5.4.2.3 and 5.4.3.2).

6.4.3 Survival of Spat

Monitoring of spatfall in natural waters is a practical and easily executed routine, and many biologists have done it, producing a profusion of data and literature. Interpretation of these data is hazardous if not futile, considering the variations of cultch, time of exposure, predation, and numerous other factors that affect the results.

Usually a high proportion of spat is lost by predation (Mackin, 1946), smothering, and competition very quickly after spatfall. Korringa (1941) estimated 90% losses on tiles before removing seed oysters. By comparing weekly and seasonal shell bags, Loosanoff and Engle (1940) found survival rates from 0 to 14% in 2 months. Using the same favorable shell bag comparison, Andrews (1948) reported survival rates from 2 to 85%, with higher rates associated with light spatfalls and low salinities. At Gloucester Point, Virginia, newly set spat on suspended shells are destroyed (90%) in a few days by the flatworm *Stylochus,* also newly set from the plankton and no larger than the oyster spat. Predation by adult *Stylochus* is discussed by Landers and Rhodes (1970).

Seed oyster areas for *C. virginica* are often located in low-salinity waters where most predators are excluded. In high-salinity areas some spat survive when intensive sets of oysters and other sedentary organisms occur upon which predators may feed voraciously. South of Delaware Bay on the Atlantic coast is a series of high-salinity lagoons with high spatfalls, and oysters survive mostly intertidally. Setting often is so intensive that crowding and poor growth become problems (Mackin, 1946; Lunz, 1954).

Survival of spat in natural waters depends upon conditions that change with season, place, and circumstances. Predation, competition, disease, and physical factors such as silting, salinity, and temperature are among the many variables that must be considered in each situation (Mac-Kenzie, 1970a,b). The development of hatcheries to control larval environment and nurseries to prevent excessive losses of spat are obvious steps to improve the wild seabeds used in oyster culture. Often the most important step is to remove young spat from the bottom where so many dangers and problems confront them. The Japanese have done this with raft culture (Cahn, 1950). A recent development in the United States is the production of cultchless seed oysters, whereby large numbers are floated or suspended in trays until a size resistant to bottom predators and hazards is reached (see Section 6.3.2.5). The feasibility of this type of mariculture is now being explored by private and public operations to determine the costs of equipment and handling. The rearing of oyster larvae in hatcheries has become routinely successful, but management of spat nurseries remains a problem with excessive costs if grown to usual seed oyster size and excessive losses if planted too soon. The problems are to avoid increased fouling, predation, and disease. Rapid growth, early marketing, and high prices seem to be essential requirements of such endeavors to be competitive with natural sets and bottom culture.

REFERENCES

Alderman, D. J. (1979). Epizootiology of *Martelia refringens*. *Mar. Fish. Rev.* **41**, (1), 67–69.

Andrews, J. D. (1948). The 1947 oyster strike in the James River. *Proc. Natl. Shellfish Assoc. 1948* pp. 61–66.

Andrews, J. D. (1971). Climatic and ecological settings for growing shellfish. *Proc. Conf. Artif. Propag. Commer. Valuable Shellfish, Coll. Mar. Stud., Univ. Delaware* pp. 77–108.

Banse, K. (1956). Uber den Transport von Meroplanktischen Larven aus dem Kattegat in die Kieler Bucht. *Ber. Dsch. Wiss. Komm. Meeresforsch.* **14**, 147–164.

Bardach, J. E., Ryther, J. H., and McLarney, W. O. (1972). "Aquaculture; the Farming and Husbandry of Freshwater and Marine Organisms," pp. 674–742. Wiley (Interscience), New York.

Baughman, J. L. (1948). "An Annotated Bibliography of Oysters, with Pertinent Material on Mussels and Other Shellfish and an Appendix on Pollution." Texas A & M Res. Found., College Station.

Bayne, B. L. (1965). Growth and the delay of metamorphosis of the larvae of *Mytilus edulis* L. *Ophelia* **2**, 1–47.

Bayne, B. L. (1969). The gregarious behavior of the larvae of *Ostrea edulis* L. at settlement. *J. Mar. Biol. Assoc. U.K.* **49**, 327–356.

Bousfield, E. L. (1955). Ecological control of the occurrence of barnacles in the Miramichi estuary. *Natl. Mus. Can. Bull.* **137**, 1–69.

Butler, P. A. (1954). Selective setting of oyster larvae on artificial cultch. *Proc. Natl. Shellfish Assoc.* **45**, 95–105.

Cahn, A. R. (1950). Oyster culture in Japan. *U.S. Fish Wildl. Serv. Fish. Leafl.* No. 383, 1–80.

Carriker, M. R. (1951). Ecological observations on the distribution of oyster larvae in New Jersey estuaries. *Ecol. Monogr.* **21**, 19–38.

Carriker, M. R. (1961). Interrelation of functional morphology, behavior, and autecology in early stages of the bivalve *Mercenaria mercenaria. J. Elisha Mitchell Sci. Soc.* **77**, 168–242.

Carter, H. H. (1967). A method for predicting broodstock requirements for oyster (*C. virginica*) production areas with application to the Manokin River. *Chesapeake Bay Inst. Johns Hopkins Univ.* Special Rept. 13.

Cerruti, A. (1941). Osservazioni ed esperimenti sulle cause di distrizione delle larve d'ostrica nel Mar Piccole e nel Mar Grande di Toronto. *Arch. Oceanogr. Limnol.* **1**, 165–201.

Coe, W. R. (1953). Influences of association, isolation and nutrition on the sexuality of snails of the genus *Crepidula. J. Exp. Zool.* **122**, 5–19.

Cole, H. A. (1938). The fate of the larval organs in the metamorphosis of *Ostrea edulis. J. Mar. Biol. Assoc. U.K.* **22**, 469–484.

Cole, H. A. (1941). The fecundity of *Ostrea edulis. J. Mar. Biol. Assoc. U.K.* **25**, 243–260.

Cole, H. A., and Knight-Jones, E. W. (1939). Some observations and experiments on the setting behavior of larvae of *Ostrea edulis. J. Cons. Perm. Int. Explor. Mer.* **14**, 86–105.

Cole, H. A., and Knight-Jones, E. W. (1949). The setting behavior of larvae of the European flat oyster, *O. edulis* L. *Min. Agr., Fish. Food, Fish Invest. London* **17**, 1–39.

Crisp, D. J. (1965). Surface chemistry, a factor in the settlement of marine invertebrate larvae. Botanica Gothoburgensis III. *Proc. Mar. Biol. Symp, 5th Goteborg* pp. 51–65.

Crisp, D. J. (1967). Chemical factors inducing settlement in *Crassostrea virginica* Gmelin. *J. Anim. Ecol.* **36**, 329–335.

Crisp, D. J., and Barnes, H. (1954). The orientation and distribution of barnacles at settlement with particular reference to surface contour. *J. Anim. Ecol.* **23**, 142–162.

Curtin, L. (1971). Oyster farming in New Zealand. *N.Z. Mar. Dep. Fish. Tech. Rep.* No. 72.

Davis, H. C. (1969). Shellfish hatcheries—Present and future. *Trans. Am. Fish Soc.* **98**, 743–750.

Davis, H. C., and Guillard, R. R. (1958). Relative value of ten genera of microorganisms as food for oyster and clam larvae. *U.S. Fish Wildl. Serv. Fish. Bull.* **58**, 293–304.

Dean, B. (1890). The present methods of oyster-culture in France. *Bull. U.S. Bur. Fish.* **10**, 363–388.

Dinamani, P. (1974). Reproductive cycle and gonadal changes in the New Zealand rock oyster, *Crassostrea glomerata. N.Z. J. Mar. Freshwater Res.* **8**, 39–65.

Dupuy, J. L., Windsor, N. T., and Sutton, C. E. (1977). "Manual for Design and Operation of an Oyster Seed Hatchery for the American Oyster, *Crassostrea virginica.*" Virginia Inst. Marine Sci., Gloucester Point, Va. Spec. Rep. Appl. Sci. and Ocean Eng. No. 142.

Galtsoff, P. S. (1964). The American oyster, *Crassostrea virginica* Gmelin. *U.S. Fish Wildl. Serv. Fish. Bull.* **64,** 1–480.

Galtsoff, P. S., Prytherch, H. F., and McMillin, H. C. (1930). An experimental study in production and collection of seed oysters. *U.S. Bur. Fish. Bull.* **46,** 197–263.

Giese, A. C. (1959). Comparative physiology: Annual reproductive cycles of marine invertebrates. *Annu. Rev. Physiol.* **21,** 547–576.

Giese, A. C., and Pearse, J. S. (1974). Introduction: General principles. *In* "Reproduction of Marine Invertebrates" (A. C. Giese and J. S. Pearse, eds.), Vol. 1, pp. 1–49. Academic Press, New York.

Gunter, G. (1950). The generic status of living oysters and the scientific name of the common American species. *Am. Midl. Nat.* **43,** 438–449.

Gunter, G. (1951). The species of oysters of the Gulf, Caribbean and West Indian region. *Bull. Mar. Sci. Gulf. Caribb.* **1,** 40–45.

Haskin, H. H. (1964). The distribution of oyster larvae. *Proc. Symp. Exp. Mar. Ecol., Grad. Sch. Oceanogr. Univ. Rhode Island* Occas. Publ. No. 2, pp. 76–80.

Hidu, H. (1967). Inshore settlement of several marine invertebrates at the Cape May shore of Delaware Bay, New Jersey, with special reference to the American oyster, *Crassostrea virginica* Gmelin. Ph.D. Thesis, Rutgers Univ., New Brunswick, New Jersey.

Hidu, H. (1969). Gregarious setting in the American oyster *Crassostrea virginica* Gmelin. *Chesapeake Sci.* **10,** 85–92.

Hidu, H., and Haskin, H. H. (1971). Setting of the American oyster related to environmental factors and larval behavior. *Proc. Natl. Shellfish Assoc.* **61,** 35–50.

Hidu, H., and Richmond, M. S. (1974). Commercial aquaculture in Maine. *Maine Sea Grant Bull.* **2,** 1–59.

Hidu, H., Drobeck, K. G., Dunnington, E. A., Jr., Roosenburg, W. H., and Beckett, R. L. (1969). "Oyster Hatcheries for the Chesapeake Bay Region," Spec. Rep. No. 2, Maryland Nat. Res. Inst. Solomons.

Hollis, P. J. (1963). Some studies on the New Zealand oysters. *Zool. Publ. Victoria Univ. Wellington* **31,** 1–28.

Hopkins, A. E. (1935). Attachment of the Olympia oyster, *Ostrea lurida,* to plane surfaces. *Ecology* **16,** 82–87.

Hopkins, A. E. (1937). Experimental observations on spawning, larval development, and setting in the Olympia oyster, *Ostrea lurida. Bull. U.S. Bur. Fish.* **48,** 439–503.

Imai, T., and Sakai, S. (1961). Study of breeding of Japanese oyster *Crassostrea gigas. Tohoku J. Agric. Res.* **12,** 125–171.

Imai, T., Hatanaka, M., Sato, R., Sakai, S., and Yuki, R. (1950). Artificial breeding of oysters in tanks. *Tohoku J. Agric. Res.* **1,** 69–86.

Joyce, E. A., Jr. (1972). "A Partial Bibliography of Oysters, with Annotations," Spec. Sci. Rep. No. 34. Florida Dep. Nat. Resour. St. Petersburg.

Knight-Jones, E. W. (1952). Reproduction of oysters in the rivers Crouch and Roach, Essex, during 1947, 1948 and 1949. *Min. Agr. Fish. Food, Fish. Invest. London,* **18,** 1–48.

Korringa, P. (1941). Experiments and observations on swarming, pelagic life and setting in the European flat oyster, *Ostrea edulis* L. *Arch. Neerl. Zool.* **5,** 1–249.

Korringa, P. (1952). Recent advances in oyster biology. *Q. Rev. Biol.* **27,** 266–308, 339–365.

Landers, W. S. (1968). Oyster hatcheries in the northeast. Proc. Oyster Culture Workshop. *Ga. Game Fish Comm. Contrib. Ser.* **6,** 35–41.

Landers, W. S., and Rhodes, E. W. (1970). Some factors influencing predation by the flatworm, *Stylochus ellipticus* (Girard), on oysters. *Chesapeake Sci.* 11, 55–60.

Lindsay, C., Westley, R. E., and Sayce, C. S. (1959). Prediction of oyster setting in the state of Washington. *Proc. Natl. Shellfish. Assoc.* 49, 59–70.

Longwell, A. C., and Stiles, S. S. (1970). The genetic system and breeding potential of the commercial American oyster. *Endeavour* 29, 94–99.

Longwell, A. C., and Stiles, S. S. (1973). Oyster genetics and the probable future role of genetics in aquaculture. *Malacol. Rev.* 6, 151–177.

Loosanoff, V. L. (1969). Maturation of gonads of oysters, *Crassostrea virginica*, of different geographical areas subjected to relatively low temperatures. *Veliger* 11, 153–163.

Loosanoff, V. L., and Davis, H. C. (1952). Temperature requirements for maturation of gonads of northern oysters. *Biol. Bull. (Woods Hole, Mass.)* 103, 80–96.

Loosanoff, V. L., and Davis, H. C. (1963a). Rearing of bivalve molluscs. *Adv. Mar. Biol.* 1, 1–136.

Loosanoff, V. L., and Davis, H. C. (1963b). Shellfish hatcheries and their future. *U.S. Fish. Wildl. Serv., Commer. Fish. Rev.* 25, 1–11.

Loosanoff, V. L., and Engle, J. B. (1940). Spawning and setting of oysters in Long Island Sound in 1937, and discussion of the methods for predicting the intensity and time of oyster setting. *U.S. Bur. Fish. Bull.* 49, 217–255.

Loosanoff, V. L., and Nomejko, C. A. (1951a). Existence of physiologically-different races of oysters, *Crassostrea virginica*. *Biol. Bull. (Woods Hole, Mass.)* 101, 151–156.

Loosanoff, V. L., and Nomejko, C. A. (1951b). Spawning and setting of the American oyster, *O. virginica*, in relation to lunar phases. *Ecology* 32, 113–134.

Lunz, G. R., Jr. (1954). The general pattern of setting in South Carolina. *Proc. Natl. Shellfish. Assoc.* 45, 47–51.

Lutz, R. A., Hidu, H., and Drobeck, K. G. (1970). Acute temperature increase as a stimulus to setting in the American oyster, *Crassostrea virginica* (Gmelin). *Proc. Natl. Shellfish. Assoc.* 60, 68–71.

MacKenzie, C. L. (1970a). Oyster culture in Long Island Sound 1966–69. *Commer. Fish. Rev.* Jan., 27–40.

MacKenzie, C. L. (1970b). Causes of oyster spat mortality, conditions of oyster setting beds, and recommendations for oyster bed management. *Proc. Natl. Shellfish. Assoc.* 60, 59–67.

Mackin, J. G. (1946). A study of oyster strikes on the seaside of Virginia. *Contrib. Va. Fish. Lab.* No. 25, 1–18.

Manning, J. H., and Whaley, H. H. (1954). Distribution of oyster larvae and spat in relation to some environmental factors in a tidal estuary. *Proc. Natl. Shellfish. Assoc.* 45, 56–65.

Marteil, L. (1976). La Conchyliculture Française. *Rev. Trav. Inst. Pêches Marit.* 40(2), 149–346.

Matthiessen, G. C. (1970). A review of oyster culture and the oyster industry in North America. *Woods Hole Oceanogr. Inst. Contrib.* No. 2528.

Medcof, J. C. (1955). Day and night characteristics of spatfall and of behavior of oyster larvae. *J. Fish. Res. Board Can.* 12, 270–286.

Menzel, R. W. (1955). Some phases of the biology of *Ostrea equestris* Say, and a comparison with *Crassostrea virginica* (Gmelin). *Tex. Inst. Mar. Sci.* 4, 70–153.

Menzel, R. W. (1968a). Cytotaxonomy of species of clam (*Mercenaria*) and oyster

(*Crassostrea*). *Proc. Symp. Mollusca, Cochin, Mar. Biol. Assoc India, Symp. Ser.* No. 3, Part 1, 75–84.

Menzel, R. W. (1968b). Chromosome number in nine families of marine pelecypod mollusks. *Nautilus* **82**, 45–58.

Menzel, R. W. (1974). Portuguese and Japanese oysters are the same species. *J. Fish. Res. Board Can.* **31**, 453–456.

Millar, R. H. (1964). Breeding and gonadial cycle of oysters in Loch Ryan, Scotland. *J. Cons., Cons. Perm. Int. Explor. Mer.* **28**, 432–439.

Millar, R. H., and Hollis, P. J. (1963). Abbreviated pelagic life of Chilean and New Zealand oysters. *Nature (London)* **197**, 512–513.

Millar, R. H., and Scott, J. M. (1968). An effect of water quality on the growth of cultured larvae of the oyster, *Ostrea edulis* L. *J. Cons., Cons. Perm. Int. Explor. Mer.* **32**, 123–130.

Needler, A. B. (1942). Sex reversal in individual oysters. *J. Fish. Res. Board Can.* **5**, 361–364.

Nelson, J. (1917). An investigation of oyster propagation in Richmond Bay, Prince Edward Island, during 1915. *Contrib. Can. Biol.* 1915–16, 53–78.

Nelson, T. C. (1924). The attachment of oyster larvae. *Biol. Bull. (Woods Hole, Mass.)* **46**, 143–151.

Nelson, T. C. (1957a). Some scientific aids to the oyster industry. *Am. Sci.* **45**, 301–332.

Nelson, T. C. (1957b). On the reactions of oyster larvae in relation to setting on the cape shore of Delaware Bay, N.J. Unpublished manuscript, Rutgers Univ., New Brunswick, New Jersey.

Nelson, T. C., and Allison, J. B. (1940). On the nature and action of diantlin, a new hormone-like substance carried by the spermatozoa of the oyster. *J. Exp. Zool.* **85**, 299–338.

Nelson, T. C., and Perkins, E. B. (1931). The reactions of oyster larvae to currents and to salinity gradients. *Anat. Rec.* **40**, 288.

Orton, J. H. (1909). On the occurrence of protandric hermaphroditism in the mollusc *Crepidula fornicata*. *Proc. Roy. Soc. London, Ser. B* **81**, 468.

Orton, J. H. (1926). On lunar periodicity in spawning of normally grown Falmouth oysters, *O. edulis*, in 1925, with a comparison of the spawning capacity of normally grown and dumpy oysters. *Mar. Biol. Assoc. U.K.* **14**, 199–225.

Orton, J. H. (1928). The dominant species of *Ostrea*. *Nature (London)* **121**, 320–321.

Orton, J. H. (1937). "Oyster Biology and Oyster Culture," Buckland Lectures for 1935. Arnold, London.

Pomerat, C. M., and Reiner, E. R. (1942). The influence of surface angle and of light on the attachment of barnacles and other sedentary organisms. *Biol. Bull. (Woods Hole, Mass.)* **82**, 14–25.

Pritchard, D. W. (1952). Distribution of oyster larvae in relation to hydrographic conditions. *Gulf Caribb. Fish. Inst., Univ. Miami Proc. 5th Annu. Sess.* pp. 123–132.

Pritchard, D. W. (1955). Estuarine circulation patterns. *Proc. Am. Soc. Civ. Eng.* **81**, 1–11.

Prytherch, H. F. (1929). Investigation of the physical conditions controlling spawning of oysters and the occurrence, distribution, and setting of oyster larvae in Milford Harbor, Connecticut. *U.S. Bur. Fish. Bull.* **44**, 429–503.

Prytherch, H. F. (1934). The role of copper in the setting, metamorphosis and distribution of the American oyster *Ostrea virginica*. *Ecol. Monogr.* **4**, 47–107.

Quayle, D. B. (1952) *Oyster Bull. B.C. Dep. Fish Shellfish Lab., Ladysmith* 3(1), 1–16.

Quayle, D. B. (1953). *Oyster Bull. B.C. Dep. Fish Shellfish Lab., Ladysmith* 3(8), 1–17

Quayle, D. B. (1964). Distribution of introduced marine Mollusca in British Columbia waters. *J. Fish Res. Board Can.* **21**, 1155–1181.

Quayle, D. B. (1969). Pacific oyster culture in British Columbia. *Bull. Fish. Res. Board Can.* **169**.

Ranson, G. (1939). Le provinculum de la prodissoconque de quelques Ostreides. *Bull. Mus. Hist. Nat. Paris* **11**, 318–332.

Ranson, G. (1948). Prodissoconques et classification des Ostreides vivants. *Bull. Mus. R. Hist. Nat. Belg.* **24**, 1–12.

Ranson, G. (1951). "Les Huitres, Biologie-Culture, Bibliographie." Lechevalier, Paris.

Ranson, G. (1967). Les espèces d'huîtres vivant actuellement dans le monde, définies par leurs coquilles larvaires ou prodissoconques. Etude des collections des grands Mussées d'Histoire Naturelle. *Rev. Trav. Inst. Pêches Marit.* **31**(2)(3), p. 128–199; 205–274.

Ritchie, T. P., and Menzel, R. W. (1969). Influence of light on larval settlement of American oysters. *Proc. Natl. Shellfish Assoc.* **59**, 116–120.

Sayce, C. S., and Larson, C. C. (1965). Willapa oyster studies. *Wash. Dep. Fish., Annu. Prog. Rep.* pp. 1–94.

Schaefer, M. B. (1937). Attachment of the larvae of *Ostrea gigas*, the Japanese oyster, to plane surfaces. *Ecology* **18**, 523–527.

Stafford, J. (1913). "The Canadian Oyster, Its Development, Environment and Culture." Mortimer Co., Ottawa.

Stauber, L. A. (1950). The problem of physiological species with special references to oysters and oyster drills. *Ecology* **31**, 109–118.

Stenzel, H. B. (1971). Oysters. *In* "Treatise on Invertebrate Paleontology, Part N, Vol. 3: Mollusca 6" (R. C. Moore, ed.). Geol. Soc. Am. and Univ. of Kansas, New York.

Thomson, J. M. (1950). The effects of the orientation of cultch material on the setting of the Sydney rock oyster. *Aust. J. Mar. Freshwater Res.* **1**(1), 139–154.

Thomson, J. M. (1954). The genera of oysters and the Australian species. *Aust. J. Mar. Freshwater Res.* **5**(1), 132–168.

Thorson, G. (1950). Reproductive and larval ecology of marine bottom invertebrates. *Biol Rev. Cambridge Philos. Soc.* **25**, 1–45.

Thorson, G. (1957). Bottom communities (sublittoral or shallow shelf). *Geol. Soc. Am., Mem.* **67**, 461–534.

Thorson, G. (1964). Light as an ecological factor in the dispersal and settlement of larvae of marine bottom invertebrates. *Ophelia* **1**, 167–208.

Ukeles, R. (1971). Nutritional requirements in shellfish culture. *Proc. Conf. Artif. Propag. Commer. Valuable Shellfish, Coll. Mar. Stud., Univ. Delaware* pp. 43–64.

Walne, P. R. (1963). Breeding of the Chilean oyster (*Ostrea chilensis* Philippi) in the laboratory. *Nature (London)* **197**, 676.

Walne, P. R. (1964a). Observations on the fertility of the oyster (*Ostrea edulis*). *J. Mar. Biol. Assoc. U.K.* **44**, 293–310.

Walne, P. R. (1964b). The culture of marine bivalve larvae. *In* "Physiology of

Mollusca" (K. M. Wilbur and C. M. Yonge, eds.), Vol. 1, pp. 197–210. Academic Press, New York.

Walne, P. R. (1966). Experiments in the large-scale culture of the larvae of *Ostrea edulis* L. *Min. Agr., Fish. Food, Fish. Invest. London* 25, 1–53.

Walne, P. R. (1974). "Culture of Bivalve Molluscs: 50 Years' Experience at Conway." Fishing News (Books).

Westley, R. E. (1956). Retention of Pacific oyster larvae in an inlet with stratified waters. *Wash. Dep. Fish., Res. Pap.* 1, 1–7.

Wood, L., and Hargis, W. J., Jr. (1971). Transport of bivalve larvae in a tidal estuary. *Proc. Eur. Mar. Biol. Symp. 4th* pp. 29–44.

Yonge, C. M. (1960). "Oysters." Collins, London.

AUTHOR INDEX

Numbers in italics refer to the pages on which the complete references are listed.

A

Abraham, K. C., 209, 214, *265*
Ahmed, M., 119, 120, *265*
Aiello, E. L., 185, 187, *265, 283, 288*
Aiyyar, R. J., 151, 204, *283*
Alagarswami, K., 135, 136, 151, 204, 209, 215, *265, 285*
Alderman, D. J., *335*
Allan, J., 60, *79*
Allen, J. A., 114, 130, 135, 153, 155, 165, 166, 167, 168, 169, 212, 225, 227, 233, 234, 236, 237, 238, 241, 251, 252, 253, 255, *265*
Allen, J. F., 116, 120, 127, 135, 214, *265*
Allen, R. D., 219, 221, *265*
Allison, J. B., 307, *339*
Altman, G., 193, *265*
Amirthalingam, C., 204, *265*
Anderson, E., 28, 32, 36, 39, 41, 42, 68, 79, 130, *276*
Anderson, W. A., 128, *266*
Andrews, J. D., 294, 334, *335*
Ansell, A. D., 122, 135, 156, 157, 158, 159, 179, 180, 181, 183, 195, 201, 212, 234, 235, 236, 238, 244, 251, 252, 254, 255, 256, 261, *266, 268*
Antheunisse, L. J., 187, 188, 189, 190, *266*

Antony Raja, B. T., 209, 215, *266*
Araki, G., 54, 57, *80*
Arey, L. B., 48, *79*
Arvy, L., 100, 101, *110*
Ashikaga, C., 181, *266*
Awati, P. R., 119, *266*

B

Baba, K., 5, 9, 10, 11, 12, 13, 14, 15, 17, 18, 19, 20, *23*
Bacci, G., 120, *266*
Baglioni, C., 225, *274*
Baier, C. R., 154, *266*
Baird, R. H., 201, *266*
Balasubramanyan, R., 152, *266*
Ballantine, D., 155, *267*
Banse, K., *335*
Bardach, J. E., 297, *335*
Bargeton-Couteaux, M., 178, *267*
Barnawell, E. B., 31, 41, 48, *79*
Barnes, H., 322, *336*
Barnes, J. R., 28, 32, 34, 39, 44, 45, 48, 58, 59, 61, 62, 63, 65, 66, 68, 72, 73, *79*
Battle, H. I., 178, 204, *267, 277*
Baughman, J. L., 296, *335*

Baxter, J. M., 28, 47, 52, 53, 74, 79
Bayne, B. L., 143, 155, 156, 159, 178, 179, 181, 183, 184, 185, 195, 203, 205, 224, 228, 229, 234, 236, 237, 238, 242, 243, 244, 247, 248, 249, 251, 252, 253, 254, 267, 273, 333, 334, 335
Beaumont, A. R., 204, 232, 275
Beckett, R. L., 297, 337
Beedham, G. E., 3, 23, 236, 267
Belding, D. L., 197, 201, 203, 209, 213, 236, 253, 255, 267
Benthem Jutting, T. van, 47, 79
Berg, W. E., 204, 226, 267, 285
Berner, L., 200, 202, 267, 268
Blacknell, W. M., 261, 268
Blake, N. J., 138, 148, 150, 156, 174, 175, 176, 177, 187, 188, 190, 191, 192, 193, 194, 195, 198, 207, 268, 287
Blaschko, H., 185, 268
Blegvad, H., 154, 268
Bloomer, H. H., 119, 268
Boney, A. D., 155, 281
Boolootian, R. A., 29, 31, 48, 53, 54, 79, 80, 83
Boss, K. J., 52, 79
Bourcart, C., 193, 194, 268
Bourne, N., 212, 268
Bousfield, E. L., 311, 315, 321, 335
Boyden, C. R., 135, 268
Boyle, P. J., 234, 240, 268
Boyle, R. P., 51, 79
Breese, W. P., 205, 268
Brenko, M., 135, 231, 232, 234, 246, 268
Brewin, B. I., 29, 39, 49, 58, 61, 62, 63, 65, 67, 79
Brousseau, D. J., 197, 268
Brukema, J. J., 181, 268
Bubic, S., 232, 268
Buley, H. M., 154, 268
Bull, J., 52, 79
Bullis, R. H., 214, 268
Bullock, R. C., 34, 42, 49, 50, 52, 59, 60, 65, 67, 75, 79
Bullock, T. H., 158, 269
Burch, J. B., 116, 130, 281
Butler, P. A., 151, 269, 324, 327, 336

C

Caddy, J. F., 158, 269
Cahn, A. R., 335, 336
Cahour, A., 116, 269
Calabrese, A., 116, 127, 139, 143, 196, 203, 214, 230, 232, 239, 245, 246, 247, 250, 251, 268, 269, 271
Campbell, S. A., 152, 157, 184, 209, 211, 269
Carefoot, T. H., 53, 81
Carriker, M. R., 236, 269, 312, 314, 319, 321, 336
Carter, H. H., 329, 336
Castagna, M., 228, 234, 269, 277
Catapane, E. J., 185, 275
Cerruti, A., 313, 336
Chanley, M. H., 114, 241, 259, 260, 262, 264, 269
Chanley, P. E., 114, 239, 240, 241, 245, 247, 249, 251, 254, 259, 261, 271, 279
Chari, T., 181, 290
Charnov, E. L., 32, 79
Cheng, T. C., 198, 199, 269
Chia, F.-S., 75, 79
Chipperfield, P., 135, 136, 137, 143, 148, 159, 178, 269
Christiansen, M. E., 30, 39, 47, 53, 59, 62, 65, 66, 73, 74, 80
Chung, M. A., 161, 162, 181, 182, 183, 274
Clark, G. R., 255, 270
Clark, R. B., 90, 93
Clark, W., 58, 60, 78, 80
Claus, C., 232, 268
Coe, W. R., 116, 117, 118, 119, 120, 122, 123, 124, 125, 129, 136, 154, 178, 195, 196, 200, 214, 270, 273, 301, 336
Cole, H. A., 238, 270, 302, 322, 323, 324, 331, 332, 336
Collier, J. R., 220, 228, 270, 290
Comely, C. A., 159, 181, 270
Conklin, E. G., 224, 270
Corner, E. D. S., 155, 281
Costello, D. P., 65, 66, 68, 80, 204, 211, 213, 219, 223, 270
Cowden, R. R., 36, 62, 80, 130, 270

Cox, L. R., 28, 58, 59, 62, 65, 67, 72, 73, *85*
Crampton, H. E., 106, *110*
Creek, G. A., 227, 234, 235, 236, 238, 240, 255, 256, 257, *270*
Crisp, D. J., 322, 332, *336*
Crozier, W. J., 28, 29, 30, 31, 48, 52, 61, 65, *79, 80*
Culliney, J. L., 234, *270*
Curtin, L., 297, *336*

D

Dakin, W. J., 120, 256, *271*
Dalmon, J., 212, *271*
Dan, J. C., 220, 221, *271, 290*
Daniell, R. J., 181, *271*
Dare, P. J., 181, *271*
Davenport, J. C., 36, *80*
Davenport, R., 36, *80*
Davids, C., 155, *271*
Davidson, M. E., 204, 211, 213, 219, 223, *270*
Davis, H. C., 140, 143, 202, 203, 204, 205, 219, 228, 229, 230, 232, 239, 240, 241, 244, 246, 247, 248, 249, 250, 251, 254, 264, *269, 271, 279,* 297, 305, *336, 338*
Davis, R. L., 154, *272*
Dean, B., 304, *336*
de Block, J. W., 201, 253, *272*
De Brunn, W., 181, *268*
Dell, R. K., 76, 77, 78, *80*
de Longcamp, D., 184, *272*
Demeter, D., 185, *292*
Deroux, G., 129, 259, *272*
De Schweinitz, E., 234, *272*
De Wilde, P. A. W., 159, *272*
De Zwann, A., 181, *272*
Dhamne, K. P., 215, *282*
Dickie, L. M., 201, *272*
Dimick, R. E., 205, *268*
Dinamani, P., 95, 101, *110,* 299, *336*
Dollfus, P., 199, *272*
Dolph, C. I., 32, *80*
Donato, A., 36, 37, *80*
Donovan, J. E., 220, *281*
Dornfeld, E., 128, *278*

Drew, G. A., 22, *23,* 116, 213, 227, 233, 239, 255, 259, 262, *272*
Drobeck, K. G., 297, 331, *336, 338*
Drosdowsky, M., 184, *272*
Duggan, W., 234, *269*
Dunnington, E. A., Jr., 297, *337*
Dupouy, J., 199, *272*
Dupuy, J. L., 298, 310, *336*
Durve, V. S., 151, 215, *272*
Duval, D. M., 136, 178, *272*

E

Eddy, E. M., 37, *80*
Edwards, D. B., 181, *271*
Eggers, A., 204, 211, 213, 219, 223, *270*
Ehinger, R. E., 156, 185, *272*
Engle, J. B., 334, *338*
Estabrooks, S. L., 181, *272*

F

Fahrmann, W., 187, 188, *273*
Feltham, C. B., 154, *292*
Field, I. A., 199, 204, *273*
Filatova, Z. A., 154, *272*
Firtel, R. A., 225, *273*
Fischer-Piette, E., 1, 6, 8, *23,* 28, *80*
Fish, C. J., 157, *273*
Foldes, I., 185, *292*
Fox, D. L., 154, *270, 273*
Fox, H. M., 204, 211, 213, 219, 223, *270*
Fraga, F., 181, *273*
Franc, A., 1, 6, 8, *23,* 28, *80,* 114, 118, *273*
Franklin, L. E., 220, 221, *273*
Franzén, Å., 9, 10, 11, 12, *23,* 45, *80,* 127, *273*
Fretter, V., 114, 116, 117, 118, 120, 202, 205, 209, *273*
Fuji, A., 164, 185, 186, 197, *273*
Fujita, T., 224, *273*
Fullarton, J. H., 227, *273*

G

Gabbott, P. A., 156, 159, 178, 179, 181, 183, 184, 194, 195, 224, 251, *267, 273*

Gabe, M., 36, 39, *80*, 187, 188, *273, 274*
Gabrielli, F., 225, *274*
Gage, J., 201, 213, *274*
Gainey, L. F., Jr., 95, *110*
Galtsoff, P. S., 90, *93*, 114, 120, 200, 202, 205, 206, 222, 224, 227, 239, *274*, 296, 298, 299, 301, 303, 305, 308, 322, 332, *337*
Garnault, P., 36, 39, *80*
Gasquet, M., 28, 57, 58, *85*
Gay, H., 211, *275*
Geelen, H. J., 201, 253, *272*
Geilenkirchen, W. L. M., 101, 103, 105, *110, 111*
George, C. J., 236, 252, *289*
George, R. Y., 87, 90, 91, *93*
Ghiselin, M. T., 120, *274*
Gibson, F. A., 201, *266, 274*
Giese, A. C., 29, 30, 31, 32, 34, 35, 36, 39, 41, 43, 44, 45, 48, 50, 51, 53, 54, 56, 57, 58, 59, *80, 82, 83, 84, 85*, 131, 137, 138, 151, 161, 162, 179, 180, 181, 182, 183, 202, 204, *274*, 298, *336*
Gilbert, M. A., 158, 185, *274*
Gimazane, J. P., 135, 151, 156, 213, *274*
Giusti, F., 137, *285*
Glynn, P. W., 28, 29, 30, 31, 46, 48, 49, 50, 51, 52, 53, 56, 57, 59, 61, 62, 63, *81*
Götting, K. J., 28, 31, 32, 33, 34, 36, 38, 39, 41, 42, 47, 78, *84*
Goldberg, R., 234, *287*
Golding, D. W., 187, *274*
Golikov, A. N., 201, 214, *274*
Gonor, J. J., 28, 73, 79, 138, 213, 224, 227, 228, 229, 231, *275, 279*
Goodrich, E. S., 7, *23*
Gosselin, R. E., 185, 187, *281*
Graham, A., 114, 116, 117, 118, 120, 202, 205, 209, *273*
Graham, H. W., 211, *275*
Granmo, A., 232, *275*
Grave, B. H., 41, 42, 47, 53, 61, 63, 65, 66, 68, 73, *81*, 118, 204, 236, *275*
Grave, C., 154, *275*
Gray, J. E., 211, *291*
Greenfield, L. J., 181, *275, 278*
Grice, G. D., 157, *275*

Grobben, C., 114, *275*
Gruffydd, L. D., 204, 232, *275*
Guillard, R. R., 247, 248, 249, 250, *271*, 297, *336*
Gunter, G., 138, 204, 206, *275*, 294, 295, 310, *337*
Gutsell, J. S., 118, 122, 197, 203, 213, *275*

H

Hagstrom, B. E., 222, *286*
Haley, J. E., 185, *275*
Haller, B., 34, 36, *81*
Hammarsten, O. D., 35, 36, 65, 69, *81*
Hanks, J. E., 143, *289*
Hansen, B., 233, 257, 259, 260, 261, 262, 264, *275*
Hargis, W. J., Jr., 319, *341*
Hart, M. A., 30, 45, 53, 54, 55, 56, *80*, 161, 162, 181, 182, 183, *274*
Harven, E. de, 221, *284*
Hashimoto, S., 220, *271*
Hashizume, M., 164, 185, 186, 197, *273*
Haskin, H. H., 311, 319, 330, *337*
Hatanaka, M., 297, *337*
Hatano, H., 179, 180, 181, *288*
Hatschek, B., 234, *275*
Heath, H., 1, 2, 6, 8, 9, 10, 11, 12, 13, 14, 17, 18, 19, 21, 22, *23*, 29, 35, 41, 42, 48, 53, 58, 59, 60, 61, 65, 66, 68, 69, 72, 73, 76, 77, *81*
Hedgpeth, J. W., 138, *275*
Heider, K., 109, *111*
Henley, C., 65, 66, 68, *80*, 204, 211, 213, 219, 223, 224, *270*
Hennick, D. P., 212, *275*
Hessler, R. R., 207, *286*
Hidu, H., 232, *271, 275*, 297, 312, 319, 327, 330, 331, 332, *337, 338*
Higley, R. M., 35, *81*
Hill, C. L., 259, *275*
Himmelman, J. H., 28, 45, 47, 48, 50, 51, 53, 55, 61, 62, 63, *81*
Hirschhorn, G., 212, *276*
Hodgkin, E. P., 126, 135, 140, 141, 142, 204, 216, 217, 218, *291*

Hoffmann, H., 28, 39, 47, 60, 65, 66, 78, 81

Hoffman, S., 4, 7, 8, 23

Hollis, P. J., 310, 321, 337, 339

Hopkins, A. E., 302, 322, 324, 327, 337

Hornell, J., 204, 276

Houtteville, P., 116, 194, 195, 276

Howard, A. D., 258, 259, 276

Hubner, E., 130, 276

Hubrecht, A. A. W., 1, 23

Hughes, R. N., 155, 185, 202, 203, 213, 276

Hull, A. F. B., 41, 42, 59, 65, 66, 72, 76, 81

Humphrey, D. G., 32, 80

Humphreys, W. J., 130, 221, 223, 236, 276

Hyman, L. H., 1, 2, 8, 23, 28, 35, 52, 65, 72, 81

I

Idler, R. D., 184, 276

Ihering, H. von, 39, 45, 84

Imai, T., 297, 337

Ingle, R. M., 214, 268

Iredale, T., 76, 81

Isham, L. B., 252, 276

Ishida, C., 32, 83

Ito, S., 221, 276

Iwata, K. S., 204, 205, 219, 276

J

Jacobson, M. K., 201, 277

James, T. W., 155, 287

Jenner, C., 114, 277

Jensen, A., 154, 277

Jørgensen, C. B., 153, 154, 155, 277

Johns, P. M., 29, 30, 32, 34, 39, 41, 44, 45, 49, 50, 52, 61, 62, 63, 64, 82

Johnson, A. C., 240, 241, 258, 259, 260, 289

Jones, A. M., 28, 47, 52, 53, 74, 79

Jones, G. F., 214, 233, 258, 260, 262, 277

Joyce, E. A., Jr., 296, 337

K

Kass, P., 64, 82

Keen, M., 103, 111

Kennedy, A. U., 178, 277

Kennedy, V. S., 228, 277

Kidder. G. M., 224, 225, 277

Kinne, O., 138, 204, 206, 229, 231, 277

Kirby-Smith, W. W., 136, 154, 155, 156, 159, 214, 277

Knight-Jones, E. W., 322, 323, 324, 331, 332, 336, 337

Knorre, H. von, 36, 76, 78, 82

Knox, G. A., 63, 82

Knudsen, J., 260, 261, 277

Kofoid, C. A., 259, 275

Korringa, P., 127, 204, 278, 296, 298, 306, 309, 312, 313, 314, 320, 322, 324, 326, 334, 337

Korschelt, E., 109, 111

Kowalevsky, A., 41, 42, 59, 60, 64, 65, 69, 70, 72, 77, 78, 82, 96, 106, 107, 111

Krishnaswamy, S., 181, 288

Kubo, M., 220, 271

Kuenzler, E. J., 155, 278

Kussakin, O. G., 41, 42, 47, 53, 59, 75, 77, 82

L

Lacaze-Duthiers, F. J. H., 96, 97, 98, 100, 101, 102, 103, 107, 108, 109, 110, 111

Lammens, J. J., 122, 123, 127, 130, 134, 135, 136, 152, 158, 159, 178, 196, 204, 209, 211, 278

Lander, K. F., 156, 179, 180, 183, 266

Landers, W. S., 297, 334, 337, 338

Landon, W. A., 154, 292

Lane, C. E., 181, 278

Lankester, E. R., 114, 116, 120, 153, 257, 258, 259, 278

Larson, C. C., 314, 340

Lavallard, R., 268

Lawrence, A. L., 35, 53, 54, 82

Lawrence, J. M., 35, 53, 54, 56, 57, 82, 179, 278

Lebour, M. V., 211, 212, 213, 233, 236, 257, 258, 259, 262, 278
Leloup, E., 5, 23, 47, 76, 82
Lemche, H., 87, 88, 89, 92, 93
Lent, C. M., 201, 278
Leonard, U. K., Jr., 214, 278
Leuchtenberger, C., 221, 276
Lewis, J. B., 41, 42, 48, 49, 59, 82
Lillie, F. R., 224, 278
Lindsay, C., 323, 338
Longo, F. J., 128, 221, 278
Longwell, A. C., 298, 338
Loosanoff, V. L., 122, 123, 126, 127, 129, 130, 131, 135, 136, 138, 139, 140, 143, 148, 151, 155, 157, 159, 178, 195, 202, 203, 204, 205, 208, 209, 213, 219, 228, 230, 232, 239, 240, 241, 242, 244, 245, 247, 248, 249, 250, 251, 254, 264, 278, 279, 297, 303, 305, 307, 334, 338
Loosmore, F. A., 156, 179, 180, 183, 266
Lopez-Benito, M., 181, 279
Lough, R. G., 213, 224, 227, 228, 229, 231, 247, 279
Lovén, S., 60, 65, 78, 82
Lubet, P., 116, 122, 123, 126, 127, 129, 130, 135, 143, 157, 178, 184, 187, 188, 189, 190, 193, 194, 195, 205, 206, 212, 268, 272, 276, 279, 280
Lucas, A., 116, 135, 195, 269, 280
Lummel, L. v., 5, 23
Lunetta, J. E., 216, 280
Lunz, G. R., Jr., 334, 338
Lutz, R. A., 234, 272, 331, 338
Lyngnes, R., 36, 39, 41, 47, 82

M

MacBride, E. W., 224, 227, 280
McCrary, A., 114, 277
MacKenzie, C. L., 335, 338
Mackin, J. G., 334, 338
McLarney, W. O., 297, 335
McLean, K., 225, 280
McMillin, H. C., 332, 337
McMullen, J. C., 212, 280
McWhinnie, M. A., 138, 280
Magne, A., 30, 31, 47, 60, 62, 65, 78, 82

Mane, U. H., 135, 181, 183, 187, 188, 189, 190, 193, 282
Mann, H., 200, 280
Manning, J. H., 314, 329, 334
Marshall, N., 154, 157, 201, 209, 213, 272, 280, 281
Marteil, L., 303, 338
Martens, E., 76, 82
Martinez, J.-C., 199, 272
Martoja, M., 28, 57, 82, 187, 280
Mason, J., 201, 212, 280
Masson, M., 247, 280
Matthews, F. G. C., 28, 29, 30, 31, 32, 34, 35, 36, 39, 42, 44, 47, 52, 53, 59, 60, 61, 63, 65, 66, 69, 72, 78, 82
Matthiessen, G. C., 297, 338
Matveeva, T. A., 211, 233, 258, 280
Medcof, J. C., 309, 314, 325, 338
Meisenheimer, J., 224, 280
Menge, B. A., 75, 82
Menzel, M. Y., 120, 281
Menzel, R. W., 120, 281, 298, 303, 304, 310, 331, 338, 339, 340
Menzies, R. J., 87, 90, 91, 92, 93
Merrill, A. S., 116, 130, 281
Metcalf, M. M., 58, 59, 60, 65, 83
Metz, C. B., 220, 281
Mihursky, J., 228, 277
Mikheyev, V. P., 155, 281
Mileikovsky, S. A., 204, 211, 281
Millar, R. H., 297, 321, 339
Millemann, R. E., 205, 268
Miller, R. L., 28, 31, 48, 53, 64, 65, 83
Miller, W. S., 239, 242, 244, 245, 247, 248, 249, 279
Milton, A. S., 185, 187, 281
Milton, S., 185, 268
Minaur, J., 215, 251, 281
Miyazaki, I., 233, 258, 261, 262, 282
Monroe, H. C., 31, 48, 53, 83
Monroy, A., 220, 221, 222, 225, 273, 281
Montalenti, G., 119, 120, 281
Moore, D. R., 137, 148, 211, 281
Moore, H. B., 214, 292
Moore, J. K., 201, 281
Moore, K. E., 187, 281
Mooseker, M. S., 220, 289
Mori, K., 181, 288
Mori, S., 201, 281

Morton, J. E., 90, *93*, 114, 120, 155, *267*, *281*

Murakami, T. H., *287*

Murdoch, M. B., 158, *291*

Murti, K. G., 28, 29, 41, 49, 58, 59, 60, 62, 65, 67, *83*

N

Nagabhushanam, R., 28, 29, 41, 49, 58, 59, 60, 62, 65, 67, *83*, 135, 152, 181, 187, 188, 189, 190, 193, 195, 215, *282*

Naidu, K. S., 116, 122, 127, 131, 135, 197, 202, 204, 209, 213, *282*

Nair, N. B., 152, 179, 180, 181, 204, 215, *282*

Narasimham, K. A., 203, 204, 209, 215, *282*, *285*

Narchi, W., 214, *282*

Nayar, K. N., 209, 215, *282*

Needler, A. B., 301, *339*

Nelson, D. A., 232, *269*

Nelson, J., 312, *339*

Nelson, T. C., 202, 205, *282*, 297, 307, 312, 319, 322, *339*

Newell, R. C., 153, 154, *282*, *283*

Nierstrasz, H. F., 1, *24*

Nimitz, M. A., 32, 34, 36, 39, 41, 43, 44, 45, 53, 54, 56, *83*

Nishikawa, S., 32, *83*

Nomejko, C. A., 303, 305, 307, *338*

Norman, K. D., 202, *284*

O

Ockelmann, K. W., 120, 128, 153, 154, 156, 159, 206, 209, 210, 211, 232, 233, 258, 259, 261, 262, 263, 264, *283*

Odhner, N., 6, *24*

Odlaug, T. O., 200, *283*

Okuda, S., 42, 48, 58, 59, 60, 61, 65, 67, 72, *83*

Oldfield, E., 117, 120, 128, 205, 211, 213, 233, 257, 258, 259, 260, 261, 262, *283*

Olsen, A. M., 202, *283*

Orton, J. H., 118, 119, 138, 202, *283*, 294, 297, 306, *339*

Owen, G., 153, 154, *283*

P

Palombi, A., *283*

Panikkar, N. K., 151, 204, *283*

Paparo, A., 185, 187, *283*

Pasteels, J. J., 219, 221, *283*, *284*

Pathansali, D., 215, *284*

Paul, M. D., 151, 204, *284*

Pearse, J. S., 28, 29, 31, 42, 44, 45, 46, 49, 50, 51, 53, 63, *80*, *83*, 131, 137, 138, 151, 202, *274*, 298, 337

Pelseneer, P., 1, *24*, 28, 30, 34, 36, *83*, 116, 257, 258, 262, 264, *284*

Perenyi, L., 185, *292*

Perkins, E. B., 319, *339*

Perlmann, P., 222, *286*

Personne, P., 128, *266*

Pfeffer, G., 76, *82*

Pfitzenmeyer, H. T., 148, 150, 211, *284*

Plate, L., 28, 32, 35, 75, 76, 77, *83*

Pomerat, C. M., *339*

Popham, M. L., 153, *284*

Porter, H., 148, 209, 213, *284*

Posgay, J. A., 116, 201, 202, *284*

Posner, G. S., 181, *278*

Potts, F. A., 153, *284*

Pratt, D. M., 157, *284*

Prenant, M., 36, 39, *80*

Pritchard, D. W., 315, 330, *339*

Prosser, C. L., 158, *284*

Pruvot, G., 1, 2, 6, 8, 9, 10, 11, 12, 13, 15, 17, 19, 21, *24*

Prytherch, H. F., 319, 322, 332, 337, *339*

Puppi, A., 185, *284*

Purchon, R. D., 114, 115, 116, 120, 152, 153, 195, *284*

Q

Quayle, D. B., 122, 123, 127, 130, 136, 139, 211, 212, 227, 236, 238, 240, 255, 256, 257, *268*, *284*, 297, 305, 309, 312, 314, 320, 321, 327, *340*

R

Rae, J. G., III, 135, *284*
Rahaman, A. A., 181, *284, 285*
Rai, H. S., 119, 214, 215, *266, 285*
Rancurel, P., 187, 188, *274*
Ranson, G., 295, 296, *340*
Rao, K. P., 155, *285, 287*
Rao, K. S., 135, 136, 152, *285*
Rao, K. V., 203, 204, 209, 215, *285*
Rasch, E., 135, *288*
Rattenbury, J. C., 226, *285*
Raven, C. P., 100, 101, 103, 104, 105, 109, *111*, 130, 217, 219, 221, 222, 223, 224, 225, 226, 227, 228, 232, 234, 235, 239, 255, 256, *285*
Rebhun, L. I., 135, 221, *285, 288*
Reddiah, K., 116, 127, 135, 196, 212, *285*
Rees, C. B., 211, 213, *285*
Reid, R. G. B., 154, 183, *285*
Reiner, E. R., *339*
Reish, D. J., 137, 148, 211, *281*
Renzoni, A., 137, *285*
Retzius, G., 9, *24*, 45, *83*, 127, *285*
Reverberi, G., 96, 100, 105, *111*, 130, 221, 227, *285, 286*
Rhodes, E. W., 232, *269, 334, 338*
Richmond, M. S., 297, *337*
Richter, H.-P., 28, 31, 32, 33, 34, 36, 38, 39, 40, 41, 42, 47, 68, 78, *83, 84*
Risbec, J., 29, 35, 39, 41, 42, 59, 65, 66, 67, 69, 72, *81, 84*
Ritchie, T. P., 331, *340*
Rokop, F. J., 101, 103, *111*, 131, 207, *286*
Ropes, J. W., 116, 127, 129, 135, 137, 140, 143, 148, 150, 202, 204, 209, 211, 214, *286*
Rosa, K., 185, *292*
Rosenburg, W. H., 228, *277*
Rosewater, J., 87, *93*
Rossenburg, W. H., 297, *337*
Rothschild, Lord, *84*
Rowe, G. T., 87, 90, 91, *93*
Runnström, J., 35, 36, 65, 69, *81*
Runnström, S., 138, 222, *286*
Ryther, J., 157, *286*, 297, *335*

S

Sabatier, A., 29, 39, *84*
Sagara, J., 205, *286*
Sakai, S., 297, *337*
Sakshaug, E., 154, *277*
Salanki, J., 185, *292*
Saleuddin, A. S. M., 120, *286*
Salvini-Plawen, L. von, 2, 3, 5, 7, 8, 9, 19, 21, *24*, 71, *84*
Sanders, H. L., 207, *286*
Saraswathy, M., 152, 179, 180, 181, 204, 215, *282*
Sastry, A. N., 117, 119, 120, 129, 130, 131, 132, 133, 134, 135, 136, 137, 138, 139, 143, 145, 146, 147, 148, 149, 150, 151, 154, 156, 157, 158, 159, 161, 164, 169, 170, 171, 172, 173, 174, 175, 176, 177, 193, 194, 195, 197, 198, 201, 202, 203, 204, 207, 209, 213, 214, 215, 217, 219, 227, 234, 235, 236, 237, 238, 239, 240, 244, 251, 252, 253, 255, 256, 257, *287*
Sato, R., 297, *337*
Savage, N., 234, *287*
Sawada, N., 219, *287*
Sayce, C. S., 314, 323, *338, 340*
Scarlato, O. A., 201, 214, *274*
Schaefer, M. B., 324, *340*
Schechter, V., 229, *287*
Scheer, B. T., 104, *111*
Scheltema, A. H., 3, 5, 6, 7, 8, *25*
Schinske, R. A., 154, *288*
Schwabl, M., 3, 5, 7, 9, *25*
Schweikart, A., 36, 39, *84*
Scott, J. M., 297, *339*
Segal, E., 151, 155, 204, *287*
Sellmer, G., 200, 233, 257, 258, 260, 264, *287*
Selwood, L., 28, 29, 32, 34, 36, 37, 38, 42, 49, *84*
Shaw, W. N., 127, 129, *287*
Simpson, R. D., 28, 41, 49, 65, 75, 76, *84*
Simroth, H., 28, *84*
Sindermann, C. J., 198, *287*
Smayda, T., 157, *287*
Smith, A. G., 27, 28, 76, 77, *84*

Smith, A. M., 161, 162, 181, 182, 183, 274
Smith, J., 118, 275
Smith, P. B., 239, 242, 244, 245, 247, 248, 249, 279
Soot-Ryen, T., 233, 288
Sorokin, Y. J., 155, 281
South, G. R., 199, 288
Southwick, W. E., 65, 68, 84
Sparks, A. K., 119, 120, 265
Sprague, L., 199, 288
Srinivasan, V. V., 181, 288
Stafford, J., 322, 340
Stasek, C. R., 90, 93
Stauber, L. A., 309, 340
Stefano, G. B., 185, 275, 288
Stenzel, H. B., 295, 340
Stephens, G. C., 154, 288
Stephenson, A., 31, 36, 39, 42, 49, 61, 63, 84, 202, 206, 288
Stevenson, R. N., 199, 288
Stickney, A. P., 120, 122, 127, 129, 135, 137, 148, 150, 211, 214, 219, 228, 230, 244, 245, 286, 288
Stiles, S. S., 298, 338
Stoll, E., 201, 288
Sugiura, Y., 116, 137, 148, 205, 288
Sullivan, C. M., 211, 288
Sutton, C. E., 298, 310, 336
Sweeney, D., 288
Swift, H., 135, 288

T

Takahashi, K., 181, 288
Takatsuki, S. I., 126, 288
Tamura, T., 184, 276
Tanaka, S., 179, 180, 181, 288
Tang, S. I., 130, 204, 288
Theede, H., 155, 288
Thiele, J., 1, 11, 12, 18, 19, 25, 32, 84
Thompson, R. J., 156, 181, 183, 184, 267, 289
Thompson, T. E., 3, 10, 11, 12, 13, 15, 16, 17, 18, 19, 20, 21, 22, 25, 45, 84
Thomson, J. M., 295, 340
Thorpe, S. R., 29, 41, 58, 59, 60, 61, 62, 63, 64, 65, 66, 72, 84

Thorson, G., 47, 74, 78, 84, 116, 202, 206, 211, 232, .233, 237, 252, 255, 258, 259, 261, 262, 289, 298, 308, 313, 330, 340
Tierney, J. Q., 252, 276
Tiffany, W. J., 201, 289
Tilney, L. G., 220, 289
Timmermans, L. P. M., 101, 103, 105, 110, 111
Tranter, D. J., 118, 119, 122, 123, 124, 126, 127, 128, 129, 130, 131, 132, 135, 136, 137, 139, 148, 196, 209, 215, 217, 289
Trevallion, A., 156, 157, 158, 159, 179, 180, 266, 289
Trueman, E. R., 3, 23
Tucker, J. S., 29, 31, 48, 54, 56, 58, 59, 80, 84
Turner, E., 28, 75, 76, 77, 85
Turner, H. J., 116, 120, 123, 129, 136, 143, 178, 196, 236, 252, 270, 289
Turner, R. D., 234, 240, 241, 258, 259, 260, 268, 270, 289
Tyler, A., 84, 104, 111, 220, 289

U

Ukeles, R., 247, 249, 289, 290, 297, 340
Umiji, S., 188, 190, 290
Uzmann, J. R., 199, 200, 290

V

van Dam, L., 155, 290
van Weel, P. B., 153, 290
Vassallo, M. T., 178, 184, 290
Vasu, B. S., 54, 85
Venkataraman, R., 181, 290
Verdonk, N. H., 96, 101, 103, 104, 105, 106, 107, 110, 111
Vernberg, F. J., 138, 158, 290
Verwey, J., 154, 155, 201, 290
Vicente, N., 28, 57, 58, 85
Volz, P., 47, 82
Von Oertzen, J. A., 139, 290

W

Wada, K., 120, *290*
Wada, R., 219, *290*
Wada, S. K., 65, *85*, 122, 205, 219, 220, 221, 222, 223, 224, 225, 227, 234, 235, 236, *271, 290*
Wainai, T., 184, *276*
Wagner, R., 28, *85*
Walne, P. R., 233, 239, 247, 250, *290, 291*, 297, 305, 306, 310, 321, *340, 341*
Watanabe, J. M., 28, 58, 59, 62, 65, 67, 72, 73, *85*
Wells, H. W., 211, *291*
Welsh, J. H., 188, *291*
Werner, B., 236, *291*
Westley, R. E., 311, 323, *338, 341*
Whaley, H. H., 314, 329, *338*
Wheeler, B. M., 157, *280*
White, K. M., 120, 121, 122, *291*
Wiborg, K. F., 116, *291*
Widdows, J., 156, 224, 251, *267, 291*
Wilbur, K. M., 255, *291*
Williams, C. S., 179, 181, 200, *291*
Williams, R. B., 158, *291*
Wilson, B. R., 126, 135, 140, 141, 142, 152, 204, 216, 217, 218, *291*
Wilson, D. P., 237, 252, 255, *291*
Wilson, E. B., 104, 105, 106, *111*

Windsor, N. T., 298, 310, *336*
Wingstrand, K. G., 88, 89, 92, *93*
Winter, J., 153, 155, *291, 292*
Wirén, A., 1, 9, *25*
Wood, L., 319, *341*
Woods, F. H., 123, 224, *292*
Woodward, J., 135, *288*
Wright, P. B., 214, *292*

Y

Yakovleva, A. M., 28, 77, *85*
Yamamoto, G., 143, 202, *292*
Yentsch, C. S., 157, *286, 292*
Yonehara, K., 220, *271*
Yonge, C. M., 90, *93*, 95, *111*, 206, *292*, 294, 296, 298, 305, 322, *341*
Yoshida, H., 227, *292*
Young, R. T., 205, 229, *292*
Yuki, R., 297, *337*

Z

Zandee, D. I., 181, *272*
Zion, H. H., 228, *277*
Zobell, C. E., 154, *292*
Zs Nagy, I., 185, *292*

SUBJECT INDEX

Entries in this index which are from a table are indicated by a "t" following the page number; page numbers for entries which are from a figure are set in boldface type.

A

Acrosomal vesicle, 128
Acrosome
  absence in chitons, 43
  in Pelecypoda, 127, 128
Acrosome reaction, in Pelecypoda, 220
Adriatic Sea, spawning periods in, 47t
Age
  gametogenesis and, 52, 53t, 195–198
  at sexual maturity, 196, 197
Alaska
  breeding periods in, 10t
  reproductive cycles in, 212t
Annulate lamellae, Polyplacophora, 39
Anus, Aplacophora, **15**
Antarctica, brooding in, 75, 76t
Apical tuft
  Aplacophora, **14, 15, 16, 17**
  Pelecypoda, 224, 235
  Polyplacophora, 68, 69, **70,** 72
  Scaphopoda, 105, 109
Archenteron
  Pelecypoda, 225–227 *passim*
  Scaphopoda, 106
Ash content and gametogenesis, 179, 180
Atrium, Aplacophora, 20
Australia
  breeding period in, 209, 214t, 216t, 217
  brooding in, 77t
  reproductive cycles in, 136, 140
  spawning period in, 49t, 63

B

Baltic Sea, reproductive cycles in, 139
Barbados, spawning period in, 48t, 49t
Bermuda
  sex ratios from, 31t
  spawning periods in, 48t
Biochemical composition and gametogenesis, 54, 160, 161, **162, 163, 175,** 179–185
Blastocoel
  Aplacophora, 13
  Pelecypoda, 225
Blastopore
  Aplacophora, **14**
  Pelecypoda, 225
Blastula
  Aplacophora, 13, 14
  Pelecypoda, 224, 225
  Scaphopoda, 106, **107**
Blastulation, Pelecypoda, 224–225
Body weight and gametogenesis, 159, **160**
Brazil, breeding periods in, 216t
Breeding season
  Aplacophora, 9, 10
  Pelecypoda, 209, 210t–216t
British Columbia
  brooding in, 77t
  reproductive cycles in, 51, 139
  spawning periods in, 47t, 48t, 305
Brood chamber, Aplacophora, **6**

Brooding
  animal size and, 75
  Aplacophora, **6**, 11, 12, 17, 18t
  egg size and, 78
  Ostreidae, 306
  Pelecypoda, 257–264
  Polyplacophora, 29, 74–78

C

California
  breeding period in, 211t, 214t
  brooding in, 76t, 77t
  reproductive cycle in, 51, 57, 137
  sex ratios from, 31t
  spawning period in, 46, 48t, 58, 64
Canada, breeding periods in, 211t, 212t
Captacula, Scaphapoda, 110
Carbohydrate levels in reproduction, 54,
  55t, 56, 57, 179–185 *passim*
Caribbean Sea, spawning in, 63
Cerebral ganglion, development of,
  Aplacophora, **16, 20**
  Pelecypoda, 227
  Polyplacophora, **70**
Chemical induction of spawning, 205–206
Chesapeake Bay
  breeding periods in, 211t, 214t
  development in species from, 230
  fecundity in species from, 302
  larval distribution in, 314, 315
Chile, brooding in, 77t
Chorion
  Aplacophora, 8
  Polyplacophora, 39, **40**, 42
Chromosome numbers, Polyplacophora,
  32
Cleavage
  Aplacophora, 13, 18t
  Ostreidae, 308
  Pelecypoda, **222**, 223–225
  Polyplacophora, 66t, 67t, 68
  Scaphopoda, 104, 105
Cloaca, Aplacophora, 20
Cloacal caecums, 8
Coeloblastula, 224, 225
Coelom, in Aplacophora, 7
Coelom formation, Polyplacophora, 69

Coelomoducts, **4**, 5, **6**, 7, 20
Coelomoduct glands, 8
Coelomopore, **4**, 7, 8
Connecticut, breeding periods in, 209,
  213t
Copulation, Aplacophora, 11, 12
Copulatory spicules, Aplacophora, 8
Cortical granules
  Pelecypoda, 221
  Polyplacophora, 39, 68
Costa Rica
  Neopilina in, 87
  reproductive cycles in, 57
Cytophores, Pelecypoda, 128, 129
Cytoplasmic localization of morphogenetic
  factors, 105–106

D

Denmark
  breeding periods in, 211t
  larval growth rate in, 243
Development
  direct, 233, 257
  lecithotrophic, 233, 264
  planktotrophic, 233
Development times, Polyplachora, 66, 67
Digestive gland
  Aplacophora, 5
  Pelecypoda, 227
  Polyplacophora, 53
Dioecism
  Aplacophora, 9
  Pelecypoda, 114, 115, 116, 119, 120
  Polyplacophora, 28
Dispersion of larvae, 313–320
Dorsal plates in Aplacophora, 21

E

Eggs
  Aplacophora, 8–12, 18t
  Monoplacophora, 91
  Polyplacophora, 39, **40**, 41–44
  Scaphopoda, 101
Embryo
  Aplacophora, **6**, 8, 11, **14, 15, 16**, 19
  Polyplacophora, **70**

Embryonic development
 Aplacophora, 13–20
 Ostreidae, 307–308
 Pelecypoda, 217–232
 Polyplacophora, 68–71, **70**
Embryonic induction, 228
Endocrines, in gametogenesis, 57
Endoderm, Aplacophora, **16**, 17, 20
Energy allocation for gamete production,
 185, **186**
England
 breeding period in, 211t, 212t, 213t
 reproductive cycle in, 10t, 30t, 31t, 47t,
  180
 spawning in, 47t, 60, 63
Epidermal papillae, Aplacophora, **14**
Europe, brooding species in, 76t, 77t

F

Fecundity, 301–303
Fertilization
 Aplacophora, 12
 external, 22
 internal, 11, 12
 Pelecypoda, 220–223, 229
 Polyplacophora, 65–68
 Scaphopoda, 103, 104, 105
Florida
 breeding period in, 209, 213t, 214t,
  217
 reproductive cycles in, 137, 143, 148,
  150, 157
 spawning in, 203
Follicles and follicle cells
 Aplacophora, 9
 Monoplacophora, 91
 Pelecypoda, 120–**127**, 129–131, 136
 Polyplacophora, 32, 34
Food, gametogenesis and, 53, 153–159,
 171
Foot groove, 3
Foregut, Aplacophora, 20
France
 breeding periods in, 213t
 sex ratios in, 30t, 31t
 spawning periods in, 47t

G

Gamete fusion, 20
Gametogenesis
 Aplacophora, 9
 factors affecting, 51–58, 137–200
 Monoplacophora, 90–91
 Pelecypoda, 126–131
 Polyplacophora, 35–45
Gastrula
 Aplacophora, 13, **14**, 15, 20
 Pelecypoda, 225–226
 Polyplacophora, 66t, 67t
 Scaphopoda, 106, **107**
Gastrulation
 Aplacophora, 13, 14, 15, 16, 17
 delamination, 13, 14
 emboly, 225
 epiboly, 13, 14, 225
 immigration, 13
 invagination, 13, 14, 106, 225, 226
 Pelecypoda, 225–226
 Polyplacophora, 68, 69
Genital pore, Pelecypoda, **124**
Genital primordium, Polyplacophora,
 36
Geographical range of Ostreidae, 303–304,
 309
Georges Bank, spawning in, 202
Germ cell origin, Polyplacophora, 35–36
Germany, spawning periods in, 47t
Germinal epithelium, Monoplacophora,
 91
Germinal vesicle,
 Monoplacophora, 90
 Pelecypoda, 217, 219
 Polyplacophora, 68
 Scaphopoda, 101, 103
Gonads
 Aplacophora, **4, 5, 6, 7**
 Monoplacophora, 88
 Pelecypoda, 120–122, **124, 125**, 158–
  167, 174
 Polyplacophora, 32, 35
 Scaphopoda, 96, **97, 98**, 110
Gonad index
 Pelecypoda, **132**–135, 144t, 145t, **158**,
  159–161, **164**, 171, **174**
 Polyplacophora, 45, 55

Gonoducts
  Aplacophora, 5
  Monoplacophora, 89
  Pelecypoda, 114, 115, 120, **121, 122,
    124, 125**
  Polyplacophora, 35
Gonopericardial ducts, 4, 6, 7
Gulf of Suez
  breeding cycle in, **46**
  sex ratios, 31t
  spawning period in, 49t, 63

H

Hatching
  Aplacophora, 21
  Pelecypoda, 227, 261–262
  Polyplacophora, 66t, 67t
Hermaphoroditism
  Aplacophora, 5
  categories of, 117
  causes of, 118
  hormones in, 118, 119
  Pelecypoda, 114, 115, 116, 117, 118,
    119, 120
  Polyplacophora, 29, 30
Hull on chiton eggs, 39, **40**, 42, 71

I

India
  breeding period in, 209, 214t, 215t
  reproductive cycles in, 136, 151, 152
  spawning period in, 49t
Induction of spawning by
  light, 62
  sperm suspension, 61
  spring plankton bloom, 62
Irish Sea, reproductive cycle in, 10
Italy, reproductive cycles in, 137

J

Japan
  breeding periods in, 214
  reproductive cycles in, 9, 10t, 185
  spawning periods in, 48t, 58

Juveniles
  Aplacophora, 12, 14, 15, 16, 18t, 21
  Ostreidae, 335
  Polyplacophora, 36, 71
  Scaphopoda, 109
Juvenile development, Pelecypoda, 255–
  264

K

Kurile Islands
  brooding in, 77t
  spawning periods in, 47t

L

Larvae
  Aplacophora, 11, 14, 15, 16, 20, 21
  Pelecypoda, 229–236 passim
  Polyplacophora, 32, 70, 71, 73, 78
Larval behavior, in Ostreidae, 311–312
Larval development, 71–79, 234–236
Larval dispersion, 313–320
Larval ecology of Ostreidae, 320–321
Larval ectoderm, 15, 68
  Aplacophora, 15
Larval eyes, development of, 68, 71, 72,
  235, 256
Larval foot
  Polyplacophora, 70, 71, 72, 73
  Scaphopoda, 109
Larval growth, 239–251, 310
Larval life in Ostreidae, 308–322
Larval musculature, 235
Larval radula sac, Polyplacophora, 70
Larval shell valves, 68–74 passim, 78
Light
  in gametogenesis, 51, 151
  induction of spawning by, 62, 63, 204–
    205
  and larval behavior, 311
Lipid levels in reproduction, 54, 55t, 56,
  57, 179–285

M

MacQuarie Island, 30t, 31t

Maine
  breeding period in, 211t, 213t
  development in, 230
  reproduction cycles in, 143
Malaysia, breeding periods in, 215t
Massachusetts
  breeding periods in, 209, 211t, 213t
  development of species from, 230
  reproductive cycles in, 126, 137, 139,
    143, 148, 149, 150, 151, 157, 161,
    163
  spawning periods in, 47t, 63, 208
Maturation
  Aplacophora, 12
  Pelecypoda, 131, 133, 217–219
  Polyplacophora, 68
Mediterranean Sea, reproductive cycle in,
  10t
Meiosis
  Pelecypoda, 130, 131
  Polyplacophora, 68
Mesoderm, Aplacophora, **16**, 17, 20
Mesoderm formation
  Aplacophora, 15, **16,** 17, 20
  Pelecypoda, 226
  Polyplacophora, 68
  Scaphopoda, 106
Metamorphosis
  Aplachophora, 15, 19, 20, 21, 22
  delay of, 331–335
  Monoplacophora, 92
  Ostreidae, 321–322
  Pelecypoda, 228–233, 238–239, 240t–
    241t, 254–255
  Polyplacophora, 71–74
  role of byssal gland in, 236, 238, 321
Micropyle, Scaphopoda, 103
Midgut, Aplacophora, **16**
Mouth, Aplacophora, **16**
Mucus, spawning and, 59, 60

                    N

Nephridia, gonads and, 89, 90, 96
Nervous system, development, 68, 109,
  256
Netherlands
  breeding period, 209, 211t

  larval settlement in, 326
  spawning periods in, 47t
Neurosecretion
  and gametogenesis, 185–195
  and spawning, 189, 306
Neurosecretory cells, 57, 58, 287, 288,
  289, 290
Newfoundland
  breeding period in, 213t
  spawning in, 202
New Jersey
  breeding period in, 209, 214t
  larval distribution in, 313, 314, **317,**
    316–20, 330
  larval settlement in, 3, 12, 327, 329,
    330, 332.
  reproductive cycles in, 140
New York
  breeding period, 209, 213t, 214t
  development in, 230
  larval distribution in, 313
  reproduction cycles in, 138, 139, 143
New Zealand
  brooding in, 77t
  spawning period in, 49t, 63
Nidosome, 60
North Carolina
  breeding period in, 209, 211t, 213t,
    214t, 217
  reproductive cycle in, 133, 134, 136,
    139, 147–148, 149, 150, 151, 157,
    158, 161, 163, 169
  spawning in, 202
North Sea
  breeding period in, 212t, 213t
  sex ratios in, 31t
North Wales, larval growth rate in, 242,
  244
Norway
  sex ratios in, 30t
  spawning times in, 47t
Nova Scotia, larval settlement in, 326
Nuage in oogenesis, 39
Nucleic acid in reproduction, 57,
  101
Nucleolus, in oogenesis, 200, 217, 219
Nutrient uptake and distribution,
    Pelecypoda, 154, 155, 156, 165,
    166t, 167t, 168t, 169t, 175, **176,**
    177, 178

O

Oogenesis
  Aplachohora, 8
  Monoplacophora, 91
  nutrient uptake and, **176,** 177
  Pelecypoda, **128,** 129–131
  Polyplacophora, 36–44
  RNA synthesis in, 36, 37
  Scaphopoda, 100, 103
  synchrony of, 130, 131
Oogonia
  Pelecypoda, 130, 131
  Polyplacophora, 36
Oregon
  breeding period in, 212t, 213t
  spawning periods in, 48t
Organogenesis
  Aplacophora, 19
  Pelecypoda, 255
Origin of germ cells
  Aplacophora, 9
  Pelecypoda, 122–126
  Polyplacophora, 35–36
Ovarian folds, in Polyplacophora, 32, **34**
Ovary
  Aplacophora, **5**
  Monoplacophora, 90
  Pelecypoda, **121**
  Polyplacophora, **33**
  Scaphopoda, 96, 97, **102**
Oviduct
  Pelecypoda, **122**
  Polyplacophora, 29, **33,** 35, 59
  Scaphopoda, 101
Ovoviparity, Polyplacophora, 74–78

P

Pallial groove
  brooding and, 74, 257
  spawning and, 65
Panama
  reproductive cycles in, 51
  sex ratio and, 31t
  spawning period in, 49t, 50
Parasites and gametogenesis, 52, 198–200

Parasitic males, 114
Parthenogenesis, lacking in Polyplaco-
    phora, 28
Pedal groove in Aplacophora, **14, 15**
Pediveliger, **237**–239, 321, 333
Pericalymma, 19, 21
Pericardial coeloms and gonads, 89
Pericardium, 4, 5–7
pH effect on development, 232
Photoreceptors, 51
Phytoplankton and gonad growth, 157,
    158
Polar bodies
  Aplacophora, 12
  Pelecypoda, 217, 219, **222**
  Polyplacophora, 68
  Scaphopoda, 104
Polar lobes
  Aplacophora, 13
  Pelecypoda, **222,** 223, 231
  Scaphopoda, 104–106
Polyembryony, lacking in Polyplacophora,
    28
Polyspermy
  Ostreidae, 308
  Scaphopoda, 104
Predation on larvae, 313
Primordial germ cells
  Pelecypoda, 122, 123, 126, 130
  Polyplacophora, 36, 44
Proctodeum, Aplacophora, **16,** 20, 21
Prodissoconch stage
  Pelecypoda, 233
  Scaphopoda, **108**
Pronuclear fusion, 223
Protandry
  Aplacophora, **5**
  Ostreidae, 299
  Pelecypoda, 117–119, 195, **196**
Protein levels in reproduction, 54, 55t, 56,
    57, 179–185
Proterogyny, 117, 118
Protonephridia, Pelecypoda, 227, 236
Prototroch
  Aplacophora, **14–16,** 19, 20
  Pelecypoda, 227
  Polyplacophora, 66t, 67t, 69, **70,** 72, 74
Puerto Rico
  reproductive cycles in, 51

Puerto Rico (*cont.*)
 sex ratios in, 31t
 spawning times in, 46, 48t–50

R

Red Sea
 reproductive cycles in, **46**
 spawning period in, 50
Reproductive behavior and spawning
 and aggregation, 200–201
 Aplacophora, 11–12
 and migration, 201–202
 Ostreidae, 303–307
 Polyplacophora, 58–65
 Scaphopoda, 103
Reproductive cycles
 Aplacophora, 9, 10t
 Ostreidae, 306
 Pelecypoda, 131–137, 144t, 145t, 149t
 Polyplacophora, 45–51
Reproductive rhythms, 61–63, 307
Reproductive system, anatomy
 Aplacophora, 6–8
 Monoplacophora, 88–90
 Ostreidae, 298–301
 Pelecypoda, 120–122
 Polyplacophora, 32–35
 Scaphopoda, 96–98
Rhode Island
 breeding period in, 209, 213t
 reproductive cycles in, 136, 185

S

Salinity
 and development, 229–231, 230t, **231**
 in gametogenesis, 151, 152
 in spawning, 204
Scotland
 breeding periods in, 212t
 spawning periods in, 47t
Self-fertilization, 117, 301
Seminal receptacles, Aplacophora, **4, 7,**
 11
Seminal vesicles

Aplacophora, **4, 7**
Pelecypoda, 114, 115
Polyplacophora, 35
Settlement
 Aplacophora, 18t
 distribution at, 324–331, 328t
 and gregariousness, 322, 331–334
 Ostreidae, 312, 322–335
 Pelecypoda, 236–238, 251–254
 Polyplacophora, 66t, 67t, 71–75
 prediction of, 323, 324
Sex determination
 Ostreidae, 299
 Pelecypoda, 115–120
 Polyplacophora, 29, 30
Sex ratio
 Ostreidae, 299–301
 Polyplacophora, 29, 30
Sex reversal, 117, 118, 195, 196, 301
Sexual dimorphism
 Aplacophora, 5
 Monoplacophora, 88
 Polyplacophora, 28, 29
 Scaphopoda, 98, 99
Shell gland development
 Aplacophora, 5, 7, 8
 Pelecypoda, 225, 227
Slime sacs, 35
South Africa
 brooding in, 76t, 77t
 reproductive cycles in, 152
South America, brooding in, 76t, 77t
Spawning
 Aplacophora, 11, 18t
 Monoplacophora, 91–92
 Ostreidae, 302–303
 Pelecypoda, 117, 133, 135–173, 202–
 209
Polyplacophora, 30, 47t–49t, 50, 51,
 55–65
 Scaphopoda, 98, 101–103
 synchrony, 206–209
Spermatocyte
 Pelecypoda, 127, 129
 Polyplacophora, 35, 45
Spermatids
 Aplacophora, 6
 Pelecypoda, 127–129
 Polyplacophora, 45

Spermatogenesis
  Aplacophora, 9
  Monoplacophora, 91
  Pelecypoda, 127–129
  Polyplacophora, 44–45
  Scaphopoda, 98–100
Spermatogonia
  Pelecypoda, **124**, 127
  Polyplacophora, 35, 44
Spermatozoa
  Aplacophora, 9, 22
  Monoplacophora, 91
  Pelecypoda, 127–129
  Polyplacophora, 45
  Scaphopoda, 98, 100
Sperm agglutination, 65, 220
Sperm balls, 305
Sperm chemotaxis, in Polyplacophora, 65, 68
Sperm ducts, 35
Spermiogenesis, Monoplacophora, 91
Sperm motility, 128
Sperm penetration, 65, 222
Spiral cleavage, 13, 223, 224
Starvation and gametogenesis, 169–173
Statocysts, 235
Stereoblastula, 224, 225
Stomadeum in development
  Aplacophora, **14, 16**, 19, 22
  Pelecypoda, 225, 227
  Polyplacophora, 68, 72
  Scaphopoda, 106
Style, crystalline, 227, 235
Substrate selection, 73, 253
Sugar glands, 90
Sweden
  reproductive cycle in, 10t
  spawning in, 60

T

Tasmania, brooding in, 75, 76t, 77t
Telotroch, Aplacophora, **14–16**, 19–22
Temperature
  and development, 228, **231**, 230t
  in gametogenesis, 51, 138–151, 170t, 171–173t, **218**
  and larval growth, 239–245, 311

  in spawning, 139, 140, 143, 146t, 202–204
Test cell larva, 21
Test cells, Aplacophora, **6, 15**–17, 19, 20
Testes
  Aplacophora, 5
  Pelecypoda, **121**
  Polyplacophora, 28, 29, 34, **44**
  Scaphopoda, 96, 97, **100**
Tides
  and gametogenesis, 52, 152
  and larval movement, 312–320 *passim*
  and reproductive rhythm, 63, 64t
Trefoil stage
  Pelecypoda, 223
  Scaphopoda, 104
Trochophores
  Aplacophora, 12, **14**, 18t, 19, 20
  Ostreidae, 308
  Pelecypoda, 226–228, **231**, **234**
  Polyplacophora, 37, 71, 72, 78
  Scaphopoda, 106–109

U

Urogenital papilla, in Pelecypoda, 122

V

Veliconcha, 236
Veliger
  Aplacophora, 19, 21
  Ostreidae, 308
  Pelecypoda, 224–264 *passim*
  Scaphopoda, 106–109
Velum, 109, 110, 227, 234–236
Virginia, larval attachment in, 327, 328t, 332
Vitelline membrane
  Aplacophora, 8, 19
  Pelecypoda, 221, 225
  Polyplacophora, 42
Vitellogenesis
  Monoplacophora, 91
  Pelecypoda, 130, 146, 171, 174

Vitellogenesis (*cont.*)
  Polyplacophora, 36–39, 42, 43
  Scaphopoda, 100, 101
Viviparity, 233

W

Washington, larvae in, 311, 314
White Sea, breeding periods in, 211t

Y

Yolk
  Pelecypoda, 130
  Polyplacophora, 39

Z

Zygotes, Aplacophora, 7, 12, 18t

TAXONOMIC INDEX

Entries in this index which are from a table are indicated by a "t" following the page number; page numbers for entries which are from a figure are set in boldface type.

A

Abra alba, 156, 181
Acanthochitona
  curiosus, 42t
  discrepans, 31t, **36,** 47t, 57, 58, 61, 65, 69
  hemphilli, 31t, 49t
Acanthochitonidae, 41t
Acanthochitonina, 27, 28, 67t
Acanthopleura
  gemmata, 31t, 42t, 49t, 50, 61, 63
  granulata, 31t, 42t, 49t, 50, 51, 53t, 56, 57, 63
  haddoni, 31t, 42t, 44, **46,** 49t, 50
Adula californiensis, 213t, 224, 227–229, **231,** 247
Amussium, 256
Amygdalum glaberrimum, 135, 140, **142,** 215t, 217, **218**
Anadara granosa, 215t
Anadonta, 224, 225, 264
  cygnea, 119
Anomia, 120, 256
  simplex, 116, 123, **125,** 240t
  squamata, 211t
Arca, 256
  glacialis, 210t
  transversa, 240t
  vivipara, 257, 258t
Argopecten (Aequipecten)
  gibbus, 214t

  irradians, 117–255 *passim,* **121, 133, 134, 146, 147, 158, 160, 163, 174, 175, 176, 177**
  irradians concentricus, 227, 234, 238, 240t, 256, 257
Astarte, 114, 120, 233
  borealis, 114, 210t, 258t, 261
  crenata, 210t, 261
  elliptica, 114, 210t, 258t, 261
  montagui, 210t, 261
  sulcata, 114, 139, 212t, 258t
Axinopsis orbiculata, 210t

B

Bacciger bacciger, 199
Bankia, 123
  fimbriatula, 240t
  setacea, 118, 227, 240t
Barnea, 123
  candida, 136, 178, 219, 221
  truncata, 116, 240t
Bathyarca, 207
Boreochiton marginatus, 28
Brachidontes, 120
  recurvus, 116, 127, 128, 188, 214t
  cf *variabilis,* 135, 140, **142,** 204, 215t, 217, **218**
Bucephalus
  haimeanus, 199
  mytili, 199

C

Cadulus
    californicus, 101, 103
    fusiformis, 96, 98
Callistochiton
    asthenes, 77t
    leei, 77t
    mawlei, 77t
    viviparus, 65, 75, 77t
Callistoplacidae, 77t
Callochiton
    crocinus, 77t
    inornata, 77t
Callochitonidae, 77t
Cardita, 262
    bailyi, 233, 258t, 260
    barbarensis, 233, 258t, 260
    ventricosa, 214t, 233, 258t, 260
Cardium, 117, 199
    edule, 117
    norwegicum, 200
    tuberculatum, 199
Centrostephanus coronatus, 51
Ceramium rubrum, 253
Cerastoderma (Cardium)
    ciliatum, 210t
    edule, 117, 135, 151, 199, 213t, 227,
        234, 236, 238, 240t, 255, 257
    elegantum, 233, 258t
    exigium, 258t
    glaucum, 135
    tuberculatum, 199
Ceratozona squalida, 44
Cercaria
    adrenocerca, 200
    milfordensis, 200
    myae, 199
    noblei, 199
Cercaria (Gymnophallus) fulbrighti, 199
Cetoconcha, 117
Chaetoderma, 3, 5–7, 9, 22
    attenuata, 2
    erudita, 10t
    nanula, 2
    nitilidum, 5, 9, 10t, 11
    scabra, 2
Chaetodermatoidea, 2–4, 8, 10t, 11, 12,
    22
Chaetopleura

apiculata, 41t, 42, 47t, 51, 53t, 61, 63,
    65, 66t, 68, 73, 78
    gemma, 44
Chaetopleuridae, 66t
Chama, 220
Chiton, 77t
    barnesii, 75, 77t
    burmanus, 42t
    glaucus, 60
    granoradiatus, 41t, 49t, 59, 60, 65, 67t
    marmoratus, 31t, 41t, 48t, 65
    nigrovirescens, 75, 77t
    olivaceus, 37, 47t, 57, 58, 65, 72
    pelliserpentis, 29, 30, 33, 41t, 43, 49t,
        50, 51, 63, 64
    septentriones, 32–34, 36, 37, 42t, 49t,
        50
    squamosus, 31t, 48t, 65
    stokesii, 31t
    torri, 77t
    tuberculatus, 29–31t, 42t, 48t, 52, 53t,
        56, 57, 59, 61, 62, 65
Chitonidae, 31t, 41t, 67t, 77t
Chlamys, 117, 256
    distorta, 135, 178, 212t, 295
    furtiva, 116, 178, 212t
    herica, 178, 184
    opercularis, 117, 204
    septemradiata, 156, 181
    striata, 116, 212t
    tigerina, 116, 178, 196, 212t
    varia, 122, 127, 135, 178, 187, 188,
        195, 196, 205, 206, 212t
Chlorella, 249, 250
Chytriodiopsis mytilovum, 199
Craspedochiton violaceus, 35
Crossostrea, 220, 224, 293–296, 302–308,
    310, 312, 322
    angulata, 178, 298, 304
    cucullata, 119
    gigas, 224, 225, 297, 298, 303, 304, 312
    virginica, 118, 139, 151, 159, 178, 187,
        189, 193, 202, 219, 224, 239, 247,
        296, 298, 300t, 301–309, 311, 326,
        327, 330, 332, 334
Crepidula, 224, 296, 301
Cryptochiton stelleri, 31t, 35, 42t, 48t,
    53t, 54, 56–61, 65, 67t, 68, 71, 72
Cryptoconchus porosus, 29, 39, 42t, 49t,
    51, 63, 65, 67t

Crystallophrisson, 3
Cumingia tellinoides, 204, 213t, 217
Cuspidaria, 153
   cuspidata, 114, 165, 167t
Cyanoplax
   dentiens, 29, 30, 41t, 48t, 58, 76t
   hartwegii, 41t, 42, **43**
Cyclas, 224, 225
Cyclinella tenuis, 214t
Cyprina islandica, 127, 135–37, 139, 155,
   156, 202, **208**, 209, 213t

D

Dentalium, 95, 96, 98, 101, 103–**109**
   antillarum, 96, 103, 104
   dentale, 101, 103, 105
   entale, 100
   entalis, 104
   neohexagonium, **97**, 98, **99**, 100, **102**
   occidentale, 100
   vulgare, **100**
Donax, 199, 201
   cuneatus, 135, 136, 151, 181, 215t
   faba, 135, 136, 151, 215t
   gouldii, 116, 200
   trunculis, 135
   variabilus, 240t
   vitatus, 199
Dosina, 127
Dreissena polymorpha, 155, 189
Dreissensia, 224, 225
Dunaliella enchlora, **248**

E

Egerina radiata, 152
Ensis directus, 219, 240t
Entocolax, 3
Entovalva, 153, 257, 258t
Epimenia, 21, 22
   verrocosa, 5, 10t, 11–15, 17, 18t, 19–
   21

F

Falcidens, 5

G

Galeomma, 257, 258t, 262
Gemma gemma, 200, 233, 258t, 260, 264
Glycymeris glycymeris, 127, 135, 195

H

Haliotis, 3
Halomenia, 14, 22
   gravida, **6**, 8, 11–13, 17–19
Hanleyidae, 30t, 41t, 76t
Hedophyllum sessile, 73
Hemiarthrum setulosum, 30t, 41t, 49t,
   65, 75, 76t
Hiatella, 234
   arctica, 210t

I

Ischnochiton, 77t
   acomplus, 41t, 59, 65, 66t
   albus, 41t
   constanti, 75, 76t
   imitator, 75, 76t
   radians, 41t
   subviridis, 76t, 78
Ischnochiton (*Ovatoplax*) *mayi*, **74**, 76t
Ischnochitonidae, 30t, 41t
Ischnochitonina, 27, 28, 66t, 76t
Isochrysis galbana, 174, 242, 248–250,
   310

K

Katelysia opima, 135, 181, 187, 188–**190**,
   193, 215t
Katharina tunicata, 29, 31t, 32, 39, 41t,
   43, 48t, 50, 51, 53t, 54, **55**, 56, 57,
   59, 61, 62, 68, 73
Kellia, 120, 257, 262
   suborbicularis, 117, 213t, 258t, 262
   symmetros, 128

L

Laevicardium mortoni, 240t
Lampsilis, 114

Lasaea, 117, 120, 257, 260, 262
    rubra, 155, 213t, 233, 257, 258t, 262
Leida, 153
Lepidochitona, 77t
    cinereus, 28, 30t, **33**, 34, **38**, 39, **40**, 41t,
        42, 47t, 51–53t, 57, **59**, 60, 61, 63,
        65, 66t, 68, 69, 72, 74, 76t, 78
Lepidopleuridae, 30t
Lepidopleurina, 27
Lepidopleurus
    asellus, 30t, 39, **47t**, 53t, 59, 62, 65,
        66t, 78
    cajetanius, **47t**
Lepidozona, 29
    cooperi, 61
    mertensii, 41t, 61
Lepton parasiticum, 258t
Limifossar talpoides, 6
Liocyma fluctuosa, 233
Lithophaga, 220
Lithophyllum, 73
Lithothamnium, 73
Loripes lacteus, 233, 258t, 262
Lyinsella, 153
Lyonsia arenosa, 210t
Lyrodus
    massa, 258t
    pedicellatus, 258t

M

Macoma, 153
    balthica, 116, 122, 123, 127, 134, 135,
        136, 152, 154, 158, 159, 178, 181,
        185, 196, 209, 211t
    calcaria, 210t
    loveni, 261
    moesta, 210t, 261
    nasuta, 135
    secta, 135
    torelli, 261
Mactra, 220
    veneriformis, 219, 221
Martesia striata, 152, 181, 234, 240t
Mercenaria, 327
    campechensis, 120, 244
    mercenaria 117, 118, 120, 122, 126,
        127, 129, 131, 135, 137, 138, 140,
        143, 157, 178, 179, 180, 182, 188,

        195, 202, 203, 209, 213t, 219, 228,
        229, 230, 232, 239, 240t, **242**, 244,
        245, **246**, 247, **248**, 249, **250**, 251,
        252, 254
Meretrix casta, 151, 187, 188, 214t
Micromenia fodiens, 59
Middendorffia polii, 41t, 47t, 64, 65, 69,
    70, 77t, 78
Minchinia
    chitonis, 52
    nelsoni, 319
Modiola, 256
Modiolaria
    discors, 258t
    discors laevigatata, 210t
    nigra, 210t, 259t
Modiolus, 123
    barbaratus, 116
    demissus, 116, 155, 179, 188, 201, 204
    modiolus, 116, 155, 156, 234
    striatulus, 215t
Monas, 297
Monochrysis lutheri, 175–177, 192, 242,
    248–250, 310
Montacuta, 117, 120, 257, 262
    ferruginosa, 117, 128, 201, 213t, 259t,
        262, 264
    percompressa, 114, 241t, 259t, 260,
        262, 264
    phascolonis, 259t
    substriata, 117, 128, 213t, 259t, 262
    tenella, 128
Mopalia, 64
    ciliata, 41t, 48t, 59, 60, 62, 63, 66t, 68
    hindsii, 31t, 41t, 46, 48t, 54, 57, 62
    imporcata, 41t
    laevior, 48t, 62
    lignosa, 41t, 48t, 58, 59, 62, 65, 67t,
        68, 72, 73
    lowei, 41t
    muscosa, 31t, 32, 36, 41t, 42, **43**, **44**,
        46, 48t, 53, 58, 61, 62, 65, 67t, 68
    porifera, 41t
Mopaliidae, 31t, 66t
Mulinia lateralis, 116, 127, 139, 143, 188,
    196, 203, 214t, 225, 228, 230t, 232,
    239, 241t, 245, 246, 247t, 251
Musculus
    discor, 233, 261, 262
    niger, 233, **261**, 262

Mya
  arenaria, 116, 121–123, 129, 135, 137, 148, 150, 155, 165, 166t, 179, 197, 199, 209, 211t, 219, 228, 230, 234, 241t, 244, 246
  truncata, 210t
Mysella, 262
  bidentata, 129, 259t
Myticola
  intestinalis, 200
  orientalis, 200
Mytilidae, 140, 142
Mytilus, 120, 123, 199, 220, 223, 225, 256
  californianus, 116, 119, 120, 154, 155, 199, 205, 229
  crassitesta, 200
  edulis, 116–254 passim, **122,** 168t, 211t, **226, 237,** 241t, **243, 244, 253,** 333
  edulis planatus, 135, 140, **141,** 215t, 217, **218**
  galloprovincialis, 135, 136, 143, 178, 189, 194, 199, 205, 231
  perna, 128, 188, 215t
  viridis, 135

N

Nannochloris atomus, 249
Nausitoria hedleyi, 152, 179, 181, 195, 215t
Nematomenia, 9, 21, 22
  banyulensis, 8, 10t, 11, 12, 13, 17, 18t, 19, 21
  banyulensis norwegica, 6
Nematopsis, 199
Neoloricata, 27
Neomenia, 9, 10t, 21, 22
  carinata, 9–13, **15–21**
Neomenioidea, 2–4, 8, 10t, 11, 13, 15, 17–21
Neopilina, 87–92
  ewingii, 89, 90
  galatheae, 87–92
Neoteredo knoxi, 241t
Neotia ponderosa, 241t
Notovola meridionalis, 202
Nucula, 127, 153, 227
  cancellata, 207

darella, 207
delphinodonta, 227, 233, 239, 255, 259t, 262
nucleus, 187, 233
pernula, 233
pontonia, 207
proxima, 213t, 233
sulcata, 181
tenuis, 210t
turgida, 233
Nuttalina
  cinerea, 64
  thomasi, 48t, 53t, 77t
Nuttallochiton
  hyadesi, 32
  mirandus, 32

O

Onithochiton lyelli, 31t, 42t, 49t, 51, 63
Ostrea, 225, 256, 257, 259t, 262, 293–295, 301, 302, 305–310, 322
  chilensis, 310, 321
  denselamellosa, 304
  edulis, 118, 119, 238, 239, 247, 264, 296, 297, 301, 303, 306, 310, 311, 322, 324
  equestris, 118, 304, 310
  frons, 304
  lurida, 118, 178, 301, 303, 304, 315
  lutaria, 310, 321

P

Pandora glacialis, 210t
  inaequivaluis, 227, 233, 234, 236, 238, 291t, 253, 255
Paphia, 123
  latevisculca, 215t
Parvatrema borealis, 200
Patinopecten
  caurinus, 212t
  yessoensis, 143, 165, 181, 185, **186,** 197, 201, 214t
Pecten, 117, 120, 123, 220, 223, 256, 295
  circularis aequisulcatus, 214t
  islandicus, 210t
  jacobeus, 181

Pecten (*cont.*)
  latiauritus var *monotimeris*, 117
  latiavitus, 214t
  maximus, 117, 122, 127, 135, 181, 201, 204, 212t, 232
  opercularis, 227
  tenuicostatus, 227
Petricola pholadiformis, 116, 123, **124**, 241t
Phaeodactylum, 165, 168t
  tricornutum, 249
Philobrya, 257
  setosa, 259t
Pholas candida, 199
Phyllomenia austrian, 7
Pinctada, 127, 217, 223, 224, 295
  albina, 119, 122–**124**, 126–129, 132, 135, 136, 195, **196**, 209, 215t
  fucata, 122, 135, 137, 195, 209, 215t
  margaritifera, 122, 135, 136, 148, 195, 209, 215t
  martensii, 180, 181, 219
  maxima, 215t, **222**, 241t
Pinnotheres, 119
  pisum, 200
Pitar calliocardia morrhuana, 204, 241t
Placopeten magellanicus, 116, 122, 127, 135, 156, 184, 197, 199, 201, 202, 209, 213t, 234
Placuna, 120
  placenta, 148, 151, 215t
Plathymenia branchiosa, 5
Platymonia, 248
Plaxiphora
  aurata, 31t, 41t, 49t
  primordia, 29, 35, 39, 41t, 59, 65, 67t
Pododesmus cepio, 214t
Poromya, 117
Portlandia, 153
  artica, 210t
Postmonorchis donacis, 200
Prochaetoderma californicum, 7
Proneomenia hawaiiensis, 2
  sopita, 6
  vagans, 2
Propeamusium
  groenlandicum, 210t
  imbriferum, 210t
  masculatus, 199
Protoeces masculatus, 199

Protothaca (= *Paphia*)
  staminea, 122, 123, 135, 136, 139, 200, 211t
Pruvotina, 11, 19
  plovidens, 18t
  praegrans, 18t
  uniperata, 18t
Prymnesium parvum, 249
Psuedokellia, 257, 259t, 260
Pseudopythina rugifera, 214t
Pteria, 223
Pycnodonte, 295

R

Radsia, 75
Rhopalomenia, 21
  aglaophenia, 11, 13, 17, 18t, 19, 21

S

Schizoplacidae, 41t, 77t
Schizoplax brandtii, 41t, 42, 47t, 53t, 75, 77t
Scioberita, 153
Scrobicularia plana, 127, 135, 155, 202, 203, 213t
Septifer
  bifurcatus, 116
  biloculoris, 135, 140, **142**, 204, 215t, 217, **218**
Siligua patula, 212t
Skeletonema costatum, 62
Solen, 199, 220
  kempi, 215t
Sphaerium, 224, 225
  striatinum, 123, 224
Spisula, 199
  sachalinensis, 201, 214t
  solidissima, 116, 122, 135, 140, 143, 188, 202, 209, 214t, 219, 220, 221, 225, 229, 241t
Spondylus, 220, 256, 295
Stenoplax heathiana, 29, 41t, **44**, 48t, 59, 60, 64, 66t, 68, **69**, 71, 73
Stichococcus, 249
Strophomenia lacazei, 10t
Stylochus ellipticus, 327, 334

Stylomenia salvatori, 10t
Synapta, 153
Syndosmya alba, 199

T

Tapes
  aereus, 199
  decussatus, 199
  pullastra, 199
  semidecussata, 199, 219, 241t
  variegata, 227
Tellina, 199
  exigus, 199
  tenuis, 156–159, 178, 180
Teredo, 117, 187, 224, 225, **234**, 256, 257, 262
  bartschi, 259t
  clappi, 259t
  diegensis, 117, 259t
  furcillatus, 152, 195, 215t, 259t
  navalis, 118, 123, 126, 195, 241t, 259t, 264
  parksi, 259t
  pedicellata, 181, 182, 233
Thracia, 117, 128
Thyasira
  flexuosa, 233
  gouldi, 210t, 261
  sarsi, 233
Tivela, 117
  stultorum, 119, 160, **162**, 179, 181, **182**
Tonica schrammi, 49t
Tonicella, 77t
  insignis, 47t, 61, 62
  lineata, 41t, 48t, 59, 61–63, 65, 66t, 68, 72, 73, 78
  marmoreus, 41t, 47t
Trachydermon raymondi, 29
Transenella tantilla, 233, 257, 259t, 260, **261**, 262, 264
Trapezium, 220
Tresus capax, 183
Tridacna, 117, 205
Turtonia, 262
  minuta, 205, 211t, 257, 259t, **261**, 262

U

Ulva, 73
Unio, 224, 225
  tumidus, 187

V

Venerupis pullastra, 212t, 236, 238, 255, 257
Venus striatula, 122, 127, 135, 155, 157, 165, 167t, 168t, 169, 195, 212t, 234, 236, 238, 255, 256
Vermetus, 3
Volsella demissus, 241t

X

Xenostrobus
  pulex, 135, 140, 141, 215t, 217, **218**
  securis, 152, 204
Xylophaga
  africana, 280t
  atlantica, 234, 260t
  bruuni, 260t
  concava, 260t
  dorsalis, 114, **115**, **116**, 195
  lobata, 260t
  panamensis, 260t, 261
  tubulata, 260t
  wolffi, 260t

Y

Yoldia, 153, 188, 227
  hyperborea, 211t
  limatula, 213t, 227, 233, 239, 255

Z

Zirphea, 220, 256
  crispata, 135, 212t